NEUROMETHODS

Series Editor
Wolfgang Walz
University of Saskatchewan
Saskatoon, SK, Canada

For further volumes:
http://www.springer.com/series/7657

Use of Nanoparticles in Neuroscience

Edited by

Fidel Santamaria

Department of Biology, The University of Texas at San Antonio, San Antonio, TX, USA

Xomalin G. Peralta

Air Force Research Laboratory 711 HPW/RHDR, Fort Sam Houston, TX, USA

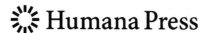 Humana Press

Editors
Fidel Santamaria
Department of Biology
The University of Texas at San Antonio
San Antonio, TX, USA

Xomalin G. Peralta
Air Force Research Laboratory
711 HPW/RHDR
Fort Sam Houston, TX, USA

ISSN 0893-2336   ISSN 1940-6045   (electronic)
Neuromethods
ISBN 978-1-4939-8528-9   ISBN 978-1-4939-7584-6   (eBook)
https://doi.org/10.1007/978-1-4939-7584-6

This Humana Press imprint is published by Springer Nature
The registered company is Springer Science+Business Media, LLC
The registered company address is: 233 Spring Street, New York, NY 10013, U.S.A.

Series Preface

Experimental life sciences have two basic foundations: concepts and tools. The *Neuromethods* series focuses on the tools and techniques unique to the investigation of the nervous system and excitable cells. It will not, however, shortchange the concept side of things as care has been taken to integrate these tools within the context of the concepts and questions under investigation. In this way, the series is unique in that it not only collects protocols but also includes theoretical background information and critiques which led to the methods and their development. Thus it gives the reader a better understanding of the origin of the techniques and their potential future development. The *Neuromethods* publishing program strikes a balance between recent and exciting developments like those concerning new animal models of disease, imaging, in vivo methods, and more established techniques, including, for example, immunocytochemistry and electrophysiological technologies. New trainees in neurosciences still need a sound footing in these older methods in order to apply a critical approach to their results.

Under the guidance of its founders, Alan Boulton and Glen Baker, the *Neuromethods* series has been a success since its first volume published through Humana Press in 1985. The series continues to flourish through many changes over the years. It is now published under the umbrella of Springer Protocols. While methods involving brain research have changed a lot since the series started, the publishing environment and technology have changed even more radically. Neuromethods has the distinct layout and style of the Springer Protocols program, designed specifically for readability and ease of reference in a laboratory setting.

The careful application of methods is potentially the most important step in the process of scientific inquiry. In the past, new methodologies led the way in developing new disciplines in the biological and medical sciences. For example, Physiology emerged out of Anatomy in the nineteenth century by harnessing new methods based on the newly discovered phenomenon of electricity. Nowadays, the relationships between disciplines and methods are more complex. Methods are now widely shared between disciplines and research areas. New developments in electronic publishing make it possible for scientists that encounter new methods to quickly find sources of information electronically. The design of individual volumes and chapters in this series takes this new access technology into account. Springer Protocols makes it possible to download single protocols separately. In addition, Springer makes its print-on-demand technology available globally. A print copy can therefore be acquired quickly and for a competitive price anywhere in the world.

Wolfgang Walz

Preface

In recent years, there has been an increase in the use of nanoparticles in various medical fields, including neuroscience. This increased use is a result of advances in our ability to control the nanoparticles' physical and chemical properties, allowing them to interact with various components of the nervous system in specific ways. Currently, there is an interest in understanding the basic principles that govern those interactions in order to take advantage of them to develop various biomedical applications. To reach that goal, there is a need to bring together the main techniques developed to quantify experiments using nanoparticles in the nervous system. These techniques include how to synthesize and functionalize nanoparticles, monitor and influence their delivery and uptake, identify and evaluate their lethal and nonlethal effects on the function and metabolic activity of the nervous system, and how to use them to affect the electrophysiological, cellular, or network activity intrinsically or through external stimuli. This book is a first step toward achieving that goal.

The book is divided into four broad areas of research:

1. Photostimulation
2. Thermal stimulation
3. Mechanical perturbation
4. Toxicity and physiological effects

The first three sections focus on methodologies where nanoparticles interact with some external source in order to perturb the neuronal system. The last one focuses on the effects resulting from having nanoparticles present in a neuronal system in the absence of any external stimuli. This division is somewhat arbitrary given that several chapters bridge two or more of these areas of research. We also want to note that although some very important topics in nanoparticle in neuroscience research may appear to be missing in this volume, such as strategies to functionalize and deliver nanoparticle, they are introduced and addressed in several chapters even if they are not the main focus of any one chapter. We encourage readers to review the contents of all chapters in order to take full advantage of all the different and complementary techniques contained within this volume.

In the following we address the basic ideas behind each area of research identified and briefly summarize the contents of the 16 chapters contained in this volume.

Photostimulation

The chapters in this section provide the techniques to utilize different wavelengths of light in combination with metallic nanoparticles to stimulate various types of neuronal cells including isolated neurons (Chapter 1), cultured neurons (Chapter 2), and brain slices (Chapters 1 and 3) with the primary goal of affecting their electrophysiological activity. In Chapter 1 the activity is monitored using voltage sensitive dyes, in Chapter 2 fluorescent imaging techniques, and in Chapter 3 whole-cell patch-clamp techniques.

Thermal Stimulation

In this section, the primary mechanism used to affect neurons is heat. In Chapters 4 and 5, heat is generated by stimulating magnetic nanoparticles with alternating magnetic fields or radiofrequency, resulting in neural modulation. In Chapter 4 the nanoparticles are synthesized using thermal decomposition, are composed of ferrites, and are coated in order to target specific cell surface receptors. In Chapter 5 the nanoparticles consist of genetically encoded ferritin iron-oxide nanocomposites which, when stimulated with radiofrequency, produce heat and affect modified temperature sensitive ion channels. These nanoparticles can also be stimulated with magnetic fields in which case they exert mechanical forces on the cell membrane that can be used for modulation, rendering this chapter relevant to Subheading 3 (mechanical perturbation) of this volume. Chapter 6 details a technique to utilize bare metallic nanoparticles for photothermal ablation of cells and organelles in brain slices. In addition, it provides a bridge to Subheading 4 of this volume by presenting a methodology to elucidate the nonlethal effects of gold nanoparticles on the firing rate of hippocampal cells.

Mechanical Perturbation

Another promising approach to affect neuronal function is to mechanically perturb the cells. As mentioned above, in Chapter 5 the authors describe a technique to exert mechanical forces on the cell membrane by stimulating magnetic nanoparticles with magnetic fields. These forces can affect specific ion channels and modulate the neuron's activity. Chapter 7 continues with the theme of stimulating superparamagnetic nanoparticles with magnetic fields. The nanoparticles can be targeted intracellularly to signaling endosomes or extracellularly to cell surface receptors and can be used to regulate growth cone motility, bridging Subheadings 3 and 4 (physiological effects) of this volume. In Chapter 8 the authors present a technique to stimulate neuroblastoma cells in culture by exciting piezoelectric nanoparticles with ultrasound waves. They monitor the electrophysiological activity with fluorescent imaging techniques. Chapter 9 is primarily concerned with affecting the uptake of chitosan-coated metallic nanoparticles by modifying the cell membrane's permeability via nanopore formation. The nanopores are formed through exposure to nanosecond pulsed electric fields and the changes in uptake are monitored using fluorescent imaging techniques.

Toxicity and Physiological Effects

Nanoparticle toxicity can be divided into lethal and nonlethal where the latter results in changes in the function and/or metabolic activity of neurons. Chapter 6 above and Chapters 10 through 14 primarily focus on the effects of nanoparticles on ionic currents as observed using electrophysiological recordings. Chapters 6 and 10 connect experimental data to computer models in order to identify which specific ionic channels in hippocampal slices (Chapter 6) and individual cultured cells (Chapter 10) are affected by bare metallic nanoparticles. The next two chapters study the effects of carbon-based nanoparticles on in vivo rat extracellular recordings (Chapter 11) and hippocampal slices monitored via the patch-clamp technique (Chapter 12). In Chapter 13, the authors present a technique to assess the effects of different types of nanoparticles (synthesized, native, and physiological) on ionic currents and neurotransmitter uptake in nerve terminals isolated from rat brain cerebral hemispheres. Chapter 14 focuses on a coculture model that can be used to assess nanoparticle uptake, length of retention within the cell, and fate following cell division. It also includes a discussion on the physicochemical characterization of the nanoparticle's corona. Although it uses magnetic nanoparticles, some of the techniques presented can be applied to nanoparticles of other compositions. Chapter 15 is the only chapter in this volume that focuses on the lethal effects of nanoparticles. It details a technique to assess toxicity of lanthanide doped nanoparticles via death distribution in cell culture. In addition, it presents a technique to assess the spatial distribution of orally administered nanoparticles in various organs, including the brain, of mice. Chapter 16 is the only chapter in this volume that focuses primarily on the physiological effects of nanoparticles. It studies the role that nanoparticles can have as nucleation sites for amyloid aggregation and presents a technique to functionalize gold nanoparticles for use in amyloid fibrillation assays.

We hope all specialized audiences and graduate students find this compilation useful for learning the techniques necessary to quantify experiments using in the nervous system.

San Antonio, TX, USA   *Fidel Santamaria*
Fort Sam Houston, TX, USA   *Xomalin G. Peralta*

Contents

Contributors

YANINA D. ÁLVAREZ • *Departamento de Física, Facultad de Ciencias Exactas y Naturales, Universidad de Buenos Aires, Ciudad de Buenos Aires, Argentina*

SATOSHI ARAI • *WASEDA Bioscience Research Institute in Singapore (WABIOS), Singapore, Singapore*

PIERRE-LUC AYOTTE-NADEAU • *CERVO Brain Research Center, Université Laval, Quebec City, Canada*

FRANCISCO BEZANILLA • *Department of Biochemistry and Molecular Biology, University of Chicago, Chicago, IL, USA*

TATIANA BORISOVA • *Department of Neurochemistry, Palladin Institute of Biochemistry, National Academy of Sciences of Ukraine, Kiev, Ukraine*

ARSENII BORYSOV • *Department of Neurochemistry, Palladin Institute of Biochemistry, National Academy of Sciences of Ukraine, Kiev, Ukraine*

WILLIAM G.A. BROWN • *Faculty of Engineering and Industrial Sciences, Swinburne University of Technology, Hawthorn, VIC, Australia*

MICHAEL BUSSE • *Systems Neuroscience and Neurotechnology Unit, Saarland University, Saarland, Germany; Neurocenter, Faculty of Medicine, Saarland University of Applied Sciences, Homburg/Saarbruecken, Germany; INM—Leibniz Institute for New Materials, Saarbruecken, Germany*

JOAO L. CARVALHO-DE-SOUZA • *Department of Biochemistry and Molecular Biology, University of Chicago, Chicago, IL, USA*

IDOIA CASTELLANOS-RUBIO • *Department of Physics, University at Buffalo, SUNY, Buffalo, NY, USA*

DIVYA M. CHARI • *School of Medicine, Institute for Science and Technology in Medicine, Keele University, Keele, Newcastle, UK*

JUTAO CHEN • *Hefei National Laboratory for Physical Sciences at Microscale and School of Life Sciences, University of Science and Technology of China, Hefei, Anhui, China*

GIANNI CIOFANI • *Smart Bio-Interfaces, Istituto Italiano di Tecnologia, Pontedera, Pisa, Italy; Department of Mechanical and Aerospace Engineering, Politecnico di Torino, Torino, Italy*

SAMANTHA K. FRANKLIN • *Department of Physics and Astronomy, University of Texas at San Antonio, San Antonio, TX, USA*

ZDZISŁAW GAJEWSKI • *Department of Large Animal Diseases with Clinic, Veterinary Research Centre, Centre for Biomedical Research, Faculty of Veterinary Medicine, Warsaw University of Life Sciences, Warsaw, Poland*

MAREK GODLEWSKI • *Institute of Physics, Polish Academy of Sciences, Warsaw, Poland*

MICHAŁ M. GODLEWSKI • *Department of Large Animal Diseases with Clinic, Veterinary Research Centre, Centre for Biomedical Research, Faculty of Veterinary Medicine, Warsaw University of Life Sciences, Warsaw, Poland; Department of Physiological Sciences, Faculty of Veterinary Medicine, Warsaw University of Life Sciences, Warsaw, Poland*

JEFFREY L. GOLDBERG • *Department of Ophthalmology, Stanford University, Palo Alto, CA, USA*

YANYAN HOU • *WASEDA Bioscience Research Institute in Singapore (WABIOS), Singapore, Singapore*

STUART I. JENKINS • *School of Medicine, Institute for Science and Technology in Medicine, Keele University, Keele, Newcastle, UK*

JAROSŁAW KASZEWSKI • *Institute of Physics, Polish Academy of Sciences, Warsaw, Poland; Department of Large Animal Diseases with Clinic, Veterinary Research Centre, Centre for Biomedical Research, Faculty of Veterinary Medicine, Warsaw University of Life Sciences, Warsaw, Poland; Department of Physiological Sciences, Faculty of Veterinary Medicine, Warsaw University of Life Sciences, Warsaw, Poland*

PAULA KIEŁBIK • *Department of Large Animal Diseases with Clinic, Veterinary Research Centre, Centre for Biomedical Research, Faculty of Veterinary Medicine, Warsaw University of Life Sciences, Warsaw, Poland; Department of Physiological Sciences, Faculty of Veterinary Medicine, Warsaw University of Life Sciences, Warsaw, Poland*

PAUL DE KONINCK • *CERVO Brain Research Center, Université Laval, Quebec City, Canada*

ANNETTE KRAEGELOH • *INM—Leibniz Institute for New Materials, Saarbruecken, Germany*

FLAVIE LAVOIE-CARDINAL • *CERVO Brain Research Center, Université Laval, Quebec City, Canada*

ATTILIO MARINO • *Smart Bio-Interfaces, Istituto Italiano di Tecnologia, Pontedera, Pisa, Italy*

VIRGILIO MATTOLI • *Center for Micro-BioRobotics, Istituto Italiano di Tecnologia, Pontedera, Pisa, Italy*

BARBARA MAZZOLAI • *Center for Micro-BioRobotics, Istituto Italiano di Tecnologia, Pontedera, Pisa, Italy*

YANYAN MIAO • *Hefei National Laboratory for Physical Sciences at Microscale and School of Life Sciences, University of Science and Technology of China, Hefei, Anhui, China; School of Life Sciences, State Key Laboratory for Biocontrol, Sun Yat-sen University, Guangzhou, China*

RAHUL MUNSHI • *Department of Physics, University at Buffalo, SUNY, Buffalo, NY, USA*

KELLY L. NASH • *Department of Physics and Astronomy, University of Texas at San Antonio, San Antonio, TX, USA*

KARINA NEEDHAM • *Otolaryngology, Department of Surgery, The University of Melbourne, East Melbourne, VIC, Australia*

JAKUB NOJSZEWSKI • *Department of Large Animal Diseases with Clinic, Veterinary Research Centre, Centre for Biomedical Research, Faculty of Veterinary Medicine, Warsaw University of Life Sciences, Warsaw, Poland; Department of Physiological Sciences, Faculty of Veterinary Medicine, Warsaw University of Life Sciences, Warsaw, Poland*

ARTEM PASTUKHOV • *Department of Neurochemistry, Palladin Institute of Biochemistry, National Academy of Sciences of Ukraine, Kiev, Ukraine*

CHIARA PAVIOLO • *LP2N-Institut d'Optique & CNRS, University of Bordeaux, Talence, France*

MARIO PELLEGRINO • *Dipartimento di Ricerca Traslazionale e delle Nuove Tecnologie in Medicina e Chirurgia, University of Pisa, Pisa, Italy*

JESICA V. PELLEGROTTI • *Centro de Investigaciones en Bionanociencias (CIBION), Consejo Nacional de Investigaciones Científicas y Técnicas (CONICET), Ciudad de Buenos Aires, Argentina*

DAVID R. PEPPERBERG • *Department of Ophthalmology and Visual Sciences, Lions of Illinois Eye Research Institute, Illinois Eye and Ear Infirmary, University of Illinois at Chicago, Chicago, IL, USA*

XOMALIN G. PERALTA • *Air Force Research Laboratory, 711 HPW/RHDR, Fort Sam Houston, TX, USA*

NATALIA POZDNYAKOVA • *Department of Neurochemistry, Palladin Institute of Biochemistry, National Academy of Sciences of Ukraine, Kiev, Ukraine*

ARND PRALLE • *Department of Physics, University at Buffalo, SUNY, Buffalo, NY, USA*

SHAHNAZ QADRI • *Department of Physics, University at Buffalo, SUNY, Buffalo, NY, USA*

TANCHEN REN • *Department of Ophthalmology, School of Medicine, University of Pittsburgh, Pittsburgh, PA, USA; McGowan Institute for Regenerative Medicine, University of Pittsburgh, Pittsburgh, PA, USA; Center for Neuroscience, University of Pittsburgh, Pittsburgh, PA, USA*

ANNA SŁOŃSKA-ZIELONKA • *Department of Large Animal Diseases with Clinic, Veterinary Research Centre, Centre for Biomedical Research, Faculty of Veterinary Medicine, Warsaw University of Life Sciences, Warsaw, Poland; Department of Physiological Sciences, Faculty of Veterinary Medicine, Warsaw University of Life Sciences, Warsaw, Poland*

NARSIS SALAFZOON • *Systems Neuroscience and Neurotechnology Unit, Saarland University, Saarland, Germany; Neurocenter, Faculty of Medicine, Saarland University of Applied Sciences, Homburg/Saarbruecken, Germany*

CHARLEEN SALESSE • *CERVO Brain Research Center, Université Laval, Quebec City, Canada*

FIDEL SANTAMARIA • *Department of Biology, The University of Texas at San Antonio, San Antonio, TX, USA*

IZABELA SERAFIŃSKA • *Department of Large Animal Diseases with Clinic, Veterinary Research Centre, Centre for Biomedical Research, Faculty of Veterinary Medicine, Warsaw University of Life Sciences, Warsaw, Poland; Department of Physiological Sciences, Faculty of Veterinary Medicine, Warsaw University of Life Sciences, Warsaw, Poland*

SARAH A. STANLEY • *Diabetes, Obesity and Metabolism Institute, Icahn School of Medicine at Mount Sinai, New York, NY, USA*

FERNANDO D. STEFANI • *Departamento de Física, Facultad de Ciencias Exactas y Naturales, Universidad de Buenos Aires, Ciudad de Buenos Aires, Argentina; Centro de Investigaciones en Bionanociencias (CIBION), Consejo Nacional de Investigaciones Científicas y Técnicas (CONICET), Ciudad de Buenos Aires, Argentina*

MICHAEL B. STEKETEE • *Department of Ophthalmology, School of Medicine, University of Pittsburgh, Pittsburgh, PA, USA; McGowan Institute for Regenerative Medicine, University of Pittsburgh, Pittsburgh, PA, USA; Center for Neuroscience, University of Pittsburgh, Pittsburgh, PA, USA*

DAVID R. STEVENS • *Department of Physiology, Faculty of Medicine, Saarland University, Homburg/Saarbruecken, Germany*

PAUL R. STODDART • *ARC Training Centre in Biodevices, Swinburne University of Technology, Hawthorn, VIC, Australia*

DANIEL J. STRAUSS • *Systems Neuroscience and Neurotechnology Unit, Saarland University, Saarbrücken, Saarland, Germany; Neurocenter, Faculty of Medicine, Saarland University of Applied Sciences. Homburg/Saarbruecken, Germany; INM—Leibniz Institute for New Materials, Saarbruecken, Germany*

MADOKA SUZUKI • *WASEDA Bioscience Research Institute in Singapore (WABIOS), Singapore, Singapore; Organization for University Research Initiatives, Waseda University, Tokyo, Japan*

SUMEYRA TEK • *Department of Physics and Astronomy, University of Texas at San Antonio, San Antonio, TX, USA*

JEREMY S. TREGER • *Department of Biochemistry and Molecular Biology, University of Chicago, Chicago, IL, USA*

BRANDY VINCENT • *Department of Physics and Astronomy, University of Texas at San Antonio, San Antonio, TX, USA*

MING WANG • *Hefei National Laboratory for Physical Sciences at Microscale, School of Life Sciences, University of Science and Technology of China, Hefei, Anhui, China*

LONGPING WEN • *Hefei National Laboratory for Physical Sciences at Microscale, School of Life Sciences, University of Science and Technology of China, Hefei, Anhui, China*

ZHUO YANG • *College of Medicine, State Key Laboratory of Medicinal Chemical Biology, Key Laboratory of Bioactive Materials, Ministry of Education, Nankai University, Tianjin, China*

JIAWEY YONG • *Environment and Biotechnology Centre, Faculty of Life and Social Sciences, Swinburne University of Technology, Hawthorn, VIC, Australia*

XIAOCHEN ZHANG • *College of Medicine, State Key Laboratory of Medicinal Chemical Biology, Key Laboratory of Bioactive Materials, Ministry of Education, Nankai University, Tianjin, China*

HAN ZHAO • *Hefei National Laboratory for Physical Sciences at Microscale, School of Life Sciences, University of Science and Technology of China, Hefei, Anhui, China*

Chapter 1

Optocapacitance Allows for Photostimulation of Neurons without Requiring Genetic Modification

Joao L. Carvalho-de-Souza, Jeremy S. Treger, David R. Pepperberg, and Francisco Bezanilla

Abstract

Optocapacitance is a novel technique that combines much of the power of optogenetics without requiring any transfection or genome modification in the organism or tissue of interest. It functions via the same principles as infrared neural stimulation, but uses cell-targeted gold nanoparticles to transduce incident light into local heating that stimulates excitable cells by altering their membrane capacitance. We have demonstrated this technique in both isolated neurons and brain slices, and in this chapter we describe in detail the methods used in these studies. Overall, optocapacitance is a technique that should be widely applicable to many cell-types and tissues, it can be performed using commercially available preparations with no modification, and this can complement optogenetic techniques in situations where the cost and difficulty of genetic modification are not justified.

Key words Optocapacitance, Photostimulation, Dorsal root ganglion neurons, Gold nanoparticles, Brain slice, Membrane capacitance

1 Introduction

Targeted photostimulation of specific classes of neurons is an exciting new subfield of neuroscience that has provided numerous insights into brain function which were difficult or impossible to achieve using prior techniques [1–5]. This targeted photostimulation is usually accomplished by employing optogenetics, a technique wherein light-sensitive ion channels are exogenously expressed in a tissue or animal under the control of a desired promotor sequence [6, 7]. The choice of promotor sequence will then dictate which populations of excitable cells will express the ion channel and thus can be stimulated with light. While optogenetics is a remarkable technique that has thus far proven very successful, it requires genetic modification of the organism or tissue under study. This process can be varyingly difficult, time-consuming, and expensive depending on what organism/tissue is being used.

Fidel Santamaria and Xomalin G. Peralta (eds.), *Use of Nanoparticles in Neuroscience*, Neuromethods, vol. 135,
https://doi.org/10.1007/978-1-4939-7584-6_1, © Springer Science+Business Media, LLC 2018

These limitations currently pose significant constraints on the use of optogenetic techniques in humans for therapeutic applications.

One alternative method for photostimulation of neurons that has been known for many years is neural stimulation at infrared wavelengths that are strongly absorbed by water (e.g., 2–4 μm) [8]. Recently, it was shown that this type of infrared neural stimulation (here termed INS) relies on changing the plasma membrane capacitance of the target neurons in response to rapid changes in temperature rather than functioning by creating holes in the membrane or by stimulating specific temperature-sensitive ion channels [9]. Thus, infrared light-induced heating of the aqueous medium in the INS method enables optical stimulation of virtually any excitable cell without causing damage or requiring genetic modification. However, INS suffers from two main limitations. First, the wavelengths used are directly absorbed by water, meaning that INS can only be performed on cells very near the surface of a tissue. Second, because it is water itself that is heated in INS, all excitable cells in the region of illumination will be equally stimulated. Thus, INS lacks the cell-type selectivity that makes optogenetics such a powerful technique.

In this chapter we describe technical details for a method of neuronal photostimulation that employs functionalized gold nanoparticles (AuNPs) to sensitize neurons to visible light [10]. This approach offers many of the advantages of optogenetics and INS while avoiding limitations such as the need to transform cells genetically (optogenetics) and the strong absorption of light by water (INS). To this end, gold nanoparticles are coupled to an antibody or other biomolecule that specifically binds to a given antigen or receptor protein of the targeted neuron's plasma membrane. When these conjugated AuNPs are introduced to a tissue or a heterogeneous population of excitable cells, they bind only to cells expressing the antigen or receptor protein. Following delivery of these cell-targeted AuNPs, the tissue is illuminated with a wavelength that is well-absorbed by the AuNPs, which transduce this light into heat. This heat diffuses through the immediately surrounding external aqueous medium to the plasma membrane, and the increase in temperature increases the membrane's electrical capacitance. This change in the membrane capacitance (C) generates a depolarizing capacitive current which is proportional to dC/dt, where t is time. If this current is sufficient to drive the membrane potential to suprathreshold values, it activates (opens) nearby voltage-gated sodium channels (Navs) and thereby triggers an action potential (Fig. 1). A critical aspect of this capacitance change is that, to produce a significant current, it must be fast because the induced current is equal to $(V - V_S)dC/dt$, where V is the membrane potential and V_S is the surface potential difference of the membrane. The coupling between the AuNPs' absorbed light energy and the resulting

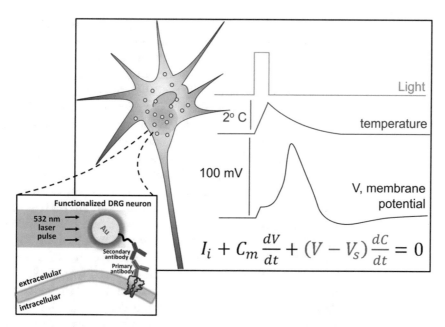

Fig. 1 The mechanism of optocapacitance that generates action potentials. The schematics show that a pulse of light, through the action of membrane-bound AuNPs (inset), causes a small (less than 2 °C) but rapid change in membrane temperature that transiently alters membrane capacitance. The differential equation represents the cell under zero current clamp conditions where I_i are the ionic currents, V is the membrane potential, V_s is the net surface membrane potential, and C_m is the membrane capacitance. Excitation occurs because the rapid capacitance change produces a depolarizing capacitive current (highlighted in red in the equation) sufficient to depolarize the membrane, opening sodium channels that initiate an action potential. (Inset) A cartoon depicting the linkage between a secondary-antibody-conjugated AuNP, a primary antibody, and a membrane target protein (in our experiments, either the P2X3 receptor or TRPV1)

change in the membrane capacitance has led us to term this light-induced process as "optocapacitance."

Optimization of the optocapacitance technique requires the tuning of three parameters. *First*, the light used to induce action potential generation must be of wavelength matched to the plasmon absorbance band exhibited by the size and structure of the AuNPs) used. In this regard it is also important to note that spherical AuNPs larger than 20 nm in diameter display substantial light scattering in addition to absorption, thus reducing the portion of the incident light energy that is converted into heat. *Second*, the AuNPs must reside in close proximity to the membrane of the targeted neuron, since heat diffusion through water is a delayed process. That is, with increasing distance of the cell-attached AuNP from the plasma membrane, the delayed nature of heat diffusion slows the rate of change of temperature sensed by the membrane, thus diminishing the capacitive current and resulting membrane depolarization. Treating the neurons with primary antibodies that are directed against a membrane protein of the intended neuron type, followed by treatment with AuNPs that are functionalized with a secondary antibody, can meet this need for AuNP proximity. That is, the primary antibody/secondary antibody linker that tethers the AuNP to the cell membrane is sufficiently short. Primary antibodies enable a broad selection of targets, and many primary antibodies are commercially available. Alternatively, the AuNPs can be attached to the cell with the use of an ion channel-binding toxin or membrane receptor ligand as the membrane-anchoring component. *Third*, the membrane depolarization must be fast enough to avoid accommodation of the sodium channels common in slow depolarization processes. To satisfy this requirement, sufficient light energy must be delivered to the AuNP within a brief (often 1 ms or less) period. Beyond providing an alternative to optogenetics that provides for selective stimulation of specific classes of excitable cells without requiring genetic modification, optocapacitance allows for stimulation of cells that are much deeper in tissues or surrounding media than can be achieved with INS. This can in principle be achieved by selection of an AuNP shape for which the plasmon absorption band is at wavelengths in the near-infrared window (700–900 nm) associated with relatively deep tissue penetration. Gold nanorods (AuNRs)) display a plasmonic absorption range from 550 to 2100 nm according to the ratio between their length and diameter.

The present text describes technical details of experiments we have conducted with dorsal root ganglion (DRG) neurons and hippocampal brain slices, using preparations of AuNPs that have been conjugated with differing cell-targeting agents. Based on our results [10], we predict that this method should be workable with many types of excitable cells, including neurons, endocrine cells and muscle cells. Importantly, data obtained using artificial lipid bilayers show that membrane proteins are not required for generation of a membrane capacitance change by the illuminated AuNPs. However, as described above, a given cell membrane protein selected for targeting by the functionalized AuNPs plays an essential role in anchoring the AuNP to the membrane via, e.g., a bridge that incorporates a primary antibody targeted to the membrane protein.

2   Materials

2.1   Gold Nanoparticles: Shape, Size, and Functionalization

The first important decision one must make when using targeted gold nanoparticles is the geometry and functionalization of gold nanoparticles. The size and geometry will affect the absorption and scattering properties of the particles, as well as their stability as a suspension of individual nanoparticles in aqueous medium, i.e., their avoidance of clumping and settling (precipitation). Increasing the particle size, in general, shifts the AuNP's plasmon absorbance band toward longer wavelengths, and AuNPs with a plasmon band peak in the near-infrared offer the advantage of relatively deep tissue penetration by the near-infrared light. Furthermore, increasing the AuNP size typically increases the particle's molar absorptivity, which can allow for lower light energy. However, increasing AuNP size also can dramatically increase light scattering, a disadvantage in applications requiring high spatial resolution of the photoactivating light to be delivered to a target tissue. Furthermore, large nanoparticles can gravitate out of solution within a few minutes, limiting their utility. Taking these facts into consideration, we have employed spherical AuNPs of 20 nm diameter and 532 nm light, a wavelength near the peak of the plasmon absorbance band for these particles. Whether or not settling (precipitation) in aqueous medium is problematic will depend on each particular application. A second crucial consideration is that, beyond the size/shape of the AuNP itself, the coating of the AuNP—that is, a polymeric coating applied by the manufacturer or by the investigator, and/or properties of the functionalizing molecules that enable cell-targeting of the AuNP—must allow stable suspensions of the AuNPs in physiological media such as regular Tyrode's or Ringer's solution. In our experience, bare citrate-stabilized gold nanoparticles immediately precipitate out of solution once they are placed into physiological media. We have found that functionalized nanoparticle preparations provided by Nanopartz, Inc. [Loveland, CO, USA/www.nanopartz.com] (which are coated with a proprietary polymer) are much more stable in physiological media. Furthermore, they can be purchased with a wide variety of attached functional groups including secondary antibodies and chemically reactive moieties. To maximize stability of the AuNPs, we arranged with Nanopartz for their preparation of AuNPs in the specific working solution (bath solution, see Table 1) used in our experiments, thereby avoiding the need to perform a solution exchange on the particles. We purchased AuNPs functionalized with either anti-IgG secondary antibodies or with streptavidin, and both performed well. Many other functionalization options are available, and the specific choice for a particular application will depend on one's choice of targeting molecule.

2.2 Targeting Molecules

Gold nanoparticles can be used without any functionalization to anchor them to the cell membrane. In this case, the AuNPs can simply be added to the solution to obtain the optocapacitance effect by their proximity to the membrane. However, if the solution is exchanged or if these nonfunctionalized particles diffuse away from the membrane the effect quickly disappears; therefore a method of anchoring the particles to the cell membrane is desirable in most cases. The anchoring can be made general to any cell membrane or specific to a particular cell type. The most broadly applicable choice of targeting molecule is to simply use primary antibodies against a neuronal membrane protein. We tested commercially available antibodies against two different targets (the

P2X3 receptor and TRPV1), and both were used successfully. To attach the antibodies to our nanoparticles, we used both native primary antibodies in conjunction with AuNPs functionalized with secondary antibodies, as well as biotinylated primary antibodies bound to streptavidin-coated AuNPs. As an alternative to antibody as the targeting component, we also tested a biotinylated scorpion toxin, Ts1 [11], which binds to voltage-gated sodium channels without pore blockade, in combination with streptavidin-functionalized nanoparticles; in this case, all cells with Na channels that bind the Ts1 toxin will be targeted. The fact that all tested combinations of targeting molecule and linkage functionality worked well supports the generality of the optocapacitance approach for multiple types of targeting molecule and linker structure. For many situations, the most convenient choice is likely to be treatment with primary antibody, followed by treatment with secondary-antibody-functionalized nanoparticles. This combination uses only materials than can be directly purchased and does not require additional chemistry to be performed.

2.3 Isolated Neurons and Hippocampal Slices

For our experiments with isolated neurons in vitro, we used neonatal rats (1–3 days old) as the source of dorsal root ganglia (DRG) neurons. We chose this age range as older animals' ganglia, in our experience, require more elaborate digestion procedures. For our experiments with hippocampal slices, 350-μm thick sagittal slices of hippocampus were cut from the brains of adult (65–102 days old) C57BL/6 mice. The slices were stored in artificial cerebrospinal fluid (aCSF, see Table 1) bubbled with carbogen (95% O_2/5% CO_2) and were used immediately after their preparation.

Table 1
Salt solutions utilized for this technique

	EBSS (mM)	Bath solution (mM)	Pipette solution (mM)	aCSF (mM)
NaCl	132.4	132	10	125
KCl	5.3	4	145	2.5
$MgCl_2$	–	1.2	4.5	1.5
$CaCl_2$	–	1.8	–	2.5
Hepes	10	10	10	–
$NaHCO_3$	–	–	–	26
NaH_2PO_4	1.01	–	–	1.25
Glucose	–	5.5	–	10
EGTA	–	–	9	–

All solutions were adjusted to a physiologic pH of 7.4

2.4 Voltage-Sensitive Fluorescent Dye

We monitored neuronal activity in hippocampal brain slices by staining the tissue with indocyanine green (ICG), typically about 5–10 min before positioning the slice in the chamber for optical recording. ICG is an FDA-approved infrared fluorescent dye that stains plasma membranes and changes its fluorescence intensity in response to changing trans-membrane potential. It has been shown to work well in rat isolated DRG neurons, hippocampal slices, and isolated cardiomyocytes [12], as well as in whole rabbit hearts [13]. This dye provides a noninvasive option to monitor cellular excitation as an alternative to traditional electrodes.

2.5 Experimental Setups

For cultures of DRG neurons, optical stimulation is performed on a Zeiss inverted microscope with a 40×, NA 0.6 objective using a 532 nm DPSS laser (Fig. 2). The microscope was originally equipped with fluorescence optics, and we used a dichroic mirror to deflect the 532 nm laser beam. The emission filter was removed to allow visualization of the laser spot through the eyepieces via leak through the dichroic mirror, since we were not attempting to record fluorescence data. The system includes a micromanipulator that holds the patch pipette and patch amplifier headstage, and an independent micromanipulator that positions a double-barreled microperfusion pipette. The 532 nm laser beam is gated by an acousto-optic modulator (AOM) and attenuated by neutral density filters prior to entering the back optical port of the microscope.

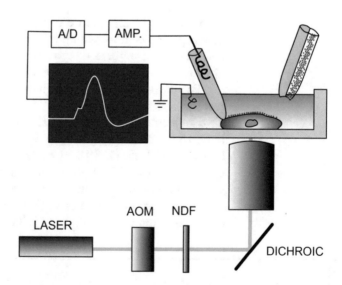

Fig. 2 Diagram showing the setup used for light stimulation of DRG neurons and electrical recording of the membrane voltage. Under current clamp, and with membrane-attached functionalized AuNPs that have previously been delivered though a nearby microperfusion pipette, a DRG neuron fires an action potential (schematically illustrated voltage trace) when presented with a laser flash of millisecond duration. *AOM* acousto-optic modulator, *NDF* neutral density filters, *AMP* patch amplifier, *A/D* analog-to-digital converter

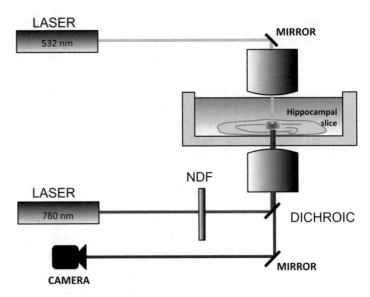

Fig. 3 Simplified schematic showing the setup used for optical stimulation and recording with hippocampal brain slices. The slice is pretreated with indocyanine green (ICG), and AuNPs are microinjected into the slice where the spots from a 532 nm laser (for AuNP photoexcitation) and a 780 nm laser (for ICG excitation) are positioned. Fluorescence emission by the ICG reports membrane voltage from cells of the tissue following presentation of a brief 532 nm flash (experimental data reported in Carvalho-de-Souza et al., 2015). *NDF* neutral density filters

Temperature changes close to the neuron under investigation for optocapacitance photostimulation are monitored by recording the resistance of a calibrated second patch pipette placed about 2 μm away from the neuronal membrane [14].

For hippocampal slices we have used a different inverted fluorescence microscope equipped with a high-speed camera (Fig. 3). Two objective lenses are used simultaneously, one above and one beneath the brain slice. The top objective (32×, NA 0.55) is used to focus the 532 nm DPSS laser that is pulsed under control of a TTL command pulse. The bottom objective (20×, NA 0.45), together with a dichroic mirror and an emission filter matched to the ICG emission spectrum, enables ICG photoexcitation from a 780 nm diode laser and the collection of ICG fluorescence images.

3 Methods

3.1 Experiments with DRG Neurons

Our experiments with DRG neurons have focused on the use of two types of AuNP conjugates. The first of these has involved conjugation of streptavidin-coated AuNPs with a biotinylated component that incorporates Ts1 toxin through click chemistry. Ts1 toxin is chemically synthesized and modified to contain an unnatural amino acid, propargylglycine (Pra), in a position that does not interfere with toxin biding affinity to Nav1.4 voltage-gated sodium channels. A biotin-PEG3-azide is joined to Ts1-Pra to produce the Ts1-biotin anchor that is attached to streptavidin-coated AuNPs [15]. The second type of conjugate has employed antibodies that target membrane proteins of DRG neurons. The ability of this type of conjugate to enable light-induced action potential firing by the DRG cell is illustrated in Fig. 4. For both approaches, neurons are dispersed from DRGs for in vitro photostimulation as described [10]. Briefly, dissected ganglia obtained from two rat pups (1–3 days old) are subjected to tryptic digestion in modified divalent cation-free Earle's Balanced Salt Solution, EBSS (see Table 1) for 15 min at 37 °C under gentle agitation. The suspension is then centrifuged, and the cellular material deposited at the bottom of

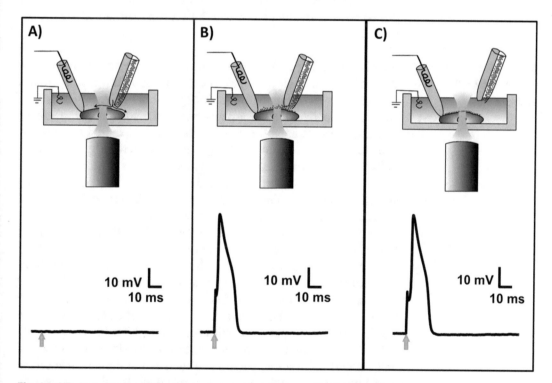

Fig. 4 AuNPs bound to the DRG cell membrane via a primary antibody/secondary antibody bridge establish DRG cell photosensitivity. Traces shown in the lower parts of the three panels are experimental data obtained from a single DRG cell upon presentation of a brief 532-nm laser pulse (green arrows). Diagrams in the upper parts of the panels schematically illustrate the prevailing experimental condition. At time **(a)**, the cell had been pretreated only with primary antibody against P2X3 receptors. In the absence of AuNP treatment, the laser flash did not alter membrane voltage. At the later time **(b)**, AuNPs functionalized with secondary antibody were being delivered to the cell from one barrel of a nearby microperfusion pipette and the laser flash elicited an action potential, indicating the attachment of AuNPs to the membrane by binding of the AuNP-joined secondary antibody to the already-attached primary antibody. At the latest time **(c)**, unbound AuNPs had been washed out by medium delivered from the second barrel of the microperfusion pipette. The similarity of the panel **c** voltage trace to that in panel **(b)** indicates the persistence of AuNP-mediated photosensitivity by AuNPs remaining bound to the membrane

the centrifuge tube is removed with a transfer pipette to a tube containing EBSS supplemented with 10% fetal calf serum, for mechanical dispersion with a set of fire-polished glass transfer pipettes of progressively decreasing tip diameter. The material is centrifuged again, the supernatant is removed, and the formed pellet is resuspended in culture medium [Dulbecco's Modified Eagle Medium (DMEM) supplemented with 5% calf serum, 100 µg/mL streptavidin, and 100 UI penicillin]. At this point in the preparation, the cells must be fully dispersed to enhance cell adhesion. Small aliquots of the cell suspension are seeded in 8–10 glass-bottom culture dishes (35 mm plastic dishes with 10 mm glass centers) that have been previously treated with poly-L-lysine. The dishes are placed for 15 min in a high-moisture 37 °C incubator (atmosphere: 95% O_2 , 5% CO_2) to allow cell adhesion. Next, the dishes are supplemented with ≈3 mL of culture medium, then placed back in the incubator where they are maintained for use for up to 7 days.

3.1.1 Labeling DRG Neurons with AuNP-Ts1 Conjugate

A dish containing a DRG cell culture is directly transferred to the stage of the microscope where the culture medium is flushed and replaced by a recording bath solution (see Table 1). A neuron is then chosen and patched. After a neuron is under current clamp and its excitability has been verified, a pair of pulses separated by 200 ms, consisting of a 1-ms depolarizing current injection by the pipette followed by a 1-ms light pulse, is applied every 3 s. The amplitude of the current injection is chosen so that virtually every pulse triggers an action potential. The laser power should be set to be just high enough for the focused laser spot to be seen at the eyepieces, which in our setup required about 10 mW. While running the paired pulses, the patched neuron is positioned at the location of the laser spot. Next, the perfusion double-barreled pipette containing both bath solution and AuNP-Ts1 is positioned about 10 µm from the neuron, and perfusion with bath solution is initiated. The laser power should now be increased to a level sufficient to excite the AuNP-Ts1-treated neurons. With our setup, 50 mW was generally sufficient to trigger action potentials, but the specific power required will depend on multiple parameters such as the numerical aperture of the microscope objective being used and will require some experimentation to determine. In our setup, the laser powers that reliably triggered action potentials in AuNP-Ts1-treated neurons had no effect on neurons in the absence of AuNPs. The perfusion with AuNP-Ts1 begins while the bath solution perfusion is shut off. At this point a supra-threshold depolarization is seen in response to every laser pulse. A good practice is to perfuse the AuNP-Ts1 for several minutes before stopping it. The perfusion with bath solution is then resumed to wash out the bulk, unbound AuNPs. The depolarizing voltage transient induced by the recurring 1-ms light pulse remains during the bath solution perfusion, indicating that AuNP-Ts1 conjugates are tightly attached to the neuronal membrane. Now the laser power and duration can be optimized as desired to reliably produce action potentials while avoiding excess depolarization by the laser pulse. Upon establishing the laser parameters (power and duration) that induce an action potential, typically no change in these parameters is needed for periods that range up to 50 min (longer periods were not tested).

3.1.2 Targeting AuNPs to DRG Neurons via Primary and Secondary Antibodies

We have found the following two-step procedure to be particularly well suited for joining AuNPs to DRG cells via a bridge that consists of a primary antibody directed against a membrane protein of the DRG cell, and secondary-antibody-coated AuNPs. The first of these steps is to treat the cells with primary antibody alone. We have chosen to use primary antibodies against TRPV1 and P2X3, known to be well-expressed in DRG neurons [16, 17]. The labeling of DRG neurons with primary antibodies against the protein antigen of choice can be made in culture medium by simply adding the antibodies to the culture dish. For this purpose, separate dishes containing the DRG cells in 2.5 mL of culture medium are supplemented with a 50-µL aliquot of each antibody at 0.8 mg/mL. Overnight incubation typically proves sufficient to label neurons with the primary antibodies. Immediately before the experiments, a culture dish containing the cells prelabeled with a given primary antibody is transferred to the stage of the microscope where the culture medium containing unbound antibodies is flushed several times and replaced by the recording bath solution. This washing step is critical to minimize the formation of AuNP clusters during the procedure's second step, which consists of delivering secondary-antibody-functionalized AuNPs to the cell under electrophysiological study. As in the case of delivery of AuNP-Ts1 to the cell (see earlier section), the preparation of AuNPs functionalized with secondary antibodies is delivered to the cell through a microperfusion pipette (Figs. 2 and 4). The secondary antibodies selectively bind to the primary antibodies, thus linking the AuNPs to the neurons, while any excess unbound AuNPs can be subsequently washed away with a perfusion of bath solution, as described above.

3.2 Using Targeted AuNPs in Mouse Hippocampal Slices

As described above, our experiments investigating the action of AuNP conjugates on hippocampal slices have employed treatment of the slice with ICG to enable optical recording of AuNP-mediated action potential generation. Hippocampal slices are prepared as described [10] and stored in aCSF (Table 1) bubbled with carbogen. The slices are used immediately after harvesting. Slices are stained for 5 min in 40 µM ICG dissolved in carbogen-bubbled aCSF. They are then extensively rinsed with fresh aCSF to remove

excess dye. For the hippocampal slices, we have used AuNPs conjugated to Ts1, identical to those used in the DRG neurons as
described above. Because the external surface of acute brain slices
consists mainly of dead cells, it is advantageous to deliver the AuNP
conjugate to the center of the slice. In addition, it is useful to code-
liver a fluorescent dye along with the AuNP-Ts1 to help locate the
injection site during the experiment. Any fluorescent dye that is
excited at 532 nm and does not significantly affect the tissue is suitable for this role; we chose to use tetramethylrhodamine-5-ma-
leimide (TMRM). Accordingly, immediately after ICG staining
and rinsing, approximately 50 nL of AuNP-Ts1 solution (containing 20 nM AuNP-Ts1 and 2 nM of TMRM) is microinjected into
the center of the slice. The slice is then mounted in the experimental setup (Fig. 3) and the 532 nm laser is focused onto the injection
site by visualizing fluorescence from the TMRM coinjected with
the nanoparticles. Experiments are performed by using continuous
780 nm illumination to visualize ICG fluorescence over time and
thus monitor electrical activity while 532 nm laser pulses stimulate
AuNP-mediated neuronal depolarization. All experiments with
brain slices are performed using carbogen-bubbled aCSF as the
recording medium.

3.3 Generality of the Optocapacitance Technique

Optocapacitance has the potential to be a universal technique for
stimulating excitable cells with light, since the basics of the mechanism that produce depolarization reside at the lipid membrane. By
anchoring AuNPs to specific membrane proteins through proper
choice of antibodies or other targeting molecules, it is possible
to place heat sources very close to the targeted cell membranes,
allowing for selective excitation of these cells while leaving other
cells too distant from the heat-generating particles to be effectively stimulated. The generality of the optocapacitance method
is exemplified by the appearance, within this past year, of multiple independent studies that have used the activation of gold-
nanoparticle-bound excitable cells by light to determine aspects of
cell physiology [18–20].

References

1. Arenkiel BR et al (2007) In vivo light-induced activation of neural circuitry in transgenic mice expressing channelrhodopsin-2. Neuron 54:205–218

2. Petreanu L, Huber D, Sobczyk A, Svoboda K (2007) Channelrhodopsin-2-assisted circuit mapping of long-range callosal projections. Nat Neurosci 10:663–668

3. Toni N et al (2008) Neurons born in the adult dentate gyrus form functional synapses with target cells. Nat Neurosci 11:901–907

4. Wang H et al (2007) High-speed mapping of synaptic connectivity using photostimulation in Channelrhodopsin-2 transgenic mice. Proc Natl Acad Sci U S A 104:8143–8148

5. Zhou XX, Pan M, Lin MZ (2015) Investigating neuronal function with optically controllable proteins. Front Mol Neurosci 8:37

6. Boyden ES, Zhang F, Bamberg E, Nagel G, Deisseroth K (2005) Millisecond-timescale, genetically targeted optical control of neural activity. Nat Neurosci 8:1263–1268

7. Deisseroth K (2015) Optogenetics: 10 years of microbial opsins in neuroscience. Nat Neurosci 18:1213–1225

8. Wells J et al (2005) Optical stimulation of neural tissue in vivo. Opt Lett 30:504–506

9. Shapiro MG, Homma K, Villarreal S, Richter C-P, Bezanilla F (2012) Infrared light excites cells by changing their electrical capacitance. Nat Commun 3:736

10. Carvalho-de-Souza JL et al (2015) Photosensitivity of neurons enabled by cell-targeted gold nanoparticles. Neuron 86: 207–217

11. Campos FV, Chanda B, Beirão PSL, Bezanilla F (2007) β-Scorpion toxin modifies gating transitions in all four voltage sensors of the Sodium Channel. J Gen Physiol 130:257–268

12. Treger JS, Priest MF, Iezzi R, Bezanilla F (2014) Real-time imaging of electrical signals with an infrared FDA-approved dye. Biophys J 107:L09–L12

13. Martišienė I et al (2016) Voltage-sensitive fluorescence of indocyanine green in the heart. Biophys J 110:723–732

14. Yao J, Liu B, Qin F (2009) Rapid temperature jump by infrared diode laser irradiation for patch-clamp studies. Biophys J 96:3611–3619

15. Dang B, Kubota T, Correa AM, Bezanilla F, Kent SBH (2014) Total chemical synthesis of biologically active fluorescent dye-Labeled Ts1 toxin. Angew Chem Int Ed Engl 53: 8970–8974

16. Hayes P et al (2000) Cloning and functional expression of a human orthologue of rat vanilloid receptor-1. Pain 88:205–215

17. Xiang Z, Bo X, Burnstock G (1998) Localization of ATP-gated P2X receptor immunoreactivity in rat sensory and sympathetic ganglia. Neurosci Lett 256:105–108

18. Eom K et al (2016) Synergistic combination of near-infrared irradiation and targeted gold nanoheaters for enhanced photothermal neural stimulation. Biomed Opt Express 7:1614–1625

19. Lavoie-Cardinal F, Salesse C, Bergeron É, Meunier M, De Koninck P (2016) Gold nanoparticle-assisted all optical localized stimulation and monitoring of Ca2+ signaling in neurons. Sci Rep 6:20619

20. Sanchez-Rodriguez SP et al (2016) Plasmonic activation of gold nanorods for remote stimulation of calcium signaling and protein expression in HEK 293T cells. Biotechnol Bioeng 113:2228–2240

Chapter 2

Nanoparticle-Assisted Localized Optical Stimulation of Cultured Neurons

Flavie Lavoie-Cardinal, Charleen Salesse, Pierre-Luc Ayotte-Nadeau, and Paul De Koninck

Abstract

Nanoparticle-assisted localized optical stimulation (NALOS) of cultured neurons is an all-optical method that allows subcellular light stimulation to investigate localized signaling in neurons. The stimulation and monitoring of localized Ca^{2+} signaling in neurons takes advantage of plasmonic excitation of gold nanoparticles (AuNPs) with infrared light. In this chapter we describe how NALOS, through its effects localized to region smaller than 10 μm^2, may be a useful complement to other light-dependent methods for controlling neuronal activity and cell signaling. We demonstrate that this technique can be applied to cultured hippocampal neurons using commercially available bare AuNPs.

Key words Photostimulation, Gold nanoparticles, Calcium imaging, Calcium signaling, Hippocampal neurons

1 Introduction

Light-induced stimulation of whole neurons or neuronal networks with optogenetic tools is a well-established and powerful approach to study circuit function [1]. With optogenetics, it is now possible to genetically control the subtype of neurons to be stimulated, to induce or inhibit neuronal activity, generally with a reduced level of invasiveness compared to electrical stimulation [2]. However, photostimulation at the nanometer to micrometer scale on neuronal membranes to investigate local signaling processes remains a challenge, primarily because light-gated channels have very small conductances. Neurotransmitter uncaging, with ultraviolet (UV) or two-photon excitation, has been used to precisely drive single synapse activation but is not optimal to activate a dendritic response at a nonsynaptic site [3]. It was shown recently that infrared (IR) short laser pulses can be used to excite neurons by affecting membrane capacitance, likely via a local heating process [4, 5]. This technique relies on the absorption of IR light by water and is

Fidel Santamaria and Xomalin G. Peralta (eds.), *Use of Nanoparticles in Neuroscience*, Neuromethods, vol. 135,
https://doi.org/10.1007/978-1-4939-7584-6_2, © Springer Science+Business Media, LLC 2018

generally performed at wavelengths between 1400 and 2200 nm to better suit the absorption spectrum of water. It is very powerful for the stimulation of larger fields but sub-cellular stimulations are not possible since all cells in the excitation volume will be stimulated through the heating of the aqueous medium [4]. In this range of wavelengths, the coupling with standard UV to visible optics that can be used for confocal imaging is very difficult and limits considerably the confinement of the excitation spot to a diffraction limited region.

Taking advantage of their property to absorb visible and IR light, gold nanoparticles (AuNPs) and nanorods (AuNRs) have been used recently to generate light-induced heat for neuronal activation or inhibition [6–9]. Excitation of AuNPs with femtosecond laser pulses; near-field enhancement, or plasma generation was previously associated with nanocavitation formation and gene transfection [10–13]. Irradiation of 20 nm AuNPs by a 1 ms light pulse at 532 nm at a repetition rate of 40 Hz could be used to trigger action potentials (APs) [7], while prolonged IR illumination was associated with neuronal inhibition [8]. Through functionalization of the AuNPs, these techniques offer a higher specificity compared to the previously described IR water-mediated stimulation techniques. However, stimulation is limited to whole cells and requires a large number of AuNPs or AuNRs bound onto the cell [7, 9].

In this chapter we introduce a technique called Nanoparticle-Assisted Localized Optical Stimulation (NALOS) for localized, diffraction unlimited stimulation of subcellular regions on cultured hippocampal neurons by the mean of plasmonic excitation of a single AuNP [14]. For this technique, IR light at 800 nm is used for off-resonance plasmonic excitation of a targeted single AuNP. The evoked increased cellular activity is monitored by Ca^{2+} optical imaging and/or electrophysiology. The monitored subcellular increase in free Ca^{2+} concentration after the illumination of a single AuNP is confined to a few μm^2. The elevation in free Ca^{2+} inside the neurons can also be widespread by raising laser intensity or the number of photostimulated NPs on the targeted cells. NALOS applied on large regions of interest (ROIs), containing more than one AuNP on the cell body, can be used to drive APs. On the other hand, NALOS applied on a dendritic compartment containing a single AuNP drives local and small currents capable of inducing localized Ca^{2+} responses. This technique does not require any genetic modification of the targeted cells, and binding of bare AuNPs onto the neuronal plasma membrane through passive sedimentation is sufficient. It is also possible to functionalize the AuNPs to target specific receptors on the membrane of the neurons [7, 9, 14, 15]. If the target receptor is cotransfected in the neuron (e.g., with a genetically encoded Ca^{2+} indicator), functionalization can help in restricting the binding of the AuNP to the fluorescent neurons. However, stability of the functionalization during storage can be limiting.

NALOS can also be used to study local signaling downstream of Ca^{2+}, by monitoring for example the spatial and temporal dynamics of the Ca^{2+}/calmodulin-dependent kinase (CaMKII) tagged with GFP in combination with Ca^{2+} imaging using a red calcium sensitive protein such as RGCaMP1.07 or NES-jRGECO1 [14]. In this configuration, two-color confocal imaging combined with infrared optical stimulation, without measurable cross talk between the three channels, is possible on most commercial confocal microscopes equipped with a two-photon excitation path. Electrophysiological recordings can also be performed simultaneously in order to characterize the membrane currents evoked by the photostimulation.

2   Materials

2.1   Gold Nanoparticles

The choice of the AuNP particle size is of great importance for NALOS, since it influences strongly the required laser intensity and the observed effect on the targeted neurons. Increasing the AuNP size generally increases light absorbance and induces a red shift of the plasmonic absorbance band, thereby increasing the amplitude of the cell response and the risks of membrane damages. On the other hand, larger AuNPs allow considerable reduction of the infrared light intensity required for photostimulation.

When suspending AuNPs in physiological media, their aggregation has to be monitored. It is important to recognize aggregates and only illuminate single AuNPs, for better reproducibility. In our hands, smaller nonfunctionalized AuNPs tend to agglomerate more easily in the culture media used compared to larger ones; we therefore favor 100 nm AuNPs.

An important consideration in the choice of the NPs is their absorption at the available wavelengths and intensities of the confocal system in hand, which needs to be compatible with the imaging application. Increasing the AuNP diameter leads to a shift to longer wavelengths of the surface plasmon absorbance (from 520 to 580 nm for 20–80 nm AuNPs) and therefore decreases the required laser intensity for an IR off-resonance plasmonic excitation [16, 17].

If the AuNPs need to be localized precisely on the cell membrane using their light scattering properties, as it is the case for NALOS, their size will also influence the required laser intensity and the choice of the wavelength [16]. With increasing AuNP size the scattering spectrum is shifted to longer wavelengths and the scattering efficiency at 600 nm is also increased [16]. We could detect easily 100 nm single AuNPs with their scattering at 633 nm, while when using 20 nm AuNPs we could detect only aggregates at this wavelength.

In our experience, AuNPs of 100 nm diameter are very good for localized photostimulation with IR light as well as their localization with red light (633 nm). This size of AuNPs enables the use of IR laser intensity well below (~1 order of magnitude) that necessary for two-photon excitation [18]. Meanwhile, 488 and 543 nm excitation (intensity below 5 μW in the back aperture of the objective) does not generate sufficient absorption by the AuNPs to cause a detectable cellular response.

2.2 Rat Hippocampal Cultures and Transfections

For the primary rat hippocampal cultures, neonatal rats of 0–2 days old were used. The dissociated cells were plated on 12 mm poly-D-lysine-coated glass coverslips at a density of 4225 cells/mm². Growth media consisted of Neurobasal and B27 (50:1), supplemented with penicillin/streptomycin (50 U/mL; 50 μg/mL) and 0.5 mM Glutamax (Invitrogen). Fetal bovine serum (2%; Hyclone) was added at time of plating. After 5 days, half of the media was replaced by media without serum and with Ara-C (5 μM; Sigma-Aldrich) to limit proliferation of nonneuronal cells. Twice a week thereon, half of the growth media was replaced with media free of serum and Ara-C. The neurons were transfected at 11–14 days in vitro and imaged 1 day after transfection.

For the transfection half of the culture media was taken out of the wells and mixed in new wells with the same amount of fresh complete growth media. The transfection was performed in the remaining media (500 μL in the case of a 24 well plate and 12 mm coverslips). For 12 mm coverslips, 0.5 μg DNA and 2 μL Lipofectamine 2000 (Invitrogen) were mixed in 100 μL Neurobasal media and added dropwise on the cells. After 3–5 h, the coverslips were transferred in the new wells containing 50/50 fresh/old growth media.

For Ca^{2+} imaging, plasmids encoding GCaMP6s [19], NES-jRGECO1 [20], or RCaMP1.07 (gift of J. Nakai) were used. Alternatively, transfections can be substituted by the incubation with cell-permeable Ca^{2+}-sensitive dyes, such as Fluo4-AM (2 μM, 30 min incubation, 30 min wash). For electrophysiology measurements, the Ca^{2+}-sensitive dye can also be loaded through the patch pipette (Fluo4, 337 nM).

2.3 Experimental Setup

NALOS can be performed on any commercial or custom-built confocal microscope. In our experiments, we used a confocal laser scanning microscope (Zeiss, LSM 510) equipped with an 80 MHz pulsed femtosecond infrared laser (Ti:Sapphire, Chameleon, Coherent). To perform photostimulation and Ca^{2+} imaging, the microscope should be equipped with both IR and visible excitation paths.

Fluorescence excitation of GCaMP6s or mGFP was performed with a 488 nm continuous wave (cw) laser (intensity in the back aperture of the objective below 5 μW) and detected using an

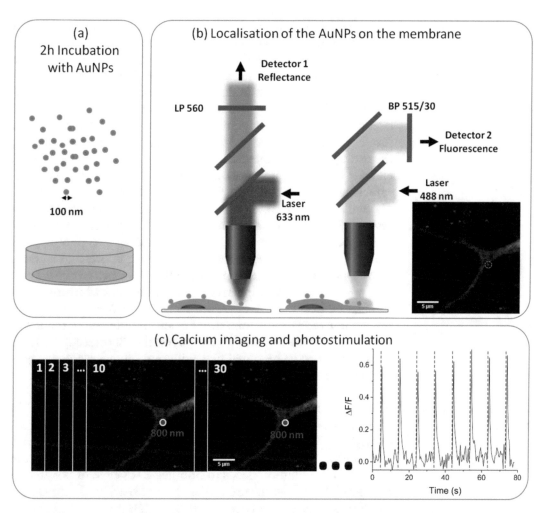

Fig. 1 Experimental procedure. (**a**) Incubation with 100 nm AuNPs for 2 h directly in the well plates. (**b**) Localization of the AuNPs on the cellular membrane in reflectance at 633 nm (red) combined with fluorescence microscopy of neurons transfected with GCaMP6s (green). (**c**) Time lapse imaging and photostimulation of the ROI marked with the white circle and DF/F signal quantified in the same ROI (dotted lines indicate times of photostimulation)

IR-blocking 500–550 nm filter (Fig. 1b). Fluorescence excitation of RCaMP1.07, NES-jRGECO, or mCherry was done with 543 nm cw laser (intensity in the back aperture of the objective below 5 μW) using an IR-blocking 565–615 nm filter. Visualization of the AuNP reflection was done with a 633 nm cw laser (laser power in the back aperture of the objective below 500 nW) and a 560 nm long pass fluorescence filter (Fig. 1b). Live-cell imaging was performed with an open perfusion chamber and a 40× 0.8NA water dipping objective.

For the photostimulation, we used a Ti-Sapphire laser at 800 nm; its power in the back aperture of the objective was tuned between 0.5 and 2 mW (0.27–1.02 MW/cm^2). For NALOS, a single scan over a ROI was performed with a pixel dwell time varying between 1.3 and 6.4 μs. In order to monitor simultaneously the AuNP positions and the cellular activity with Ca^{2+} imaging, the microscope should be equipped with at least two independent detection paths.

It is important to consider that the plasmon absorbance maximum of AuNPs is in the visible range of the spectrum and that the laser intensity used for confocal imaging should be below the threshold necessary for plasmonic excitation. In our experience, the excitation of the fluorescent proteins in the visible range with a cw laser at the imaging intensity of 0.2–0.9 kW/cm² had no measurable effect on the neurons in contact with AuNPs.

Since a femtosecond pulsed IR laser at 800 nm is used for off-resonance plasmonic excitation of the AuNPs, it needs to be tuned to intensities that are below the ones needed for two-photon excitation of the chosen fluorescent marker. Green fluorescent markers show typically a good two-photon absorption between 800 and 920 nm and therefore the laser power in the back aperture of the stimulation beam was kept below 2 mW (1.02 MW/cm²), which was on our setup about one order of magnitude below the lower threshold of two-photon excitation of the fluorescent markers used. This need to be tuned depending on the microscope objective and the fluorescent probes used.

2.4 Optical Imaging Solution

Neurons were imaged in an open chamber in HEPES-aCSF (in mM: 102 NaCl, 5 KCl, 10 HEPES, 1.2 CaCl₂, 1 MgCl₂, 10 C₆H₁₂O₆, pH 7.4, 230–240 mOsm) using a perfusion system with temperature adjusted to 29–30 °C. The osmolarity of the HEPES-aCSF solution needs to be adjusted as close as possible to the osmolarity of the culture media. In our experiments, the osmolarity was adjusted between 230 and 240 mOsm (similar to Neurobasal). For electrophysiology experiments, 0.5 μM tetrodoxin (TTX) was added to the imaging solution.

2.5 Intracellular Solutions for Patch Clamp Recordings

For patch clamp recordings in current clamp mode, the solution consisted of (in mM) 111 KMeSO₃, 10 diNa-phosphocreatine, 10 HEPES, 2.5 MgCl₂, 2 ATP-Tris, 0.4 GTP-TRIS, at pH 7.25–7.35; the osmolarity was adjusted to 210–220 mOsm/L. For patch clamp recordings in voltage clamp mode, the solution consisted of (in mM) 96 CsMeSO₃, 20 CsCl, 10 diNa-phosphocreatine, 10 HEPES, 2.5 MgCl₂, 0.6 EGTA, 4 ATP-TRIS, 0.4 mMGTP-TRIS, at pH 7.25–7.35; the osmolarity was adjusted 210–220 mOsm/L.

3 Methods

3.1 Incubation with AuNPs

Prior to imaging, incubation for 2 h of the AuNPs onto the neurons inside the incubator allowed for sufficient sedimentation (Fig. 1a). Fifty microliters of the stock solution (50 μg/mL, 100 nm diameter, Nanopartz, A11-100-CIT-100, 5.71×10^9 AuNPs/mL, $\varepsilon = 1.1 \times 10^{11}$ M^{-1} cm^{-1}, potential zeta -46 mV) was mixed with 50 μL of Neurobasal culture media and added dropwise to the cells in 1 mL growth media. The optimal ratio between the mixed AuNPs-Neurobasal solution and the growth media in the well was 1:10. This incubation protocol yielded between 20 and 40 detectable AuNPs in an area of 40×40 μm. Longer incubation times or higher AuNP concentration increased the probability of AuNP aggregation. After incubation, no washing step was necessary; the coverslip with neurons in contact with AuNPs could be transferred directly to the imaging chamber.

3.2 Nanoparticle-Assisted Local Optical Stimulation

First, a fluorescence image of a neuron expressing a Ca^{2+} indicator and a reflection image of the AuNPs were taken and overlaid (Fig. 1b). AuNPs that appear in contact with a dendrite of the transfected neuron were then identified. To avoid AuNP aggregates, only AuNPs showing a diffraction limited point spread function were chosen for photostimulation. Since the scattering of the AuNPs is shifted to longer wavelengths upon agglomeration, the intensity of the reflection PSFs of AuNPs at 543 and 633 nm can also be compared. Indeed, the monomers should show a stronger reflection at 543 nm, while clusters will reflect more at 633 nm.

The photostimulation step was performed using the ROI-bleaching module of the LSM-510 Zeiss microscope. ROIs of $1–10$ μm^2 were chosen around an AuNP for the IR-photostimulation (Fig. 1c). A single scan over a ROI was sufficient to induce a measurable local Ca^{2+} response. Increasing the number of scans over a ROI (up to ten iterations) can be used to increase the probability of successful stimulation at low laser intensity.

Using the ROI-bleaching module of the LSM510 microscope, a series of ten images were taken prior to stimulation. The IR-illumination was then performed for a given number of iterations (generally between 1 and 5) on the ROI and imaging was carried on for the desired lapse of time. Repeated photostimulation on the same AuNP was possible, while monitoring its movement via the reflection images (Fig. 1c).

Laser power below 2 mW at the back aperture of the objective was required to induce a reversible widespread Ca^{2+} response, while less than 1 mW was preferable to induce a local and reversible response [14]. At this power, no cross talk between stimulation and fluorescence excitation was observed. Higher illumination intensities lead to irreversible Ca^{2+} response suggesting damages of the plasma membrane.

3.3 Calcium Imaging

Optical imaging of Ca^{2+} can be performed alone or in combination with a second fluorescent signal, such as CaMKII tagged with a fluorescent protein (Fig. 1c). The scanning speed and size of the imaged region were adjusted to obtain sufficient temporal resolution of the Ca^{2+} signals, according to the kinetics of the chosen indicator. The pixel size was adjusted to fulfill the Nyquist criterion and the pixel dwell time varied between 1.3 and 6.4 µs. Ca^{2+} indicators with slow kinetics and bright fluorescence, such as GCaMP6s, NES-JRGECO1, and RCaMP1.07, were chosen to obtain a sufficient signal to noise ratio even for a local Ca^{2+} response confined to a few µm². Imaging rates between 2 and 4 Hz were sufficient.

The fluorescence intensity over a chosen ROI was averaged and subtracted with the averaged background intensity. The obtained fluorescence intensity (F) was corrected by the mean average intensity in this region in the ten imaged frames before stimulation (F_0) to obtain the corrected $\Delta F / F_0$ traces (Fig. 1c).

3.4 Electrophysiological Recordings

As described in Fig. 2a, we identified a fluorescent neuron that had dendrites with overlapping AuNPs, using fluorescence imaging and reflectance. For recordings, a glass pipette with a resistance of 3.5–5 MΩ was filled with intracellular solution. Alternatively, a Ca^{2+} sensitive dye such as Fluo4 could be added to the intracellular solution for recordings on nontransfected cells. Low positive pressure was applied in the pipette while in the bath, and was released upon touching a neuronal cell body, which could be seen by a slight deflection of the membrane, visually identified using a 40× water-immersion objective and infrared differential interference contrast (Zeiss Axioscop FS2). Negative pressure was then applied until resistance reached at least 1 GΩ. Pipette capacitance was automatically adjusted by the software before breaking the seal and achieving whole-cell configuration. Cell health, monitored throughout the experiment, was established by stable holding

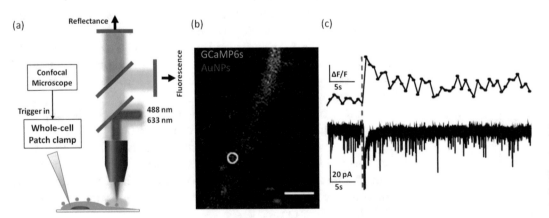

Fig. 2 Simultaneous electrophysiological and optical recordings. (**a**) Simultaneous electrophysiological and optical recording combined with photostimulation. The electrophysiological recording was triggered by the LSM510 microscope to ensure synchronization of both traces. (**b**) Ca^{2+} imaging (GCaMP6s fluorescence, green) after AuNP localization (reflectance, red); scale bar 5 µm. (**c**) Ca^{2+} signal (upper trace) is correlated to the corresponding electrophysiological trace (lower trace)

current and series resistance: $a \pm 20\%$ change was tolerated. Data acquisition (filtered at 1.8–2 kHz and digitized at 10 kHz) was performed using a Multiclamp 700B amplifier and the Clampex 10.6 software (Molecular Devices). The electrophysiological recordings were triggered by the LSM510 confocal microscope for synchronization with the Ca^{2+} imaging (Fig. 2a, b). Data were analyzed using Clampfit 10.6 (Molecular Devices) and Igor Pro (WaveMetrics).

References

1. Fenno L, Yizhar O, Deisseroth K (2011) The development and application of optogenetics. Annu Rev Neurosci 34:389–412

2. Deisseroth K (2011) Optogenetics. Nat Methods 8:26–29

3. Matsuzaki M, Ellis-Davies GC, Nemoto T et al (2001) Dendritic spine geometry is critical for AMPA receptor expression in hippocampal CA1 pyramidal neurons. Nat Neurosci 4:1086–1092

4. Shapiro MG, Homma K, Villarreal S et al (2012) Infrared light excites cells by changing their electrical capacitance. Nat Commun 3:736

5. Wells J, Kao C, Mariappan K et al (2005) Optical stimulation of neural tissue in vivo. Opt Lett 30:504–506

6. Eom K, Kim J, Choi JM et al (2014) Enhanced infrared neural stimulation using localized surface plasmon resonance of gold nanorods. Small 10:3853–3857

7. Carvalho-de-Souza JL, Treger JS, Dang B et al (2015) Photosensitivity of neurons enabled by cell-targeted gold nanoparticles. Neuron 86:207–217

8. Yoo S, Hong S, Choi Y et al (2014) Photothermal inhibition of neural activity with near-infrared-sensitive nanotransducers. ACS Nano 8:8040–8049

9. Nakatsuji H, Numata T, Morone N et al (2015) Thermosensitive ion channel activation in single neuronal cells by using surface-engineered plasmonic nanoparticles. Angew Chem Int Ed 54:11725–11729

10. Schomaker M, Heinemann D, Kalies S et al (2015) Characterization of nanoparticle mediated laser transfection by femtosecond laser pulses for applications in molecular medicine. J Nanobiotechnol 13:10

11. Baumgart J, Humbert L, Boulais É et al (2012) Off-resonance plasmonic enhanced femtosecond laser optoporation and transfection of cancer cells. Biomaterials 33:2345–2350

12. Boulais É, Lachaine R, Meunier M (2012) Plasma mediated off-resonance plasmonic enhanced ultrafast laser-induced nanocavitation. Nano Lett 12:4763–4769

13. Schomaker M, Killian D, Willenbrock S et al (2015) Biophysical effects in off-resonant gold nanoparticle mediated (GNOME) laser transfection of cell lines, primary- and stem cells using fs laser pulses. J Biophotonics 8:646–658

14. Lavoie-Cardinal F, Salesse C, Bergeron É et al (2016) Gold nanoparticle-assisted all optical localized stimulation and monitoring of Ca2+ signaling in neurons. Sci Rep 6:20619

15. Bergeron E, Boutopoulos C, Martel R et al (2015) Cell-specific optoporation with near-infrared ultrafast laser and functionalized gold nanoparticles. Nanoscale 7:17836–17847

16. Jain PK, Lee KS, El-Sayed IH et al (2006) Calculated absorption and scattering properties of gold nanoparticles of different size, shape, and composition: applications in biological imaging and biomedicine. J Phys Chem B 110:7238–7248

17. Link S, El-Sayed MA (2010) Shape and size dependence of radiative, non-radiative and photothermal properties of gold nanocrystals. Int Rev Phys Chem 19:409–453

18. Mütze J, Iyer V, MacKlin JJ et al (2012) Excitation spectra and brightness optimization of two-photon excited probes. Biophys J 102:934–944

19. Chen T-W, Wardill TJ, Sun Y et al (2013) Ultrasensitive fluorescent proteins for imaging neuronal activity. Nature 499:295–300

20. Dana H, Mohar B, Sun Y et al (2016) Sensitive red protein calcium indicators for imaging neural activity. Elife 5:pii: e12727

Chapter 3

Stimulation of Primary Auditory Neurons Mediated by Near-Infrared Excitation of Gold Nanorods

Chiara Paviolo, Karina Needham, William G.A. Brown, Jiawey Yong, and Paul R. Stoddart

Abstract

Neural stimulation plays an important role in achieving therapeutic interactions with both the central and peripheral nervous systems, and forms the basis of neural prostheses such as cochlear implants and pacemakers. The interactions are commonly based on electrical stimulation delivered by microelectrodes, which are implanted in the vicinity of the target tissue. Electrical stimulation has limited selectivity, as the resolution of the stimulus is degraded by current spread. Moreover, the implantation may cause injury to the target tissue and the host inflammatory response can reduce stability. In order to improve the performance of neural interfaces, optical stimulation is attracting increasing attention, based on techniques such as optogenetics, photoactive molecules, and infrared neural stimulation. However, optical techniques at present tend to rely on visible or infrared wavelengths that have a limited penetration in tissue. Alternatively, the near-infrared region, corresponding to the therapeutic window in tissue, can be accessed by two-photon stimulation with relatively expensive light sources, or by the introduction of extrinsic light absorbers. For the latter approach, gold nanorods have recently been shown to provide efficient stimulation in a range of cell types, when exposed to near infrared light. Given the wide range of surface functionalizations and relatively low toxicity of gold, this approach is expected to draw increasing interest in the field of neural stimulation. This Method describes experimental procedures that have been used to prepare primary auditory neurons with gold nanorods for near-infrared excitation. It is anticipated that these procedures could be adapted to a range of related neural stimulation studies.

Key words Neural stimulation, Optical stimulation, Gold nanorods, Auditory neurons

1 Introduction

For decades, electrical stimulation has been the gold standard technique for depolarising neuronal cells and generating action potentials. This approach relies on the use of a stimulating electrode that injects electrical charge and raises the membrane potential above the stimulation threshold of the target nerves [1]. Despite many advantages, such as the delivery of current in a controllable and quantifiable manner, this way of stimulating neurons has revealed several limitations. Firstly, the stimulating electrode involves

Fidel Santamaria and Xomalin G. Peralta (eds.), *Use of Nanoparticles in Neuroscience*, Neuromethods, vol. 135, https://doi.org/10.1007/978-1-4939-7584-6_3, © Springer Science+Business Media, LLC 2018

physical contact with the surrounding tissues, which can lead to toxicity or tissue damage. Intimate contact also tends to generate an inflammatory response, which changes the electrical characteristics of the electrode-tissue interface over time [2]. Secondly, the spatial selectivity of stimulation is largely dependent on the dimension of the electrode and on the spread of the electrical current in the tissue. Thirdly, electrical artefacts coexisting with the biological signal are often generated during the recording [3].

To overcome these limitations, researchers have now demonstrated that mid-infrared light (typically ca. 1450–2200 nm) could potentially replace (or complement) the traditional electrical methods. Indeed, infrared light delivery into nerve tissue via optical fibers has provided an increased spatial selectivity of stimulation without the need for any genetic modification or intimate physical contact with the tissue [4]. This technique (also called infrared neural stimulation, or INS) relies on the absorption of mid-infrared light by the water within and surrounding the target tissue [5]. The absorbed light is then converted into thermal energy, which is believed to lead to the processes responsible for the stimulation, such as (1) the generation of transient changes in cell membrane capacitance [6] and/or (2) the activation of temperature-sensitive ion channels [7]. However, in many in vivo models, the presence of extraneous absorbing layers of tissue has limited the efficiency of stimulation, resulting in the use of high laser powers to reach the action potential threshold [8]. Moreover, recently developed thermal models suggest that repetitive stimulation at multiple sites could lead to cumulative heating. This may affect the target nerves, potentially causing damage to the tissues in practical applications [9].

In response to these challenges, researchers have recognized that the use of extrinsic absorbers might improve the penetration depth and produce more localized heating effects in the nerves [10]. Ideally, the photoabsorbing chromophores should have optical absorption in the therapeutic window of biological tissues where transmission is highest (near-infrared, NIR, typically ca. 650–1400 nm [11]). Other desirable features include biocompatibility, large absorption cross section, highly efficient photon-to-heat conversion and preferably imaging capabilities. Gold nanorods (Au NRs) have been proven to satisfy all of these requirements, given their efficient photothermal conversion in the NIR [12], their strong absorption cross sections at the localized surface plasmon resonance [13], the wide range of surface functionalization options that can be used to optimize their biocompatibility [14], and their use as imaging agents [15]. Indeed, laser-exposed Au NRs were shown to increase the neurite length and generate calcium transients in NG108-15 neuronal cells [16, 17]. These findings strongly suggested that it may be possible to stimulate cells using the heat generated by the excitation of Au NRs at their plasmon resonance. Therefore cell membrane depolarization was recorded in vitro after NIR laser illumination of Au NRs in primary auditory neurons [18] and in vivo in rat sciatic nerves [19]. More recently, photothermal heat generated by NIR exposure of Au NRs was shown to activate transient receptor potential vallinoid 1 (TRPV1) channels and induce calcium influx in HEK293T cells [20]. Neural stimulation mediated by near-infrared excitation of Au NRs is indeed a promising strategy, and holds great potential to underpin the next generation of neural prostheses.

In the present paper, protocols for pulsed NIR laser stimulation of primary spiral ganglion neurons cultured with silica-coated Au NRs are outlined. Cell electrical activity was measured with the whole-cell patch-clamp recording technique. NIR irradiation was based on a fiber-coupled laser diode (LD) that allows safe operation and repeatable alignment. Heating effects associated with the NR excitation around the neurons were measured by an open-pipette method [21]. The Au NR sample preparation and laser irradiation methods can be further extended to different particle shapes and neuronal cultures, providing that the specific synthesis and culture protocols are known, respectively.

2   Materials

Use ultrapure water to prepare all of the solutions (ultrapure water is obtained by purifying deionized water to a sensitivity of 18 MΩ cm at 25 °C). All reagents can be prepared at room temperature and stored at 4 °C (unless otherwise indicated). Diligently follow all waste disposal regulations when discarding waste materials.

2.1   Preparation of Solutions

1. Poly-L-ornithine solution: decant poly-L-ornithine (500 µg/ml, Sigma Aldrich) into 5 ml volumes and store it in a sterile manner until use.

2. Laminin/Poly-L-ornithine solution: add 50 µl of mouse laminin (0.01 mg/ml, Invitrogen) to 5 ml of poly-L-ornithine solution.

3. Minimal Essential Medium (MEM): mix 490 ml of minimal essential medium (Invitrogen) with 5 ml of non-essential amino acids (10 mM, Invitrogen) and 5 ml of penicillin-streptomycin (Invitrogen). Complete medium is discarded within 2 weeks of preparation.

4. Neurobasal medium (NBM): 47.5 ml neurobasal A, 0.5 ml N2 supplement (Invitrogen), 1 ml B27 supplement (Invitrogen), 0.5 ml L-glutamine (Invitrogen), and 0.5 ml penicillin-streptomycin. Complete medium is discarded within 2 weeks of preparation.

5. Intracellular (micropipette) solution: 115 mM K-gluconate, 7 mM KCl, 10 mM HEPES, 0.05 mM EGTA, 2 mM Na_2ATP, 2 mM MgATP, 0.5 mM Na_2GTP (adjust to pH 7.3 with KOH; adjust to 295 mOsmol/kg with sucrose). To sterilize the solution, use a filter (0.2 μm) and then divide the final volume into 200 μl aliquots. These can be stored at −20 °C until the day of recording.

6. Extracellular (bath) solution: 137 mM NaCl, 5 mM KCl, 2 mM $CaCl_2$, 1 mM $MgCl_2$, 10 mM HEPES, 10 mM glucose (adjust to pH 7.4 with NaOH; adjust to 300–310 mOsmol/kg with sucrose). This solution is generally made on the day of recording.

2.2 Equipment

1. Darkfield light scattering and microspectroscopy: darkfield images of spiral ganglion neurons were taken with an inverted microscope (Eclipse Ti-U, Nikon) using an oil immersion type darkfield condenser (NA 0.9–1.45) and a 60× oil immersion iris diaphragm objective (PlanFluor, Nikon, NA 0.5–1.25) (*see* **Note 1**). The microscope was equipped with a colour camera (Digital Sight DS-Vi1 or DS-Ri1, Nikon) whose shutter was controlled by the NIS-Elements computer software (Nikon). The scattering spectra were acquired using an Isoplane SCT-320 imaging spectrometer coupled with a ProEM CCD camera (Princeton Instruments). No filter or polarization optics were inserted during the spectral acquisitions. The acquisitions were synchronized to the computer via the LightField software version 4.0. NR spectra were acquired over a wavelength range from 400 to 1000 nm using a 150 g/mm diffraction grating.

2. Electrophysiology system: Dodt gradient contrast images of spiral ganglion neurons were taken with an upright microscope (AxioExaminer D1, Carl Zeiss Pty Ltd.) using a 40× water-immersion objective lens and monochrome CCD camera (Spot RT SE18, Diagnostic Instruments). Patch-clamp recordings were made via a multi-clamp data acquisition unit (Digidata 1440A, Molecular Devices) and amplifier (Multiclamp 700B amplifier, Molecular Devices), controlled by AxoGraph software (AxoGraph Scientific). Micropipettes filled with intracellular solution were prepared from borosilicate glass (1.0 mm outside diameter, 0.58 mm inside diameter, with filament) with a laser-based electrode puller (P-2000, Sutter Instrument) to yield a tip resistance of 2–8 MΩ, and positioned via a micromanipulator (MPC-325, Sutter Instrument). The gravity-fed perfusion system was equipped with aspirator bottle, pinch valve, polyethylene tubing, inline heater for rapid heating of the solution, and a peristaltic pump to remove spent solution by suction. Recordings were digitized at 50 kHz and low pass filtered (Bessel) at 10 kHz. Series resistance was routinely compensated

online (up to 70%), and, in current clamp, micropipette capacitance neutralization and bridge balance were utilized to compensate errors due to series resistance. Corrections for liquid junction potential (12.8 mV) were made offline using JPCalcW (Prof P. H. Barry, Sydney).

3. Laser stimulation setup: a class IIIB single mode fiber-coupled laser diode (OptoTech) with output wavelength λ of 780 ± 5 nm and variable peak power (max ≥100 mW) was used throughout the experiments (*see* **Note 2**). The optical fiber (SMF-28, Corning) had a numerical aperture (NA) of 0.14, and a diameter of 125 μm (8.2 μm core) (*see* **Note 3**). Laser power was measured by a handheld laser power meter integrated with a silicon sensor (LaserCheck, Coherent Scientific) (*see* **Note 4**). Laser pulses were controlled manually by a function generator (TDS1002, Tektronix) triggered by the patch clamp data acquisition unit. The fiber was held by a micropositioner. See Fig. 1 for the schematic of the experimental setup.

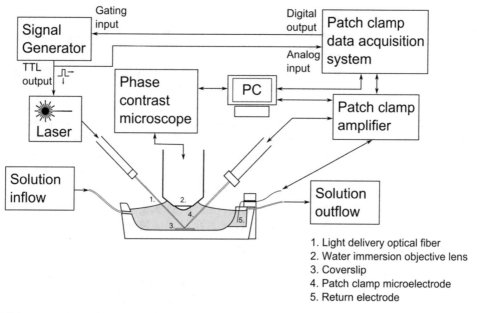

1. Light delivery optical fiber
2. Water immersion objective lens
3. Coverslip
4. Patch clamp microelectrode
5. Return electrode

Fig. 1 Schematic of the experimental setup (not to scale) (*adapted from* [23] *with permission from JoVE*)

3 Methods

3.1 Culture of Spiral Ganglion Neurons

All of the equipment used in this section has to be sterile or autoclaved unless otherwise specified.

1. Use forceps to transfer small round (e.g., 10 mm diameter) glass coverslips into individual wells of a 4-ring 35 mm petri dish or 4-well plate. Pipette 150 μl of laminin/poly-L-ornithine

solution onto the top surface of the coverslip and place in a cell culture incubator (37 °C/5% CO_2) for up to 48 h. Ensure that the coverslips remain immersed in solution (*see* **Note 5**).

2. Prepare dissociated spiral ganglion neurons from post-natal day 4–7 rat pups. As previously described, rat pups are anaesthetised on ice, decapitated and the temporal bones removed and placed into chilled MEM. Bullae are isolated, the modiolus extracted and placed into chilled NBM. Tissue is then dissociated using both enzymatic (0.025% trypsin and 0.001% DNase I in NBM for 10 min at 37 °C, arrested with 1 ml fetal bovine serum) and mechanical techniques (gentle trituration through 16–23 gauge needles) [22].

3. Aspirate any remaining laminin/poly-L-ornithine solution from the coverslips and wash briefly with NBM.

4. Add 150–200 µl of the dissociated spiral ganglion neuron suspension to the coverslips and place into a cell culture incubator (37 °C, 10% CO_2). Up to 20 coverslips can be prepared from an average litter of eight rat pups.

5. Four hours after plating neurons, aspirate the solution to remove cell debris and replace with 150–200 µl warmed fresh NBM.

6. Return coverslips to the incubator until required for electrophysiological recordings (*see* **Note 6**).

3.2 Nanoparticle Preparation for Cell Culture

1. Measure the initial optical density (OD) of concentrated Au NR solution (e.g. silica-coated Au NR) suspended in sterile ultrapure water at 780 nm via UV-Vis spectroscopy, by recording the absorption values from 300 to 1000 nm with a resolution of 0.5–2 nm (*see* **Notes 7** and **8**).

2. Prior to incubation, dilute the NR solution in NBM to an OD value (at 780 nm) of ≈ 0.18 (*see* **Note 9**).

3. Fifteen to seventeen hours prior to electrophysiological recordings, replace the standard NBM in the cultured spiral ganglion neurons with 150–200 µl of NBM-NR solution (*see* **Note 10**). Ensure that the coverslips remain immersed in solution.

4. Incubate overnight to allow NR internalization (*see* **Note 11**).

3.3 Darkfield Imaging and Micro-spectroscopy

1. Rinse the neuronal cultures once with warmed PBS to remove excess Au NRs from the NBM.

2. Fix the samples in ice-cold 100% methanol for 10 min.

3. Rinse the cells with PBS (3 × 5 min) to remove excess methanol.

4. Mount the coverslips on standard glass slides using a mounting agent (Aquatex, Merck) with a refractive index of 1.5 (*see* **Note 12**).

5. Collect darkfield images (inset in Fig. 2, *see* **Notes 1** and **13**).

6. Collect nanorod spectra until a satisfactory signal-to-noise ratio is obtained (typically a 100 ms acquisition time from a ≈5 μm slit width).

7. Correct the raw signals by subtracting the dark currents measured with the camera shutter closed. Use the following equation:

$$S(\lambda) = \frac{I_{\text{raw}}(\lambda) - I_{\text{dark}}(\lambda)}{I_{\text{white}}(\lambda) - I_{\text{dark}}(\lambda)} \tag{1}$$

where $S(\lambda)$ is the calculated scattering intensity for each wavelength, $I_{\text{raw}}(\lambda)$ is the signal intensity of a given pixel, $I_{\text{dark}}(\lambda)$ is the signal intensity of the dark current, and $I_{\text{white}}(\lambda)$ is the signal intensity of the halogen lamp.

8. Further correct the signal by subtracting the average normalized control cell background spectrum ($S_{\text{cell}}(\lambda)$) (Fig. 2, *see* **Note 14**):

$$S'(\lambda) = S(\lambda) - S_{\text{cell}}(\lambda) \tag{2}$$

3.4 Laser Stimulation and In Vitro Electrophysiology

1. Fill the appropriate container of the perfusion system with extracellular solution and adjust the flow rate to 1–2 ml/min.

2. Transfer the cultured neurons to the recording chamber of the microscope.

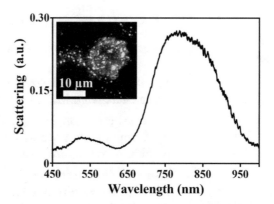

Fig. 2 Representative average scattering spectrum of silica-coated Au NRs in spiral ganglion neurons. The inset shows the corresponding dark-field image. The red dotted line indicates the line of spectral acquisition from which the spectrum was acquired (*reproduced from* [18] *with permission from Wiley-VCH*)

3. Flow in the system warmed extracellular solution to remove excess Au NRs from the cell culture medium.

4. Using a 40× water-immersion objective lens, visually locate the spiral ganglion neurons within the culture with the microscope in phase-contrast mode (*see* **Note 15**).

5. Use the micropositioner to move the output fiber until the tip is close to the target neuron in both the horizontal and vertical planes (*see* **Note 16**).

6. Once the fiber is in position, move it out by a known amount to enable straightforward positioning of the micropipette for patch clamp recording.

7. Fill the micropipette with intracellular solution and securely fit into place on the headstage of the amplifier.

8. Using tubing attached to the side of the microelectrode holder, apply a small amount of positive pressure to prevent clogging of the micropipette.

9. Using a micromanipulator, move the micropipette into position just above the target neuron. Refer to Brown et al. for the protocol on whole cell electrophysiology recordings [23] (*see* **Note 17**).

10. When a neural recording is achieved, move the optical fiber back into position next to the neuron (Fig. 3a).

11. While recording electrophysiological data in either current or voltage clamp configurations, run the NIR laser at 90 mW with pulse lengths ranging from 25 μs to 50 ms and energies of ≈0.002–5 mJ per pulse (*see* **Notes 18–20**).

12. Repeat steps 3–10 to record action potentials from different neurons. See Fig. 3b, c for representative results on optically evoked membrane currents and potentials, respectively.

3.5 Local Temperature Measurements

All of the procedures below were adapted from Yao et al. [21].

Calibration procedure (to be repeated for each micropipette):

1. Fill the appropriate container of the perfusion system with extracellular solution and adjust the flow rate to 1–2 ml/min.

2. Place a thermocouple (e.g., K type thermocouple, #409-4908, RS Components Pty Ltd.) in the solution to provide an independent temperature measurement for the calibration process.

3. Use an inline heater (e.g., Warner SH-27B) or equivalent to raise the solution temperature to approximately 40 °C as measured by the thermocouple.

4. Fill the recording micropipette with intracellular solution, attach it to the headstage and apply a positive pressure (i.e., steps 7 and 8 of Method 3.4).

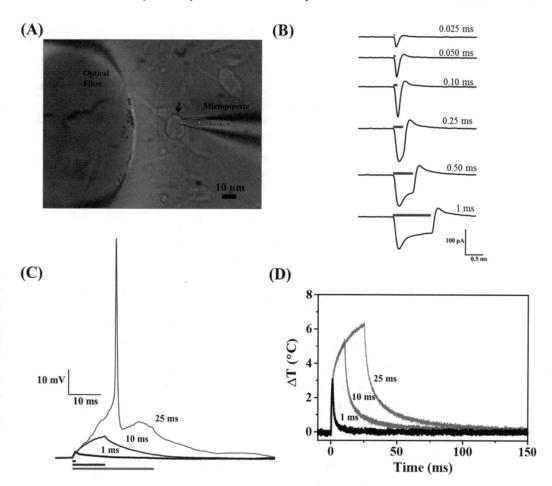

Fig. 3 Optical stimulation data. (**a**) Phase contrast micrograph showing the position of the optical fiber and microelectrode relative to the target cell. (**b**) Voltage-clamp data of an auditory neuron with Au NRs in response to laser pulses of different duration (from top to bottom: 0.025, 0.05, 0.1, 0.25, 0.5, 1 ms). (**c**) Current-clamped recordings of primary auditory neurons cultured with silica-coated Au NRs showing subthreshold membrane potentials and an action potential, evoked in response to 1, 10, and 25 ms laser pulses, respectively. All neurons were held at −73 mV. Bars indicate the onset and duration of laser pulses. (**d**) Representative temperature measurements for 1, 10, and 25 ms laser pulse durations. *Figure adapted from* [18] *with permission from Wiley-VCH*

5. Immerse the tip of the micropipette in the heated extracellular solution as close as possible to the tip of the thermocouple (*see* **Note 21**).

6. Apply small voltage pulses (e.g., 5 mV pulses of 10 ms duration) to the micropipette at a rate of at least 20 Hz.

7. Average the current response over 1–1.5 s. This value is representative of the micropipette resistance (R_p), related to the applied voltage (V) and the measured current (I) by Ohm's law:

$$R_p = \frac{V}{I} \qquad (3)$$

8. Turn off the heater and repeatedly measure the micropipette resistance as the temperature returns to room temperature. For each resistance measurement also record the corresponding thermocouple reading.

9. Calculate the relationship between the current amplitude (i.e., the resistance of the micropipette) and the temperature (T) using the following equation [21]:

$$T = \frac{1}{\dfrac{1}{T_0} - \dfrac{R}{E_a} \ln\left(\dfrac{I}{I_0}\right)} \qquad (4)$$

where R is the gas constant, T_0 and I_0 is the corresponding current measurement at room temperature, T and I are the temperature and current measurements at higher temperatures, and E_a is the Arrhenius activation energy.

10. Plot $\ln(I/I_0)$ against $(1/T_0 - 1/T)$ (Arrhenius plot) and fit the data points with a linear equation.

11. To calibrate the micropipette, derive E_a from the slope of the linear fit.

Temperature Measurement

1. Move the calibrated micropipette to a safe position to avoid accidental breakage.

2. Follow steps 4–6 from Method 3.4.

3. Position the micropipette as close as possible to the target cell without touching it (*see* **Note 22**).

4. Move the optical fiber back into position next to the target cell.

5. Apply laser illumination while measuring the current through the micropipette.

6. Calculate the change in temperature due to laser illumination by inserting the measured current into Eq. 4 along with the value of E_a obtained during calibration (see Fig. 3d for representative results).

7. Determine the mean temperature change from at least three neurons per pulse setting.

4 Notes

1. To prevent direct light from entering the objective, the numerical aperture of the objective must be inferior to the lower limit of the numerical aperture of the darkfield condenser.

2. For maximum efficiency, the laser wavelength should match the longitudinal localized surface plasmon resonance peak of the NRs.

3. The fiber tip should have a high quality surface (i.e., the tip should be perpendicular to the fiber axis and appear flat upon visual inspection). Cleave it prior to the experiments using standard techniques. Choose the appropriate methods based on the diameter of the fiber.

4. It is recommended to check the laser power every time the light delivery fiber is cleaved or a significant adjustment is made to the laser (e.g. transportation from one lab to another).

5. Laminin/Poly-L-ornithine solution is used to improve cell attachment on glass.

6. Dissociated spiral ganglion neuron cultures can be used for electrophysiological experiments 4 h after dissociation, but recordings are most reliably obtained from neurons cultured for 24–72 h after dissociation. This time should be taken into account during data acquisition. Replenish NBM every 24–48 h to avoid dehydration.

7. Au NRs can be synthesized by a number of recipes [24], or purchased from commercial vendors. Different coating materials change the surface charge [25] and the stability of the particles [26]. For example, due to the inert chemistry of the silica shell, silica-coated Au NRs can be suspended in a variety of polar organic solvents without forming aggregates [24].

8. The UV-Vis spectrum of Au NRs should exhibit two resonance bands, referring to the oscillations of the free electrons along the two axes of the rod (i.e. longitudinal and transverse axes). This is due to the two possible orientations of the rods with respect to the electric field of incident light. The electron oscillation along the transverse direction induces a weak absorption band in the visible region, while oscillations along the longitudinal direction produce a strong absorption band in the NIR. These two localized surface plasmon resonance bands are the characteristic spectral feature of Au NRs [27].

9. If needed, Au NRs can be centrifuged without altering their colloidal stability for 20 min at $7800 \times g$. Au NR solution can be prepared in advance and stored at 4 °C. Do not freeze.

10. Cytotoxicity caused by the nanoparticles varies with the coating material [26]. In previously published results, silica-coated

Au NRs showed a good preservation of the normal neuronal functions in terms of proliferation and cell membrane integrity [16].

11. In the cell culture environment, it is well known that unspecific adsorption of proteins on the nanoparticle surface might form the so-called protein corona [14]. It has been shown that this organic shell facilitates nanoparticle internalization [28].

12. Choose the mounting media carefully. To enhance the clarity of the images, it is important to match the refractive index of the mounting media as closely as possible to the refractive index of the imaging objective's immersion medium.

13. Darkfield imaging of gold nanoparticles relies on the colourful light scattered from the gold surface. The wavelength of the scattered light highly depends on the nanoparticle shape and should match the localized surface plasmon resonance peaks (inset in Fig. 2).

14. The sharpness and the position of the longitudinal plasmon peak provides information on Au NR aggregation and coupling [29]. Broadening of the peak in the cell culture environment might be an indication of particle agglomeration (Fig. 2). Protein coronas can also influence colloidal stability and produce local refractive index changes, thus affecting the position of the peak.

15. A typical spiral ganglion neuron is phase-bright, round, with a diameter of approximately 15–20 μm, and a prominent nucleus (Fig. 3a).

16. Accurate knowledge of positional variations may be required to resolve possible differences between stimulation processes. Refer to [23] for more details on fiber positioning relative to the target neuron.

17. Whole-cell patch clamp recordings were made at room temperature (≈ 21 °C). Coverslips were replaced after $\approx$3–4 h in the recording chamber.

18. Setting the repetition rate of laser pulses to 1 Hz or less may be useful for initial experiments, since it will minimize the effects of the total power delivered to the cells.

19. To eliminate the direct electric field of the laser as a possible stimulus for the cells, check the stimulatory effect of the NIR excitation in the absence of Au NRs.

20. The efficiency of stimulation is highly dependent on Au NR aggregation (i.e. on the shift of the longitudinal localized surface plasmon resonance peak). Water is a weak chromophore in the NIR (absorption coefficient of ≈ 0.005 mm^{-1} [30]), thus exhibiting poor photon-to-heat conversion compared to aqueous NRs.

21. The tip of the micropipette should be positioned as close as possible to the tip of the thermocouple to minimise the uncertainty due to spatial variations in temperature.

22. Position the micropipette as close as possible to the cell, to obtain a representative measurement of the local temperature at the cell surface. Note that contact with the cell may cause the pipette tip to become fouled (i.e. partially blocked by debris, thereby changing its resistance). If this is the case, a new pipette and calibration would be required.

References

1. Hodgkin AL, Huxley AF (1952) A quantitative description of membrane current and its application to conduction and excitation in nerve. J Physiol Lond 117(4):500–544

2. Newbold C, Richardson R, Millard R, Seligman P, Cowan R, Shepherd R (2011) Electrical stimulation causes rapid changes in electrode impedance of cell-covered electrodes. J Neural Eng 8(3):036029–036029. https://doi.org/10.1088/1741-2560/8/3/036029

3. Wells JD, Cayce JM, Mahadevan-Jansen A, Konrad PE, Jansen ED (2011) Infrared nerve stimulation: a novel therapeutic laser modality. In: Welch AJ, van Gemert MJC (eds) Optical-thermal response of laser-irradiated tissue. Springer, Netherlands, pp 915–939. https://doi.org/10.1007/978-90-481-8831-4_24

4. Richter CP, Matic AI, Wells JD, Jansen ED, Walsh JT (2011) Neural stimulation with optical radiation. Laser Photonics Rev 5(1):68–80. https://doi.org/10.1002/lpor.200900044

5. Roggan A, Friebel M, Dörschel K, Hahn A, Müller G (1999) Optical properties of circulating human blood in the wavelength range 400-2500 nm. J Biomed Opt 4(1):36–46. https://doi.org/10.1117/1.429919

6. Shapiro MG, Homma K, Villarreal S, Richter C-P, Bezanilla F (2012) Infrared light excites cells by changing their electrical capacitance. Nat Commun 3:736. https://doi.org/10.1038/ncomms1742

7. Albert ES, Bec JM, Desmadryl G, Chekroud K, Travo C, Gaboyard S, Bardin F, Marc I, Dumas M, Lenaers G, Hamel C, Muller A, Chabbert C (2012) TRPV4 channels mediate the infrared laser-evoked response in sensory neurons. J Neurophysiol 107(12):3227–3234. https://doi.org/10.1152/jn.00424.2011

8. Thompson AC, Wade SA, Brown WGA, Stoddart PR (2012) Modeling of light absorption in tissue during infrared neural stimulation. J Biomed Opt 17(7):075002–075002. https://doi.org/10.1117/1.jbo.17.7.075002

9. Thompson AC, Wade SA, Cadusch PJ, Brown WG, Stoddart PR (2013) Modeling of the temporal effects of heating during infrared neural stimulation. J Biomed Opt 18(3):035004

10. Migliori B, Di Ventra M, Kristan W (2012) Photoactivation of neurons by laser-generated local heating. AIP Adv 2(3):032154. https://doi.org/10.1063/1.4748955

11. Weissleder R (2001) A clearer vision for *in vivo* imaging. Nat Biotechnol 19(4):316–317

12. Huang X, Jain PK, El-Sayed IH, El-Sayed MA (2006) Determination of the minimum temperature required for selective photothermal destruction of cancer cells with the use of immunotargeted gold nanoparticles. Photochem Photobiol 82(2):412–417. https://doi.org/10.1562/2005-12-14-ra-754

13. Jain PK, Lee KS, El-Sayed IH, El-Sayed MA (2006) Calculated absorption and scattering properties of gold nanoparticles of different size, shape, and composition: applications in biological imaging and biomedicine. J Phys Chem B 110(14):7238–7248. https://doi.org/10.1021/jp057170o

14. Dykman LA, Khlebtsov NG (2014) Uptake of engineered gold nanoparticles into mammalian cells. Chem Rev 114(2):1258–1288. https://doi.org/10.1021/cr300441a

15. Choi WI, Sahu A, Kim YH, Tae G (2011) Photothermal cancer therapy and imaging based on gold nanorods. Ann Biomed Eng 40(2):534–546. https://doi.org/10.1007/s10439-011-0388-0

16. Paviolo C, Haycock JW, Yong J, Yu A, Stoddart PR, McArthur SL (2013) Laser exposure of gold nanorods can increase neuronal cell outgrowth. Biotechnol Bioeng 110(8):2277–2291. https://doi.org/10.1002/bit.24889

17. Paviolo C, Haycock JW, Cadusch PJ, McArthur SL, Stoddart PR (2014) Laser exposure of gold nanorods can induce intracellular calcium transients. J Biophotonics 7(10):761–765. https://doi.org/10.1002/jbio.201300043

18. Yong J, Needham K, Brown WGA, Nayagam BA, McArthur SL, Yu A, Stoddart PR (2014) Gold-nanorod-assisted near-infrared stimulation of primary auditory neurons. Adv Healthc Mater 3(11):1862–1868. https://doi.org/10.1002/adhm.201400027

19. Eom K, Kim J, Choi JM, Kang T, Chang JW, Byun KM, Jun SB, Kim SJ (2014) Enhanced infrared neural stimulation using localized surface plasmon resonance of gold nanorods. Small 10(19):3853–3857. https://doi.org/10.1002/smll.201400599

20. Nakatsuji H, Numata T, Morone N, Kaneko S, Mori Y, Imahori H, Murakami T (2015) Thermosensitive ion channel activation in single neuronal cells by using surface-engineered plasmonic nanoparticles. Angew Chem Int Ed 54(40):11725–11729. https://doi.org/10.1002/anie.201505534

21. Yao J, Liu B, Qin F (2009) Rapid temperature jump by infrared diode laser irradiation for patch-clamp studies. Biophys J 96(9):3611–3619. https://doi.org/10.1016/j.bpj.2009.02.016

22. Needham K, Nayagam BA, Minter RL, O'Leary SJ (2012) Combined application of brain-derived neurotrophic factor and neurotrophin-3 and its impact on spiral ganglion neuron firing properties and hyperpolarization-activated currents. Hear Res 291(1–2):1–14. https://doi.org/10.1016/j.heares.2012.07.002

23. Brown WGA, Needham K, Nayagam BA, Stoddart PR (2013) Whole cell patch clamp for investigating the mechanisms of infrared neural stimulation. JoVE (77). https://doi.org/10.3791/50444

24. Pérez-Juste J, Pastoriza-Santos I, Liz-Marzán LM, Mulvaney P (2005) Gold nanorods: synthesis, characterization and applications. Coord Chem Rev 249(17–18):1870–1901. https://doi.org/10.1016/j.ccr.2005.01.030

25. Hauck TS, Ghazani AA, Chan WCW (2008) Assessing the effect of surface chemistry on gold nanorod uptake, toxicity, and gene expression in mammalian cells. Small 4(1):153–159. https://doi.org/10.1002/smll.200700217

26. Hu X, Gao X (2011) Multilayer coating of gold nanorods for combined stability and biocompatibility. Phys Chem Chem Phys 13(21):10028–10035. https://doi.org/10.1039/c0cp02434a

27. Chen HY, Shao L, Li Q, Wang J (2013) Gold nanorods and their plasmonic properties. Chem Soc Rev 42(7):2679–2724. https://doi.org/10.1039/C2CS35367A

28. Walkey CD, Olsen JB, Song F, Liu R, Guo H, Olsen DWH, Cohen Y, Emili A, Chan WCW (2014) Protein corona fingerprinting predicts the cellular interaction of gold and silver nanoparticles. ACS Nano 8(3):2439–2455. https://doi.org/10.1021/nn406018q

29. Funston AM, Novo C, Davis TJ, Mulvaney P (2009) Plasmon coupling of gold nanorods at short distances and in different geometries. Nano Lett 9(4):1651–1658. https://doi.org/10.1021/nl900034v

30. Hale GM, Querry MR (1973) Optical constants of water in the 200-nm to 200-μm wavelength region. Appl Opt 12(3):555–563. https://doi.org/10.1364/AO.12.000555

Chapter 4

Nanoparticle Preparation for Magnetothermal Genetic Stimulation in Cell Culture and in the Brain of Live Rodents

Idoia Castellanos-Rubio, Rahul Munshi, Shahnaz Qadri, and Arnd Pralle

Abstract

Remote deep brain activation of specific neurons to evoke functional behavior in unrestrained subjects allows studies of how brain circuitry controls behavior. The local heating of superparamagnetic nanoparticles in alternating magnetic fields can be coupled to temperature sensitive ion channels to modulate neuronal activity in cell culture and in the behaving, awake animal. Application of synthesized ferrite nanoparticles requires pacification of their surface to achieve biocompatibility. In addition, it is often necessary to target the particles to a specific cell region or protein, which requires further conjugation with ligands or antibodies. This chapter discusses the advantages of various nanoparticle surface coating strategies.

Key words Magnetogenetics, Superparamagnetic nanoparticles, Surface pacification, Targeting of nanoparticles, Magnetic hyperthermia

1 Introduction

Magnetic nanoparticles (MNPs) based on iron oxide are studied by an increasing number of researchers as they offer new modalities to physically interface and modulate biological processes. Three effects have been used: local MNP heating in alternating magnetic fields to trigger temperature sensitive proteins [1–3]; dipole–dipole interaction in magnetic field gradients to aggregated MNP tethered surface receptors [4]; and similarly mechanical force application to individual proteins [5]. All these applications require that the MNP are targeted to specifics proteins which necessitates a surface coat containing specific ligands.

Most commonly used MNPs are composed of spinel ferrites MFe_2O_4 (where M = Mn, Fe or Co) and synthesized from the thermal decomposition of organometallic precursors at high temperatures, because they provide high quality nanocrystals which exhibiting the best magnetic properties for heat generation under alternating magnetic fields or force application in gradient fields

Fidel Santamaria and Xomalin G. Peralta (eds.), *Use of Nanoparticles in Neuroscience*, Neuromethods, vol. 135, https://doi.org/10.1007/978-1-4939-7584-6_4, © Springer Science+Business Media, LLC 2018

[6–8]. However, the resulting NPs are capped with hydrophobic ligands, such as oleic acid, making them dispersible only in organic solvents. Thus, any use of these materials in biomedical applications requires surface modification with hydrophilic and biocompatible molecules. Especially for the use of MNPs in vivo a proper surface modification it is of utmost importance due to their tendency to self-associate through dipolar interactions, which could provoke either blood thrombi or substantially change the magnetic response [9, 10]. In the last two decades several water transfer procedures have been developed, which can essentially be grouped into two main strategies [11, 12]: (1) Ligand exchange and (2) Encapsulation.

Ligand exchange consists of the removal of the native capping molecules and replacing them with bifunctional hydrophilic ligands, in which one moiety directly interacts with the nanocrystal surface while the other moiety affords the water solubility. On the contrary, the encapsulation strategy preserves the native capping agents (i.e., oleic acid) of the hydrophobic nanocrystals in order to interdigitate between them the hydrophobic long chains of amphiphilic polymers or phospholipids. The interdigitation is driven via hydrophobic interactions, while the hydrophilic moieties of the polymer form the surface of the micelle and provide water solubility.

Encapsulation can be achieved using polymers [13], or by inert inorganic materials, such as gold [5], or silica [14]. The encapsulation strategy presents two important advantages: First, the polymer coating does not affect, a priori, the surface of the inorganic material (metal, alloy, or oxide); therefore, the physicochemical properties of the core are less disturbed than in ligand-exchange procedures [15]. Second, as the hydrophobic interactions between the native capping molecules and the polymer chains are nonspecific, the same polymer can be used to coat a wide variety of nanocrystals [16]. However, encapsulation carries may generate empty micelles without enclosed nanocrystals. These empty micelles are difficult to remove and only ultracentrifugation has succeeded [17]. Also, protocols need to be optimized to avoid encapsulating more than one nanocrystal per micelle. In addition, polymer coating tends to significantly increase the hydrodynamic size of the nanoparticles, which limits their use in applications requiring small probes.

The amphiphilic polymers that have been used more widely for the encapsulation consist of hydrophobic side chains and a backbone of maleic anhydride rings, which acts as anchor for further functionalization or bioconjugation providing excellent possibilities for multifunctionality [13]. It is possible to tune the polymer by premodifying it before the water transfer process, because amine-containing molecules react spontaneously with maleic anhydride moieties to form an amide bond [18]. In most cases, the prefunctionalization reaction requires moderate heating. It should be performed in nonpolar solvents because the hydrophobic side chains destabilize the polymer in water. Alternatively, a modification of the polymer shell can be achieved in aqueous solution after the encapsulation of the hydrophobic nanoparticles. Post-modification can be easily attained by reacting primary amine molecules with carboxyl groups at the polymeric coating via EDC/NHS chemistry [19], which requires milder reaction conditions and is more suitable for some biomolecules like proteins or antibodies [20, 21].

For example, Polyethyleneglycol (PEG) molecules can be incorporated in the amphiphilic polymer by both premodification and postmodification. The hydrophilic, disordered PEG is known to increase the colloidal stability of NPs in salt-containing solutions and to reduce opsonization and nonspecific interactions with biomolecules and cells [22, 23]. Consequently, PEGylation is usually considered an initial functionalization step; however, in some cases it becomes unnecessary and is omitted for simplicity. Recently, several zwitterion moieties have been used as an alternative to PEG to gain colloidal stability minimizing the hydrodynamic size of the NPs [24]. Additionally, the insertion of fluorophore molecules in the shell of the MNPs has shown to be very useful not only as a marker but also as a temperature [1] or pH [25] probes.

Clearly, the biological fate of MNPs is highly affected by the NPs' surface chemistry. Varying the nature of the coating greatly influences several important biological parameters such as cellular uptake, NP localization inside cells, toxicity, circulation, biodistribution, and protein corona [26, 27]. The design of the surface of NPs represents a key step toward the optimal bioperformance of the MNPs. In order to accomplish specific applications nanoparticles need to be anchored onto specific cell or tissue sites. Specific targeting can be achieved by attaching specific recognition units, i.e., peptides, proteins, vitamins, antibodies, etc., on the NPs' surface [28, 29]. This stage is remarkably complex because it requires making the NP system resistant to the changes in pH, salt concentrations, protein loads as well as mechanical forces encounter when interacting with cells in culture or within an animal.

The affinity of streptavidin (SA) or NeutrAvidin (NA) for biotin is one of the strongest noncovalent intermolecular interactions known ($K_D \approx 10^{-15}$ M). Furthermore, biotin is a very small molecule (MW = 244 Da), which can easily be attached to proteins and antibodies without perturbing their function, may be expressed genetically [30], and is biocompatible as it appears naturally as vitamin B. Hence, the MNP functionalized with a biotin binding protein, such SA or NA, become a versatile tool capable to bind any biotinylated protein, antibody or cell. NeutrAvidin (NA) was engineered from SA by removing surface charges which induced nonspecific binding, and is now preferred over SA. NA and SA are tetrameric and capable of binding up to four biotins, which may cause aggregation of the MNP and the target proteins and should be avoided.

2   Materials

Prepare all solutions using analytical-grade reagents and ultrapure water (R>18 MΩ cm at 25 °C). Prepare and store all reagents at room temperature, unless indicated otherwise. Diligently follow all waste disposal regulations when disposing of waste materials.

- Poly(isobutylene-alt-maleic anhydride), PMA, (Mw = 6000 Da, Sigma #531278).

- Dodecylamine (98%, Sigma #D222208).

- Tetrahydrofuran, THF, (≥99.9%, Aldrich #186562).

- Anhydrous chloroform (EMD Millipore DriSolv® #CX1057-6).

- *N*-Hydroxysulfosuccinimide, Sulfo-NHS, (Thermo Scientific #24510). Store at 4 °C.

- 1-Ethyl-3-(3-dimethylaminopropyl)carbodiimide hydrochloride, EDC, (Thermo Scientific #22980). Store at −20 °C in a desiccator.

- NeutrAvidin, NA, Biotin binding protein (Thermo Scientific #31000). Store at 4 °C.

- Alexa Fluor 647 Succinimidyl Ester (Thermo Scientific #A20006). Store at −20 °C in a desiccator. Protect from light.

- Dimethyl sulfoxide, DMSO (Sigma #D8418).

- Hydroxylamine (Thermo Scientific #JG123590A).

- EZ-Link Sulfo-NHS-LC-Biotin (Thermo Scientific #21335). Store at −20 °C in a desiccator.

- Anti-A2B5 antibody (abcam #ab53521). Store short time at +4 °C, long term at −20 °C.

3   Methods

1. *Synthesis of the amphiphilic polymer.*

    (a) Dissolve 1 g (6.4 mmol monomer) of poly(isobutylene-alt-maleic anhydride) polymer and 1.1 ml (4.8 mmol) docecylamine in 30 ml of THF, use a round flask (*see* **NOTE** below).

    (b) Stir the mixture magnetically and maintain it under reflux (at 60 °C) overnight.

    (c) Evaporate the solvent in the rotary evaporator and redissolve the resulting pale yellow solid in 13 ml of chloroform in order to have a stock solution of 0.5 M (based on monomer). Store the stock solution in the fridge.

NOTE: The amphiphilic polymer is prepared by grafting dodecylamine molecules to the poly(isobutylene-alt-maleic anhydride) polymer (PMA), 75% of the anhydride groups of the polymer are reacted with dodecylamine [13]. A wide variety of amphiphilic polymers with different side chains and different grafting percentage have been used for encapsulation [11]. Docecylamine grafted (75%) poly(isobutylene-alt-maleic anhydride) has been shown to be one of the most stable in a wider pH range and at higher salt concentration [16]. In this step PEGylation in premodification process could be carried out [31]. Some of the MNPs used have a diameter of around 13 nm and are sufficiently stable to not require PEGylation.

2. *Polymer coating.*

   (a) In order to transfer to water 1 mg of MNPs (MFe_2O_4; M = Mn, Fe, Co), of around 13 nm in diameter, add 167 µl of PMA from 50 mM stock solution in $CHCl_3$ to 5 ml solution of MNPs (0.2 mg/ml) in $CHCl_3$ (*see* **NOTE** below)

   (b) Shake the mixture for 20 min and evaporate the solvent in the rotary evaporator.

   (c) Add 5 ml of $CHCl_3$ to the dry film, shake the mixture until the NPs are redispersed, and evaporate the solvent again (repeat the process twice).

   NOTE: This method follows the previously published polymer coating [13] with a few modifications. The amount of monomer units added per nm^2 of NP surface is reduced to 50 in order to reduce the amount of empty micelles. The NPs dispersion is diluted to 0.2 mg/ml in order to prevent encapsulating multiple nanocrystals in one micelle.

3. *Hydrolysis of anhydrides maleic anhydride groups.*

   (a) Add 5 ml of sodium borate buffer (50 mM, pH 9) to the dry film.

   (b) Sonicate the resulting mixture until the nanoparticles are properly dispersed in the aqueous solution (see NOTE)

   NOTE: In general it takes around 2 h to completely redisperse the NPs in the aqueous solution, but depending on the characteristics of the sample and the amount of monomer added per nm^2 of NPs surface, longer sonication may be required.

4. *Cleaning.*

   (a) To remove possible large aggregates, filter the nanoparticle solution through a syringe filter with a 0.2 µm pore size.

   (b) Perform gel electrophoresis or ultracentrifugation to eliminate the unbound polymer or empty micelles (*see* **NOTE** below).

   (c) Make a stock solution of around 5 mg/ml in distilled water and store in the fridge.

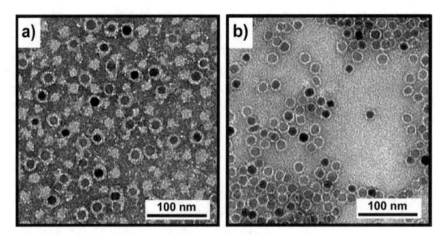

Fig. 1 Negative stain TEM to visualize the polymer coating around the MNPs and the empty micelles, (**a**) after gel electrophoresis and (b) after ultracentrifugation

NOTE: In principle both gel electrophoresis [19, 32] and ultracentrifugation [33] may be used to remove empty micelles. However, gel electrophoresis can only process small amounts of NPs, and often does not remove all empty micelles (Fig. 1). In some applications, empty micelles do not pose a problem, but when targeting the NP to a limited number of receptors, or when the aim is a very high density of NP, then empty micelles occupy receptors or space not available to the MNPs. Therefore, we perform ultracentrifugation to completely remove empty micelles [34]. Excessive ultracentrifugation should be avoided to prevent inducing NP aggregation.

High speed centrifugation can be used as (1) zonal separation by viscosity/density gradient [35, 36] or as (2) sedimentation-based separation [37, 38]. If the sample is homogenous in size and shape, the excess of free ligand can be successfully removed by pelleting the NPs down (centrifugation speed ant time will depend on the NPs features). Then remove the supernatant, redisperse the NPs in distilled water, and repeat the process twice.

5. *Labeling Neutravidin with Fluorophore.*

   (a) Prepare a stock solution of 2 mg/ml NA in PBS (1×) (*see* **NOTE** 1 below). Store it in the fridge.

   (b) Prepare a stock solution of 10 mg/ml Alexa Fluor 647NHS Ester in DMSO (see NOTE 2 below). Store it at −20 °C.

   (c) For a labeling ratio of NA:Dye = 1:5, pipette 200 µl of NA stock solution in an Eppendorf tube and add 4.2 µl of Alexa Fluor 647 NHS Ester stock solution on top.

   (d) Shake the mixture gently for 2 h.

(e) Use centrifuge filter of 10 kDa until the filtrate clears out (around 4 runs at 8000 r.c.f. between 5 and 10 min).

(f) Top the filter up (400 µl) each time with PBS (1×).

(g) Collect the retentate and resuspend in 100 µl PBS in order to have 4 mg/ml NA-Dye stock solution. Store it in the fridge.

NOTE 1: Allow only gentle shaking and wait for NA to dissolve completely.

NOTE 2: Avoid excessive pipetting and protect from light at all stages hereon.

6. *Activation of the carboxylic groups in the PMA coating of the MNPs.*

(a) Transfer 100 µg (20 µl) of MNPs from MNP@PMA water stock solution (5 mg/ml) into an Eppendorf tube.

Prepare 10 mg/ml distilled water stocks of EDC and Sulfo-NHS (see NOTE 1).

(b) Add to the 20 µl MNP solution: 81 µl of PBS, 5 µl of EDC solution and 14 µl Sulfo-NHS solution (final volume of 100 µl where [EDC] = 0.5 mg/ml and [Sulfo-NHS] = 1.4 mg/ml).

(c) Shake the mixture gently for 18 min.

(d) Clean the sample using centrifuge filter of 300 kDa, two runs at 8000 r.c.f. for 3 min (see NOTE 3).

(e) Resuspend the retentate in about 50 µl water so that the [MNPs] = 2 mg/ml (see NOTE 3).

NOTE 1: The solutions must be used fresh; they cannot be stored.

NOTE 2: Do not top the volume up, add about 50–100 µl water for each run.

NOTE 3: There should be no pellet formation at this stage. All retentate should get suspended in distilled water. Make sure not to wait longer than 30–35 min before the following step.

7. *Covalent functionalization of activated MNP with Neutravidin.*

(a) Add the previously prepared 100 µl solution of dye labeled NA (4 mg/ml) to the 50 µl activated MNP suspension dropwise, with gentle shaking between each drop. Add ten drops of 10 µl and wait 1 min between drops (*see* **NOTE** 1).

(b) Wait 2 min and quench the reaction by adding 3 µl hydroxylamine 500 mM and pipette mix.

(c) Centrifuge the suspension until the filtrate is clear. Use 300 kDa filter at 8000 r.c.f. for 6 min and in each run top the filter volume with distilled water (*see* **NOTE** 2).

Note 1: The amount NA required to cover the entire surface of the MNP@PMA depends on the MNP size. We use MNP with a magnetic core diameter of 13 nm, which after coating results in a hydrodynamic diameter of 20 nm. These require 64 NA per MNP, which equates to 1 mg NA per 1 mg MNPs. To ensure a proper functionalization fourfold excess is used, so 400 μg NA-Dye for 100 μg MNPs. Equivalent NA:MNP-surface ratios have been used to functionalize semiconductor nanorods with NA [20].

Note 2: Higher r.c.f. or longer time can cause otherwise stable colloidal MNPs to precipitate. Spend enough time on this step. Reduce run times and adjust speeds depending on MNP stability in water. Consider changing the filter if already in use for over 30 min.

8. (A) *Targeting of Functionalized MNP to cell surface receptor via biotinylated antibody*

   (a) Note: many antibodies may be purchased already biotinylated. To biotinylate your antibody of choice use EZ-Link Sulfo-NHS-LC-Biotin following the manufactures protocol:

   - If antibody is in amine-free buffer at pH 7.2–8.0, it may be used as it, otherwise exchange buffer to PBS, use antibody at 1 mg/ml

   - Immediately before use, prepare a 10 mM solution of the biotin reagent in distilled water.

   - To achieve 4–5 biotin groups per antibody molecule, add 20-fold molar excess of biotin reagent, i.e., add 1.5 μl biotin reagent (10 mM) to 100 μl of 1 mg/ml antibody solution.

   - Incubate reaction on ice for 2 h or at room temperature for 30 min.

   - Remove excess nonreacted and hydrolyzed biotin reagent using desalting or dialysis. We recommend microcon centrifugation using a 100 kDa filter.

   (b) Add biotinylated antibody to cells grown on coverslips in imaging buffer or PBS at a dilution of 1:200 (or the manufacture recommended dilution for the chosen antibody), at room temperature. Allow 10 min bind to antigens. Longer time may lead to endocytosis of labeled surface proteins. Do not add antibody to unfiltered conditioned media as it may interfere with antibody binding.

   (c) Wash antibody containing buffer solution completely and replace with MNP-NA suspension. Allow 10 min for biotin–avidin binding at final MNP concentration of 10–20 μg/ml.

(d) Wash unbound MNP away and image.

8. (*B*) *Targeting of Functionalized MNP to biotinylated cell surface receptor*

Note: An interesting alternative is to genetically engineer the cells to express a surface protein, which encodes a biotinylation site and also engineer the biotinylation enzyme into the cell [30]. This approach works very well in most cell culture cells, but less so in neurons. Recently, a wide range of protein targeting tools have been developed, such as halo- and SNAP-tags, nanobodies and Click-chemistry; for specifics we refer to the specific literature [31, 39, 40].

9. *Application of Alternating Magnetic Field during fluorescence microscopy*

(a) The heating of the MNP requires the application of alternating magnetic fields (AFM) of 100 G and higher. We use a custom-built water-cooled copper coil (5 mm diameter, five turns, see Fig. 2).

(b) The coil is driven by a 7.5 kW AC power supply (MSI Automation), and placed directly around the sample chamber on the microscope stage.

(c) As the AMF induces eddy currents in any metal, the metallic microscope parts need to be replaced with non-conducting elements. Still, parts of the objective lens may heat up, possibly causing focus shift and aberrations, which need be corrected in real time. We recommend using a fast, laser interferometer based, analog feedback piezoelectric auto-focus system (Motion X Corporation, Fig. 1).

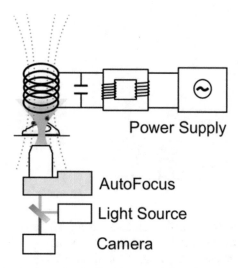

Fig. 2 Schematic of microscope with AMF. An inverted fluorescence microscope is modified to hold the objective lens on a fast autofocus. The water-cooled coil is mounted above

10. *Fluorescence microscopy measurement of heating*

    (a) The temperature response of the fluorophore used to label the NA in the MNP coating should be calibrated after attaching it to the MNP. Ideally, the fluorescence lifetime of a dilute suspension of labeled MNPs is measured as function of temperature.

    (b) Alternatively, the temperature response of the fluorescence intensity may be measured in the microscope, as long one takes care to minimize bleaching and measures both directions, up and down in the temperature.

4   Conclusions

Polymer encapsulated MNPs have been successfully targeted to the membrane of neurons in culture and in vivo [1, 41]. The integration of the fluorophore Dylight550 into the protein coat permits recording the temperature rise during the application of the AMF and to correlate this to the stimulation of the neurons, which is recorded by the calcium indicator GCaMP6f. Figure 3 shows

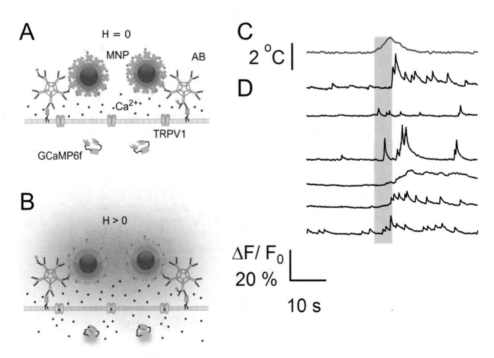

Fig. 3 Magnetothermal genetic neurostimulation. (**a**) Schematic showing NA-polymer-coated MNP bound via antibodies to the cell membrane of neurons expressing TPRV1 and the calcium indicator GCaMP6f. (**b**) Application of the alternating magnetic field causes MNP to heat opening the TRPV1 channels and letting calcium flow into the cells, which is detected by GCaMP6f and causes action potential generation. (**c**) Membrane temperature profile recorded via Dylight fluorescence during a 5 s field application (gray). (**d**) GCaMP6f responses of several neurons exposed to the same magnetic field stimulus (after 41)

sample data for temperature increase and neuronal stimulation on cultured hippocampal neurons.

The summarized methods and protocols on MNP encapsulations, local temperature measurement and neuronal stimulation are widely adaptable to different NP systems, cell types and experiments. Further modification of the coating is easily possible. In vivo experiments injecting NP into the body often require the attachment of polyethylene (PEG) to prevent aggregation of NP and immune response of the host [22, 42, 43] .

Acknowledgment

This work was supported by NIH grants R01MH111872 (A.P.) and R01MH094730 (A.P.), HFSP grant RGP0052/2012 (A.P.) and a fellowship from the Basque Government, the Becas posdoctoralesto, to I.C.-R.

References

1. Huang H, Delikanli S, Zeng H et al (2010) Remote control of ion channels and neurons through magnetic-field heating of nanoparticles. Nat Nanotechnol 5:602–606

2. Chen R, Romero G, Christiansen MG et al (2015) Wireless magnetothermal deep brain stimulation. Science (New York, NY) 347: 1477–1480

3. Stanley SA, Gagner JE, Damanpour S et al (2012) Radio-wave heating of iron oxide nanoparticles can regulate plasma glucose in mice. Science (New York, NY) 336:604–608

4. Mannix RJ, Kumar S, Cassiola F et al (2008) Nanomagnetic actuation of receptor-mediated signal transduction. Nat Nanotechnol 3: 36–40

5. Seo D, Southard KM, Kim J et al (2016) A mechanogenetic toolkit for interrogating cell signaling in space and time. Cell 165:1–12

6. Chen R, Christiansen MG, Sourakov A et al (2016) High-performance ferrite nanoparticles through nonaqueous redox phase tuning. Nano Lett 16:1345–1351

7. Castellanos-Rubio I, Insausti M, Garaio E et al (2014) Fe$_3$O$_4$ nanoparticles prepared by the seeded-growth route for hyperthermia: electron magnetic resonance as a key tool to evaluate size distribution in magnetic nanoparticles. Nanoscale 6:7542–7552

8. Lee J-H, Jang J-T, Choi J-S et al (2011) Exchange-coupled magnetic nanoparticles for efficient heat induction. Nat Nanotechnol 6:418–422

9. Ovejero JG, Cabrera D, Carrey J et al (2016) Effects of inter- and intra-aggregate magnetic dipolar interactions on the magnetic heating efficiency of iron oxide nanoparticles. Phys Chem Chem Phys 18:10954–10963

10. López A, Gutiérrez L, Lázaro FJ (2007) The role of dipolar interaction in the quantitative determination of particulate magnetic carriers in biological tissues. Phys Med Biol 52: 5043–5056

11. Palui G, Aldeek F, Wang W et al (2014) Strategies for interfacing inorganic nanocrystals with biological systems based on polymer-coating. Chem Soc Rev 44:193–227

12. Bohara RA, Thorat ND, Pawar SH (2016) Role of functionalization: strategies to explore potential Nano-bio applications of magnetic nanoparticles. RSC Adv 6:43989–44012

13. Lin C-AJ, Sperling RA, Li JK et al (2008) Design of an amphiphilic polymer for nanoparticle coating and functionalization. Small 4:334–341

14. Zhang H, Huang H, He S et al (2014) Monodisperse magnetofluorescent nanoplatforms for local heating and temperature sensing. Nanoscale 6:13463–13469

15. Tomczak N, Liu R, Vancso JG (2013) Polymer-coated quantum dots. Nanoscale 5: 12018–12032

16. Quarta A, Curcio A, Kakwere H et al (2012) Polymer coated inorganic nanoparticles: tailoring the nanocrystal surface for designing nanoprobes with biological implications. Nanoscale 4:3319

17. Zrazhevskiy P, Dave SR, Gao X (2014) Addressing key technical aspects of quantum dot probe preparation for bioassays. Part Part Syst Charact 31:1291–1299

18. Wang W, Ji X, Bin Na H et al (2014) Design of a multi-dopamine-modified polymer ligand optimally suited for interfacing magnetic nanoparticles with biological systems. Lang Des 30:6197–6208

19. Sperling RA, Liedl T, Duhr S et al (2007) Size determination of (bio) conjugated water-soluble colloidal Nanoparticles: a comparison of different techniques. J Phys Chem C 111:11552–11559

20. Lippert LG, Hallock JT, Dadosh T et al (2016) NeutrAvidin functionalization of CdSe/CdS quantum nanorods and quantification of biotin binding sites using biotin-4-fluorescein fluorescence quenching. Bioconjugate Chem 27:562–568. https://doi.org/10.1021/acs.bioconjchem.5b00577

21. Parolo C, De Escosura-mun A, Polo E et al (2013) Design, preparation, and evaluation of a fixed-orientation antibody/gold-nanoparticle conjugate as an IMMUNOSENSING label. ACS Appl Mater Interfaces 5:10753–10759

22. Gref JMR, Domp A, Quellec P, Blunk T, Muller RH, Verbavatz RL (1995) The controlled intravenous delivery of drugs using PEG coated sterically stabilized nanospheres. Adv Drug Deliv Rev 16:215

23. Xu C, Shi S, Feng L et al (2016) Long circulating reduced graphene oxide-iron oxide nanoparticles for efficient tumor targeting and multimodality imaging. Nanoscale 8:12683–12692

24. Wang W, Ji X, Kapur A et al (2015) A multifunctional polymer combining the imidazole and zwitterion motifs as a biocompatible compact coating for quantum dots. J Am Chem Soc 137:14158–14172

25. Huang J, Ying L, Yang X et al (2015) Ratiometric fluorescent sensing of pH values in living cells by dual-fluorophore-labeled i-motif Nanoprobes. Anal Chem 87:8724–8731

26. Chen K, Xu X, Guo J et al (2015) Enhanced intracellular delivery and tissue retention of nanoparticles by mussel-inspired surface chemistry. Biomacromolecules 16:3574–3583

27. Saha K, Rahimi M, Yazdani M et al (2016) Regulation of macrophage recognition through the interplay of nanoparticle surface functionality and protein corona. ACS Nano 10:4421–4430. https://doi.org/10.1021/acsnano.6b00053

28. Ma Y, Huang J, Song S et al (2016) Cancer-targeted Nanotheranostics: recent advances and perspectives. Small 12:4936–4954

29. Tonga GY, Saha K, Rotello VM (2014) 25th anniversary article: interfacing nanoparticles and biology: new strategies for biomedicine. Adv Mater 26:359–370

30. Howarth M, Takao K, Hayashi Y et al (2005) Targeting quantum dots to surface proteins in living cells with biotin ligase. Proc Natl Acad Sci U S A 102:7583–7588

31. Yu WW, Chang E, Falkner JC et al (2007) Forming biocompatible and nonaggregated nanocrystals in water using amphiphilic polymers. J Am Chem Soc 129:2871–2879

32. Sperling R a, Pellegrino T, Li JK et al (2006) Electrophoretic separation of nanoparticles with a discrete number of functional groups. Adv Funct Mater 16:943–948

33. Bonaccorso F, Zerbetto M, Ferrari AC et al (2013) Sorting nanoparticles by centrifugal fields in clean media. J Phys Chem C 117:13217–13229

34. Zhang Q, Castellanos-rubio I, Munshi R et al (2015) Model driven optimization of magnetic anisotropy of exchange-coupled core-shell ferrite nanoparticles for maximal hysteretic loss. Chem Mater 27:7380–7387

35. Akbulut O, Mace CR, Martinez RV et al (2012) Separation of nanoparticles in aqueous multiphase systems through centrifugation separation of nanoparticles in aqueous multiphase systems through centrifugation. Nano Lett 12:4060–4064

36. Prantner AM, Chen J, Murray CB et al (2012) Coating evaluation and purification of monodisperse, water-soluble, magnetic nanoparticles using sucrose density gradient ultracentrifugation. Chem Mater 24:4008–4010

37. Khoury LR, Goldbart R, Traitel T et al (2015) Harvesting low molecular weight biomarkers using gold nanoparticles. ACS Nano 9:5750–5759

38. Nagaraja AT, You Y, Choi J et al (2016) Journal of colloid and Interface science layer-by-layer modification of high surface curvature nanoparticles with weak polyelectrolytes using a multiphase solvent precipitation process. J Colloid Interface Sci 466:432–441

39. Kolb HC, Finn MG, Sharpless KB (2001) Click chemistry: diverse chemical function from a few good reactions. Angew Chem Int Ed Engl 40:2004–2021

40. Rothbauer U, Zolghadr K, Tillib S et al (2006) Targeting and tracing antigens in live cells with fluorescent nanobodies. Nat Methods 3:887–889

41. Munshi R, Qadri S, Zhang Q, et al (2017) Magnetothermal genetic deep brain stimulation of motor behaviors in awake, freely moving mice. eLife 2017;6:e27069

42. Jokerst JV, Lobovkina T, Zare RN et al (2011) Nanoparticle PEGylation for imaging and therapy. Nanomedicine 6:715–728

43. Otsuka H, Nagasaki Y, Kataoka K (2012) PEGylated nanoparticles for biological and pharmaceutical applications. Adv Drug Deliv Rev 64:246–255

Chapter 5

Genetically Encoded Nanoparticles for Neural Modulation

Sarah A. Stanley

Abstract

Regulating the activity of genetically and/or anatomically defined neurons is invaluable in trying to determine the physiological role of specific neural populations. Existing tools for neural activation and inhibition such as optogenetics and chemogenetics have transformed our understanding of neural circuits but they are not universally applicable. This chapter describes a system for remote, bidirectional regulation of neural activity in vivo controlled by noninvasive electromagnetic signals (Stanley et al. Nature 531:647–650, 2016). Using targeted expression of modified temperature sensitive channels and genetically encoded ferritin to generate intracellular nanoparticles, it is possible to activate or inhibit genetically targeted neurons using radiofrequency (RF) or magnetic fields. This technology provides an additional technique for rapid neural regulation in freely moving organisms.

Key words Radiogenetics, Magnetogenetics, Transient receptor potential vanilloid 1 receptor, Ferritin, Nanoparticles

1 Introduction

A greater understanding of the physiological roles of specific cells and circuits requires detailed knowledge of the component cell populations and their anatomical connections. But the definitive test of a neuron's involvement relies on switching cell activity on or off and determining the effects on physiological functions, ideally in a freely moving animal with as little disturbance to its natural behavior as possible.

In recent years, numerous tools for studying how specific neural populations contribute to physiological functions have become available. Initial functional studies relied on assessing the effects of nonspecific, irreversible, neural ablation in anatomically defined CNS regions [1]. However, this technique also damaged fibers passing through the ablated area and produced variable sized lesions. Subsequent protocols used nonspecific cell activation tools, such as glutamate uncaging [2] in vitro or direct electrical stimulation via implanted electrodes [3] in vivo. Direct electrical

Fidel Santamaria and Xomalin G. Peralta (eds.), *Use of Nanoparticles in Neuroscience*, Neuromethods, vol. 135, https://doi.org/10.1007/978-1-4939-7584-6_5, © Springer Science+Business Media, LLC 2018

stimulation activates both cell bodies and fibers, is highly variable, causes significant tissue damage and requires a permanent implant [4]. Despite their limitations, these ablation and electrical activation studies provided significant information about the circuitry and roles of specific CNS regions. However, they have been largely superseded by targeted tools for neural modulation that allowed the investigation of cell-type specific neural populations. There is now an array of genetically encoded tools ranging from light [5] and ligand-gated ion channels [6] to modified G-protein coupled receptors which modulate neural activity in defined populations [7]. These tools have transformed our understanding of specific circuits but each has its own advantages and limitations.

Rapid modulation of neural activity has been achieved by technologies based on ion channels. These have several advantages; ion channel structure and function are usually well described, they have a rapid time course of activation and a wide range of channels exist that can be employed to achieve neural modulation. Although ligand-gated ion channels have been used [8], optogenetics, using the light-activated cation channel channelrhodopsin (ChR2), have been most extensively employed to investigate neural function [9]. ChR2 can be genetically targeted to defined neural populations which are then rapidly activated when treated with blue light [5]. Further advances have seen the development of variants which allow more rapid control, prolonged depolarization for coordinated firing or neural silencing [10]. By examining the effects of selected optogenetic stimulation and/or silencing, it has been possible to identify neural populations regulating complex behaviors including feeding [11], reward [12], depression [13] and sleep [10].

These optogenetic tools have provided an important step in investigating neural function but they are not universally applicable. Light penetration into CNS tissue is less than a millimeter [14] and so activation in vivo requires fiber optic implants which modulate only local neural populations. Permanent implants can both damage tissue and possibly interfere with behavior. Optogenetics is also unsuitable for regulating cell-specific neural populations that are found in several CNS regions as this would require multiple fiber-optic implants. In addition, optical fibers for light delivery can be difficult to use in regions such as the spinal cord, whose function is perturbed by insertion of a fiber, or in the developing CNS. Finally, long-term studies could be limited by the potential toxicity of or immune response to nonmammalian opsin proteins [15].

Cell-type specific populations can extend across large areas of the CNS but the physiological roles of dispersed cell populations have been less well studied, in part, because they are less accessible to optogenetic manipulation. An alternative system for cell-specific activation involves chemogenetics: the use of engineered ion channels [6, 8] or G-protein coupled receptors [7, 16] (e.g., Designer receptors exclusively activated by designer drugs (DREADDs)) activated by otherwise inert small molecules. Chemogenetic systems can be targeted to specific neural populations (though some require a specific transgenic mouse line to reduce background effects) and can modulate cell populations throughout the CNS. These tools have been used to modulate cell-type specific populations both locally and those found in multiple CNS regions to determine their roles [17]. Agonists for chemogenetic systems can be applied locally, e.g., by uncaging [18] in vitro or by direct injection into the tissue in vivo [6] and under such circumstances, the onset of action is rapid but agonist application is invasive. However, for systems that use peripherally administered agonists the onset of action is typically of the order of minutes and the effects may last for hours to days which can complicate correlations between neural activation and behavior [7]. In addition, at the moment, the agonists for chemogenetic systems need to be delivered by handling and injecting the animal immediately before assessing behavior. A system for rapid, specific, noninvasive modulation of neural activity might be helpful for assessing the roles of neural populations in specific behaviors.

1.1 Radiogenetics and Magnetogenetics

Noninvasive neural modulation requires a signal that can penetrate tissue and this can be achieved using radiofrequency [19] (RF) or magnetic fields. Low-frequency RF penetrates tissue with minimal absorption [20] and is already in clinical use for noninvasive reprogramming of cardiac pacemakers and for cochlear implants [21]. Similarly, magnetic fields are also used in medicine for noninvasive imaging (magnetic resonance imaging) and treatment (transcranial magnetic stimulation) [22]. In contrast to tissue, both RF and magnetic fields interact with metals and particularly, with metal nanoparticles. Ferrous and nonferrous metal nanoparticles absorb RF energy converting it into heat through various mechanisms including Neel rotation and eddy currents [23] and this, in turn, results in Brownian motion [24]. Magnetic fields also freely penetrate tissue and can exert a mechanical force on metal nanoparticles. The combination of magnetic fields and nanoparticles have been used for many ex-vivo purposes including improved cell transfection and exerting mechanical forces on cells to alter their properties. However, the energy transferred from radiowaves and magnetic fields to nanoparticles can also be harnessed for cell modulation.

Several groups have used extracellular nanoparticles for noninvasive cell activation [25, 26]. These systems used a combination of three components (1) low frequency radiowaves (465 kHz), (2) iron oxide nanoparticles, and (3) a temperature sensitive ion channel. Target cells are modified to express a temperature sensitive cation channel, transient receptor potential vanilloid 1 receptor (TRPV1) [27]. This may be altered with an epitope tag incorporated into its first extracellular domain [25]. Iron oxide nanoparticles, either coated with anti-tag antibody or functionalized to reduce internalization [26], are injected and decorate the surface of the target cells. In the presence of radiowaves, the cell surface nanoparticles absorb energy which in turn opens the multimodal TRPV1 resulting in cation entry. This signal can then be used to depolarize endocrine cells [25] or neurons [26], or switch on a reporter gene in the targeted cells [28]. These systems also have limitations. Nanoparticle injection is invasive, its effectiveness is time-limited as a result of nanoparticle internalization, probably even with functionalization, and because the injected particles only decorate cells locally, the system cannot be used to activate cells over larger areas. However, some of these problems can be overcome with a system using transgenic expression of either a chimeric ferritin protein [29, 30] or an iron-sulfur cluster assembly protein 1 [31] to continuously generate nanoparticles within the targeted cells.

The genetically encoded system described here makes use of metal nanoparticles formed within a GFP-tagged ferritin complex. Ferritin forms a 24 subunit protein shell with ferroxidase activity [32, 33]. It accumulates Fe ions from the cytoplasm and converts them into an iron oxide nanoparticle of approximately 5–8 nm that forms within the protein shell [34, 35]. The iron core of endogenous ferritin is heterogenous in composition and consists of magnetite/maghemite, hematite, and ferrihydrite crystals. It can also vary in size, morphology, and structure depending on the composition of the ferritin shell and the cellular environment. As the iron concentration in the environment falls the proportion of superparamagnetic magnetite/maghemite increases [36, 37]. Superparamagnetic nanoparticles increase their internal energy when treated with radiofrequency fields and are also susceptible to magnetic fields. Transgenic expression of ferritin in vitro confers superparamagnetic properties and applying an external magnetic field can be used to introduce energy. In the current system, a green fluorescent protein (GFP) tag is added to the surface of the ferritin multimer and the temperature sensitive TRPV1 is modified by the addition of an N-terminal GFP binding domain [38]. The anti-GFP binding protein fused to TRPV1 tethers the GFP-tagged ferritin to the channel. In the presence of radiowaves, energy transfer through heat and/or motion of metal nanoparticles results in opening of the TRPV1 channel, cation ion entry into neurons and depolarization. This approach is shown in Fig. 1. TRPV1 is not only temperature sensing but also responds to mechanical stimuli. This characteristic combined with the superparamagnetic properties of the ferritin core means an external magnetic field can be used to activate neurons. In the presence of a magnetic field, iron oxide nanoparticles within the tethered ferritin align with the field [39] resulting in a mechanical force on the channel to open TRPV1 and allow ion entry and depolarization. In contrast to chemogenetic tools,

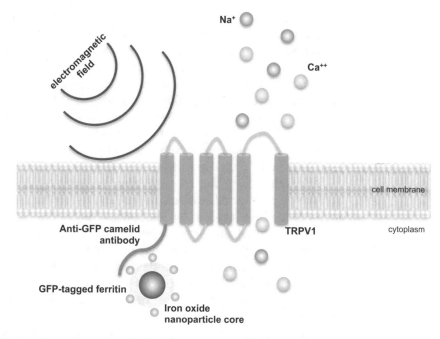

Fig. 1 Schema of remote neural activation. Electromagnetic fields such as radiowaves or magnetic fields heat/move the iron core of ferritin to open TRPV1. Cation entry depolarizes the neuron

the on- and off-rates of these responses are of the order of seconds [29] which may allow better correlation between behavioral changes and neural modulation.

In addition, it is possible to modify the TRPV1 channel to alter the channel permeability from positively charged calcium and sodium ions to chloride ions [29]. This modification, combined with genetically encoded nanoparticles and radiowaves or magnetic fields results in a tool for noninvasive neural silencing in vivo. Together, these tools provide a means for rapid neural modulation in vitro and in vivo.

1.2 Equipment, Materials, and Setup

The method described in this chapter requires a transgenic mouse expressing cre-recombinase in the cell population of interest, often using the endogenous promoter to regulate the expression of cre. There are multiple mouse lines expressing cre-recombinase under the control of specific promoters available from the sites below.

Mutant mouse resource and research center (https://www.mmrrc.org/catalog/StrainCatalogSearchForm.php?jboEvent=Search&SourceCollection=GENSAT).

Jackson Laboratories.

(https://www.jax.org/research-and-faculty/tools/cre-repository#).

Detailed instructions on developing and validating transgenic lines expressing cre-recombinase are beyond the scope of this chapter but may be found elsewhere [40, 41].

The second component required is a cre-dependent viral vector that targets the expression of the radiogenetic components to the cell type of interest. The plasmids to generate cre-dependent adenovirus are available from the corresponding authors of the publications. These can be used to generate high-titer viruses using established methods [42], through academic viral core facilities (University of Iowa, http://www.medicine.uiowa.edu/vectorcore/products, University of Pennsylvania https://www.med.upenn.edu/gtp/vectorcore/production.shtml) or commercial companies (e.g., Viraquest, Vector biolabs).

After identifying the cre-expressing line and obtaining high-titer cre-dependent adenovirus, the constructs are targeted to the appropriate CNS region by stereotactic injection. Stereotactic equipment is available from several companies (e.g., Stoetling, Kopf, World Precision Instruments) and allows precise control over the site of injection. The virus is delivered using a gas-tight syringe and manual or digital syringe pumps or microinjectors can be used to control the volume and rate of viral delivery. A detailed protocol for surgery and stereotactic injection are described below.

The targeted neurons can be modulated using either a radiowave or variable magnetic field. There are several sources to obtain a radiowave generator/amplifier of the appropriate wavelength (465 kHz) and field strength (e.g., Ultraflex, Magnetherm). These typically generate a field using a small coil (3–6 cm diameter coil) and so mice need to be anesthetized for treatment using these systems and anesthesia effects need to be taken into consideration. The RF coils are usually water-cooled and so the mice are placed in a plastic container with insulation such as Styrofoam to prevent them from coming into contact with the cold coil.

Alternatively, a variable magnetic field may be used. We used the magnetic field generated outside the electromagnetic coil of a clinical MRI machine. The freely moving animals were placed in a plastic chamber in an area where the field strength is at least 1 T. However, it is possible that lower field strengths may also be effective. The dimensions of the chamber can be modified as needed; for feeding studies, we used a chamber $37.3 \times 23.4 \times 14.0$ cm. For these studies, the chamber and all other equipment used to test behavior or other parameters (food containers, cameras, etc.) need to be MRI safe/nonmetallic.

2   Materials

2.1   Virus and Stereotactic Surgery

- Cre-expressing mouse line.

- Cre-dependent virus expressing radiogenetic construct.

- Anesthesia (injected anesthesia, e.g., ketamine/xylazine 100 mg/kg and 10 mg/kg body weight, respectively (see http://cshprotocols.cshlp.org/content/2006/1/pdb. rec702) or inhaled, e.g., isoflurane delivered by vaporizer).

- 5 µl gas-tight syringe (e.g., Hamilton syringe).
- Surgical tools (drill bit, tweezers, scissors, or scalpel, sterilized by autoclaving).
- Hair trimmer or hair removal cream.
- Ophthalmic ointment.
- Analgesia (e.g., buprenorphine or carprofen).
- Skin cleaner such as Betadine or chlorhexidine.
- Alcohol.
- 3% hydrogen peroxide.
- Cotton swabs.
- Hand-held drill or drill attached to stereotactic frame.
- Sutures or vetbond animal adhesive.
- Warming pad.
- Mouse Brain in Stereotactic Coordinates by Paxinos and Franklin.
- Iron dextran (e.g., Durvet 01 DME1030 Iron Dextran Injection).

2.2 Radiogenetic Modulation

- Anesthesia (e.g., ketamine/xylazine or inhaled isoflurane, as above).
- Radiofrequency generator.
- Mouse chamber (typically a cylinder 30 mm diameter and 115 mm in length, the dimensions will depend on the coil being used).

2.3 Magnetogenetic Modulation

- Variable magnetic field (e.g., field outside the coil of MRI machine).
- Mouse chamber (typically 37.3 × 23.4 × 14.0 cm but the dimensions will depend on the parameters being studied).

2.4 Verification of Injection Site

- Anesthesia (e.g., ketamine/xylazine or inhaled isoflurane, as above).
- Phosphate buffered saline.
- 4% paraformaldehyde or 10% formalin.
- Surgical tools.
- Vibratome.
- Paint brushes.
- 12-well plates.
- Primary antibody raised against GFP.
- Microscope slides.
- Fluorescent microscope.

3 Procedures

3.1 Virus Preparation

Adenovirus should be used according to the biosafety guidelines of the institution. The adenovirus should be stored at −80 °C until use. Adenovirus produced by commercial or academic viral facilities may be supplied in relatively large volumes (e.g., 1 ml) and should be aliquoted into smaller volumes when it is first thawed. The virus should be thawed at room temperature and remain at room temperature during use. Aliquots of no less than 50 μl are stored in sterile cryotubes (such as Sarstedt cryopure tubes) at −80 °C. The aliquots can be freeze-thawed approximately twice without significant changes in the viral titer and the titer of unthawed aliquots stored at −80 °C should be stable for approximately several years.

A gas-tight syringe is used to inject the virus. A small volume of sterile saline (approximately 1 μl) is drawn up followed by 0.5–1 μl of air then a small volume (e.g., 2 μl) of thawed virus. The amount of virus administered is measured by following the top of the meniscus of the virus in the syringe. The syringe is then placed in the holder on the arm of the stereotactic frame. There are a number of factors that should be considered when determining the virus volume to inject such as the size of the targeted structure, the spread of virus within the tissue and the extent to which cre-recombinase is expressed in the structure and adjacent regions. It is often necessary to perform pilot studies to determine the optimum injection volume for the targeted region.

3.2 Surgery

Surgery should be performed according to the approved animal protocol(s) at the institution.

Injections are usually undertaken in adult mice age 8–12 weeks using cre-expressing transgenic mice and wild-type littermate controls. Mice are weighed before surgery to ensure that they recover to within 10% of their presurgery weight. Mice are anesthetized using either intraperitoneal injection of ketamine and xylazine or inhaled anesthesia (isoflurane 2% to effect). The depth of anesthesia should be checked by ensuring the mice do not respond to a toe pinch (or as defined by the institution) and analgesia is given (e.g., Buprenorphine, 0.05 mg/kg, ip). The hair over the dorsal surface of the head is removed either with hair clippers or using hair removal cream. The skin is then wiped clean to remove the cut hair so that it does not enter the surgical site. Next, the mouse is placed in the stereotactic apparatus with nontraumatic ear bars and nosepiece. It is important to ensure there is no movement and the head is stable. The mouse should be kept warm using a heat pad and sterile ophthalmic ointment placed on each eye. The depth of anesthesia should be monitored regularly during surgery using pedal reflex and respiratory rate.

Once the mouse is stabilized in the stereotactic frame, the skin is cleaned with betadine and then alcohol before a small vertical incision is made in the skin over the back of the head. The skin is gently drawn to each side with tweezers and the skull cleaned with cotton swabs. If the skull sutures are difficult to see, the skull can be cleaned with a cotton swab soaked in 3% hydrogen peroxide, with further swabs used to clear the skull surface. These steps will help to ensure that bregma and lambda landmarks can be seen clearly.

It is important to make sure that the skull is level in both the vertical (rostrocaudal) and horizontal (medial-lateral) axes. The syringe is gently lowered until the tip of the needle is touching the skull at lambda and the coordinates are noted or the Z axis is zeroed if a digital frame is being used. The syringe is then lifted and moved until the needle tip is touching the skull at bregma. Adjust the height of the nosepiece as needed to ensure that the Z-axis reading at bregma is within 0.1 mm of that at lambda and that the skull is level in the rostrocaudal plane. The tip of the needle is then placed on the surface of the skull at bregma then moved 1 mm laterally from the midline and lowered again until the needle touches the skull. The coordinates are noted or the Z-axis zeroed as before. Next the syringe is moved 1 mm to the opposite side of the midline and lowered until the needle tip touches the skull. The difference between the Z coordinates on each side of the midline should be less than 0.1 mm and the height of the ear bars can be adjusted to ensure the skull is level in the medial-lateral plane.

The next step is to prepare the skull for injection. If a stereotactic frame with a drill arm is being used, the syringe arm is removed and the drill secured to the frame then the drill tip is lowered to touch the skull at bregma. The coordinates are noted or the X and Y axes zeroed on a digital frame. The drill is then moved to the correct coordinates for the injection site. The coordinates can be calculated using a stereotactic mouse brain atlas (such as "The Mouse Brain in Stereotactic Coordinates"), from publications targeting the same region or from the coordinates used to target defined brain regions in wild-type or cre-expressing strains by the Allen Brain Institute (http://help.brain-map.org/display/mouseconnectivity/Documentation). (Pilot studies can be done by injecting fluorescent beads at the coordinates for the targeted brain region then perfusing, fixing and sectioning of the injected brain to confirm the injection site is correct.) The drill is lowered to produce a small hole in the skull. The drill arm is then removed and the syringe arm replaced in the frame. If a handheld drill is being used, the syringe remains in the frame and is moved to the coordinates for injection. It is gently lowered to the skull surface and the position marked. Then the syringe is then raised out of the way and the drill used to produce a small hole in the skull ensuring it does not damage the brain surface. If there is bleeding, this can be controlled using cotton swabs or eye swabs.

After the skull has been drilled, the syringe arm is again positioned over bregma and the X and Y coordinates zeroed. The syringe is then moved to the correct X-Y injection coordinates and lowered through the hole in the skull until the tip touches the cortical surface. At this point the Z axis is zeroed. The syringe is then slowly lowered to the correct depth in the brain. The viral volume (typically 150 nl to 1 μl depending on the injection site and transgenic strain) is then slowly injected over 10 min or more. Digital or manual syringe pumps can be useful to make sure the injection is performed slowly and smoothly. The syringe should remain in place for at least 5 min after the end of the injection, then slowly withdrawn from the brain. If bilateral structures need to be targeted, the injection is repeated on the opposite side of the brain.

In addition to the viral delivery of the radiogenetic constructs, iron dextran is also injected into the CSF to make sure that the ferritin shell is sufficiently loaded with iron. Iron dextran (100 mg/ml, 3 μl) is injected to lateral ventricle over 10 min using the protocol above.

After the syringe has been withdrawn, the skull is dried with cotton swabs and the skin over the skull closed using either sutures or Vetbond (being careful to ensure it does not go anywhere near the eyes). The mouse is then removed from the frame and placed in a clean dry environment with a heating pad. When they are moving normally, the mice are returned to their home cage. They should be monitored regularly and weighed to make sure they regain their presurgery weight. Analgesia for 72 h after surgery is recommended.

The time taken from virus injection until recombination and expression depends on the viral vector. Construct expression with adenovirus delivery occurs relatively quickly (several days) but the ferritin shells also need to be adequately loaded with iron oxide so we usually wait a week from injection before starting studies. This period also allows the animals to fully recover, to be acclimated to handling or any novel environments or to be separated into individual cages if needed. For example, if the mice are being treated with the small RF coil and will be anesthetized with injected anesthesia, they are handled and poked with an insulin syringe for at least 3 days (preferably longer) before the study to reduce stress.

3.3   Studies Using RF Treatment

Mice are moved to the treatment room at least 2 h before the start of the study to minimize stress. Temperature and humidity are kept as stable as possible and noise is minimized as much as possible.

Mice are anesthetized using either isoflurane or ketamine/xylazine for RF studies using the small mouse coil. When fully anesthetized (15 min after injection/induction, with absent pedal

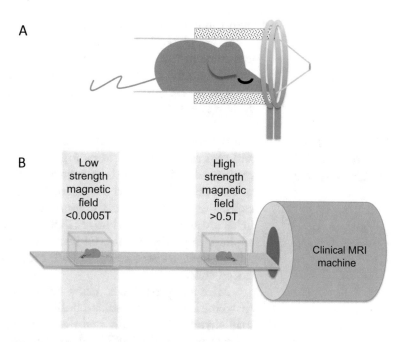

Fig. 2 Remote neural activation using radiowaves or magnetic field. (**a**) Anesthetized mice are placed in a chamber inside a water-cooled RF coil (3c, diameter) for RF treatment. (**b**) Nonuniform magnetic fields generated outside the electromagnetic coil of a clinical MRI machine can be used for magnetic field modulation of neural activity

reflexes), the mice are placed in the treatment chamber and then in the RF coil and treated with field strengths of 0–32 mT for the required period (for example, treatment periods of 10–45 min were used in studies examining blood glucose) (Fig. 2a). Crossover studies (with mice randomized to treatment or no treatment in the initial arm then switched to the opposite arm) can be helpful to improve the statistical power of studies by allowing paired comparisons.

3.4 Studies Using Magnetic Field Treatment

The mice are treated using a magnetic field generated by the electromagnetic coil of an MRI machine. A gauss meter is needed to define an area outside the coil with a minimum field strength. We used an area where the field strength on the side of the chamber furthest from the coil was at least 0.5 T at the side furthest from the coil so the mice are treated with a minimum field strength of 0.5 T (Fig. 2b). The untreated group was placed in the MRI room, but the field strength they experienced was no greater than that of the earth's magnetic field.

Mice are moved to the preparation room at least 2 h before the start of the study. Temperature and humidity are kept as stable as possible and disturbance is minimized. Mice are placed in the study chamber for a 20-min acclimation period in the preparation room then the chamber is moved to MRI room. Mice in the treatment group are placed in the area with a minimum field strength of 1 T and mice in the control group are placed in the area with a field strength equivalent to the earth's magnetic field.

The investigator will need to determine when in the light–dark cycle it is best to evaluate the behavior being studied.

3.5 Confirmation of Injection Site and Neural Activity

Immunohistochemistry is used to confirm that the correct cell type and region are targeted in radiogenetic studies. It is important to confirm that the virus is expressed in the correct region in the treated mice since it is the virus injection, in combination with cre expression, which targets the appropriate region. RF and magnetic fields are applied to the entire CNS or body and are not regionally applied. Both TRPV1 and GFP-tagged ferritin are generated from a bicistronic construct and so GFP expression should mark transduced cells.

Mice are anesthetized using ketamine/xylazine and then perfused initially with phosphate buffered saline and then 4% paraformaldehyde (PFA). The brain is then removed and post-fixed overnight in cold 4% PFA before being transferred to PBS. The brains can be sectioned on a vibratome (40–50 μm) in multiple series. The sections are can be immunostained for GFP using standard immunohistochemical protocols then mounted on slides, allowed to dry and coverslipped. Fluorescent microscopy is used to visualize GFP and confirm the site of construct expression. Stereological techniques or image analysis software may provide an estimate of the number of infected cells. Virus injection into cre mice crossed to a mouse line with cre-dependent expression of a fluorescent marker (e.g., Rosa-tdTomato (Jackson Laboratories)) can be used to confirm that the virus is only transducing the required cell population.

Immunohistochemistry for c-fos can be used as a marker for neural activation. If needed, the mice should be treated with the radiofrequency or magnetic field 90–120 min before perfusion. However, it is important to remember that anesthesia (needed for treatment with the small RF coil) can induce significant fos expression.

3.6 Troubleshooting

Radiogenetic/magnetogenetic neural modulation relies on correct and adequate expression of a cre-dependent construct in a transgenic mouse line.

A number of viral factors may produce variability. It is important to ensure that high titer viral stocks are used (we used viral stocks with 10^{12} physical particles, 10^{10} plaque forming units) and that the titer is consistent between studies. However, the cre-dependent construct expressing both the modified TRPV1 and the GFP-ferritin chimera is relatively large and packaging this construct is less efficient so it is possible that not all the viral particles are able to transduce cells. This would reduce the efficiency of viral transduction. The viruses are engineered to only infect cre-expressing cells but there may be batch-to-batch variation in the virus so that recombination may produce some viral particles that are not cre-dependent and constructs transduce cells that do not express cre. It may be sensible to test new viral lots by injecting into cre-expressing mice and comparing their ability and specificity to transduce cre-expressing cells with previous lots of virus.

Variability can also be due to the transgenic mouse line and the injection site. The ideal transgenic mouse line would express cre only in a genetically defined cell population that is in a relatively compact anatomical region that can be targeted with a single virus injection or at least with no adjacent regions expressing cre. This is seldom the case. Several adjacent brain areas may express cre and so it is vital to modify the target coordinates and possibly the injection angle to make sure the virus hits the correct region and to adjust the volume to limit spread to surrounding areas. It is obviously important that the transgenic mice are correctly genotyped but there may be ectopic sites of cre expression that result in ectopic construct expression. In addition, there may be variability in viral expression between mouse lines. It should be possible to determine the injection site and numbers of transduced neurons from GFP immunohistochemistry performed at the end of the study and any mice with injections outside predetermined parameters (site, number, specificity) can be excluded from statistical analysis.

The variability often seen in in vivo studies can also be reduced by careful acclimation to changes in housing, test chambers, handling and minimizing stressors as much as possible. However, there is still likely to be some variation and it is best to ensure that treatment and control groups in cross-over studies do not all receive the same treatment at the same time.

4   Conclusion

Radiogenetics and magnetogenetics allow remote, targeted modulation of neural populations in vivo [26, 29–31]. These technologies have distinct characteristics compared to other tools for neuromodulation. Unlike optogenetics, RF and magnetic field treatment enables neural modulation without the need for a permanent optical fiber and without tethering for light delivery during assessment. Limited light penetration also restricts optogenetic modulation to cells immediately adjacent to the optical fiber rather than across larger anatomical areas which can be achieved with radio/magnetogenetics. While chemogenetic techniques can also regulate dispersed neural populations, the faster kinetics of radiogenetics and magnetogenetics [29, 30] and the ability to regulate neural activity without needing to handle the animal and administer an agonist immediately before assessment may be advantageous for certain studies. Radiogenetics and magnetogenetics expand the repertoire of tools available to investigate the physiological roles of specific neural populations and, as these technologies continue to evolve and improve, they may be more broadly applicable for example, to regulate neural activity in the developing nervous system or modulating cell populations in the periphery.

Acknowledgments

The work by S.A.S. was supported by the National Institutes of Health (MH105941).

References

1. Sclafani A (1971) Neural pathways involved in the ventromedial hypothalamic lesion syndrome in the rat. J Comp Physiol Psychol 77:70–96

2. Sternson SM, Shepherd GM, Friedman JM (2005) Topographic mapping of VMH --> arcuate nucleus microcircuits and their reorganization by fasting. Nat Neurosci 8:1356–1363

3. Shimazu T, Fukuda A, Ban T (1966) Reciprocal influences of the ventromedial and lateral hypothalamic nuclei on blood glucose level and liver glycogen content. Nature 210:1178–1179

4. Stock G, Sturm V, Schmitt HP, Schlor KH (1979) The influence of chronic deep brain stimulation on excitability and morphology of the stimulated tissue. Acta Neurochir 47:123–129

5. Boyden ES, Zhang F, Bamberg E, Nagel G, Deisseroth K (2005) Millisecond-timescale, genetically targeted optical control of neural activity. Nat Neurosci 8:1263–1268

6. Arenkiel BR, Klein ME, Davison IG, Katz LC, Ehlers MD (2008) Genetic control of neuronal activity in mice conditionally expressing TRPV1. Nat Methods 5:299–302. https://doi.org/10.1038/nmeth.1190

7. Nawaratne V et al (2008) New insights into the function of M4 muscarinic acetylcholine receptors gained using a novel allosteric modulator and a DREADD (designer receptor exclusively activated by a designer drug). Mol Pharmacol 74:1119–1131

8. Lerchner W et al (2007) Reversible silencing of neuronal excitability in behaving mice by a genetically targeted, ivermectin-gated Cl– channel. Neuron 54:35–49

9. Williams SC, Deisseroth K (2013) Optogenetics. Proc Natl Acad Sci U S A 110:16287. https://doi.org/10.1073/pnas.1317033110

10. Jego S et al (2013) Optogenetic identification of a rapid eye movement sleep modulatory circuit in the hypothalamus. Nat Neurosci 16:1637–1643. https://doi.org/10.1038/nn.3522

11. Betley JN, Cao ZF, Ritola KD, Sternson SM (2013) Parallel, redundant circuit organization for homeostatic control of feeding behavior. Cell 155:1337–1350. https://doi.org/10.1016/j.cell.2013.11.002

12. Domingos AI et al (2011) Leptin regulates the reward value of nutrient. Nat Neurosci 14:1562–1568. https://doi.org/10.1038/nn.2977

13. Chaudhury D et al (2013) Rapid regulation of depression-related behaviours by control of midbrain dopamine neurons. Nature 493:532–536. https://doi.org/10.1038/nature11713

14. Melo CA et al (2001) Characterization of light penetration in rat tissues. J Clin Laser Med Surg 19:175–179. https://doi.org/10.1089/104454701316918925

15. Iyer SM et al (2014) Virally mediated optogenetic excitation and inhibition of pain in freely moving nontransgenic mice. Nat Biotechnol 32:274–278. https://doi.org/10.1038/nbt.2834

16. Dong S, Rogan SC, Roth BL (2010) Directed molecular evolution of DREADDs: a generic approach to creating next-generation RASSLs. Nat Protoc 5:561–573. https://doi.org/10.1038/nprot.2009.239

17. Agulhon C et al (2013) Modulation of the autonomic nervous system and behaviour by acute

glial cell Gq protein-coupled receptor activation in vivo. J Physiol 591:5599–5609. https://doi.org/10.1113/jphysiol.2013.261289

18. Zemelman BV, Nesnas N, Lee GA, Miesenbock G (2003) Photochemical gating of heterologous ion channels: remote control over genetically designated populations of neurons. Proc Natl Acad Sci U S A 100:1352–1357. https://doi.org/10.1073/pnas.242738899

19. Young JH, Wang MT, Brezovich I (1980) Frequency/depth-penetration considerations in hyperthermia by magnetically induced currents. Electron Lett 16:358–359

20. Stauffer PR, Cetas TC, Jones RC (1984) Magnetic induction heating of ferromagnetic implants for inducing localized hyperthermia in deep-seated tumors. IEEE Trans Bio-med Eng 31:235–251. https://doi.org/10.1109/TBME.1984.325334

21. Halperin D et al (2008) Pacemakers and implantable cardiac defibrillators: software radio attacks and zero-power defenses. P IEEE S Secur Priv. pp 129–142

22. Kedzior KK, Gellersen HM, Brachetti AK, Berlim MT (2015) Deep transcranial magnetic stimulation (DTMS) in the treatment of major depression: an exploratory systematic review and meta-analysis. J Affect Disord 187:73–83. https://doi.org/10.1016/j.jad.2015.08.033

23. Hergt R, Anrda W, d'Ambly CG, Hilger I, Kaiser W, Richter W, Schmidt H-G (1998) Physical limits of hyperthermia using magnetite fine particles. IEEE Trans Magn 34:3745–3754

24. Fortin JP et al (2007) Size-sorted anionic iron oxide nanomagnets as colloidal mediators for magnetic hyperthermia. J Am Chem Soc 129:2628–2635

25. Stanley SA et al (2012) Radio-wave heating of iron oxide nanoparticles can regulate plasma glucose in mice. Science 336:604–608. https://doi.org/10.1126/science.1216753

26. Chen R, Romero G, Christiansen MG, Mohr A, Anikeeva P (2015) Wireless magnetothermal deep brain stimulation. Science 347:1477–1480. https://doi.org/10.1126/science.1261821

27. Caterina MJ et al (1997) The capsaicin receptor: a heat-activated ion channel in the pain pathway. Nature 389:816–824. https://doi.org/10.1038/39807

28. Stanley S et al (2013) Profiling of glucose-sensing neurons reveals that GHRH neurons are activated by hypoglycemia. Cell Metab 18:596–607. https://doi.org/10.1016/j.cmet.2013.09.002

29. Stanley SA et al (2016) Bidirectional electromagnetic control of the hypothalamus regulates feeding and metabolism. Nature 531:647–650. https://doi.org/10.1038/nature17183

30. Wheeler MA et al (2016) Genetically targeted magnetic control of the nervous system. Nat Neurosci 19:756–761. https://doi.org/10.1038/nn.4265

31. Long X, Ye J, Zhao D, Zhang SJ (2015) Magnetogenetics: remote non-invasive magnetic activation of neuronal activity with a magnetoreceptor. Sci Bull (Beijing) 60:2107–2119. https://doi.org/10.1007/s11434-015-0902-0

32. Theil EC (2013) Ferritin: the protein nanocage and iron biomineral in health and in disease. Inorg Chem 52:12223–12233

33. Arosio P, Ingrassia R, Cavadini P (2009) Ferritins: a family of molecules for iron storage, antioxidation and more. Biochim Biophys Acta 1790:589–599. https://doi.org/10.1016/j.bbagen.2008.09.004

34. Zhang Y, Orner BP (2011) Self-assembly in the ferritin nano-cage protein superfamily. Int J Mol Sci 12:5406–5421. https://doi.org/10.3390/ijms12085406

35. Iordanova B, Robison CS, Ahrens ET (2010) Design and characterization of a chimeric ferritin with enhanced iron loading and transverse NMR relaxation rate. J Biol Inorg Chem 15:957–965. https://doi.org/10.1007/s00775-010-0657-7

36. Galvez N et al (2008) Comparative structural and chemical studies of ferritin cores with gradual removal of their iron contents. J Am Chem Soc 130:8062–8068. https://doi.org/10.1021/ja800492z

37. Garcia-Prieto A et al (2015) On the mineral core of ferritin-like proteins: structural and magnetic characterization. Nanoscale 8:1088–1099. https://doi.org/10.1039/c5nr04446d

38. Kirchhofer A et al (2010) Modulation of protein properties in living cells using nanobodies. Nat Struct Mol Biol 17:133–138. https://doi.org/10.1038/nsmb.1727

39. Gilles C, Bonville P, Rakoto H, Broto JM, Wong KKW, Mann S (2002) Magnetic hysteresis and superantiferromagnetism in ferritin nanoparticles. J Magn Magn Mater 241:430–440

40. Cho A, Haruyama N, Kulkarni AB (2009) Generation of transgenic mice. Curr Protoc Cell Biol Chapter 19:Unit 19 11. https://doi.org/10.1002/0471143030.cb1911s42

41. Haruyama N, Cho A, Kulkarni AB (2009) Overview: engineering transgenic constructs and mice. Curr Protoc Cell Biol Chapter 19:Unit 19 10. https://doi.org/10.1002/0471143030.cb1910s42

42. Luo J et al (2007) A protocol for rapid generation of recombinant adenoviruses using the AdEasy system. Nat Protoc 2:1236–1247. https://doi.org/10.1038/nprot.2007.135

Chapter 6

Two Applications of Gold Nanostars to Hippocampal Neuronal Cells: Localized Photothermal Ablation and Stimulation of Firing Rate

Fidel Santamaria and Xomalin G. Peralta

Abstract

We present a procedure to synthesize gold nanostars which are readily internalized by mouse hippocampal cells. These nanoparticles tend to localize close to the nuclei and have a surface plasmon resonance in the near infrared therefore we can stimulate their luminescence with two-photon technology while imaging structural or functional neuronal fluorescent markers. This allows us to evaluate the effect of the nanoparticles on the neuron's behavior without any other external stimuli. We found that these nanoparticles increase the firing rate by modifying the activity of the potassium channels. In addition, by increasing the intensity of the laser used for imaging, we can stimulate the one-photon surface plasmon mode of the nanoparticles to destroy single cells and organelles containing nanostars while neighboring cells remain intact.

Key words Nanoparticles, Two-photon microscopy, Photothermal ablation

1 Introduction

In recent years, there has been an increased interest in using nanoparticles in neuroscience research for various applications including neuroimaging, drug delivery, photothermal ablation, and neuronal stimulation. Given that their optical properties can be tuned by size and shape, some of these applications make use of metallic nanoparticles such as gold or silver. Metallic nanoparticles do not photobleach, they do not blink when imaged, and they can be functionalized in order to localize them at the subcellular level. Another property of metallic nanoparticles is that when a metal nanoparticle is illuminated by light, a coherent oscillation of the electrons can be excited at the interface of the metal nanoparticle and the surrounding media, known as a surface plasmon. This results in an absorbance peak in the nanoparticle's spectra. Most studies which make use of nanoparticles for biological applications select gold as the material of choice due to its inert nature. The peak

Fidel Santamaria and Xomalin G. Peralta (eds.), *Use of Nanoparticles in Neuroscience*, Neuromethods, vol. 135, https://doi.org/10.1007/978-1-4939-7584-6_6, © Springer Science+Business Media, LLC 2018

surface plasmon resonance of spherical gold nanoparticles is in the visible range of the electromagnetic spectrum [1], where tissue has a high absorptivity. This property imposes a limitation on the biological application of spherical gold nanoparticles. However, there are methods in place that can shift the peak plasmon resonance of gold nanoparticles to the near-infrared (NIR) region of the electromagnetic spectrum, where tissue (hemoglobin and water) has the highest transmissivity [2]. This can be accomplished by coating a dielectric nucleus with a thin gold layer [3] or by modifying the shape of the nanoparticle into rods [4–6], boxes [7], or stars [8, 9]. In the former, the resonant wavelength is a function of the layer's thickness and diameter of the particle [10–12] while in the latter it is directly related to the shape of the nanoparticle.

In this study, we use a seed-mediated and silver-assisted growth method to obtain a high yield of star shaped gold nanoparticles with a strong surface plasmon resonance in the NIR [13]. Some reports, and our own experiments, indicate that the internalization of nanoparticles by cells is shape dependent [14, 15]. It is also known that some nanoparticles inherently localize in different cell types or at specific locations within a cell, enabling their use without functionalizing them [16, 17]. In particular, the nanostars produced with this method were readily internalized into neurons, presumably by endocytosis [18–20], even without any biological functionalization. To fully develop any of the applications of nanomaterials to neuroscience proposed to date, there is a need to understand the mechanism through which functionalized and bare nanoparticles interfere with the normal functioning of a cell. This includes evaluating lethal effects as well as nonlethal effects which can result in changes in the neuron's behavior [19, 21, 22].

1.1 Effect of Nanoparticles on Neuronal Function

We have recently shown that star shaped gold nanoparticles are rapidly internalized by neurons across the nervous system without any surface functionalization making them ideal candidates for biomedical applications that target cells without biological functionalization [23]. By engineering the shape and size of the nanoparticles, we can stimulate their luminescence with two-photon technology. Thus, we can simultaneously visualize our nanoparticles while imaging structural or functional neuronal fluorescent markers, allowing us to study **the nonlethal effects of nanostars internalized by mouse hippocampal cells**. We found that the mere presence of the nanoparticles within the neurons has an effect on the firing rate of action potentials highlighting the need for more studies on the nontoxic functional effects of nanoparticles on neurons.

1.2 Photothermal Ablation for Nanosurgery

Photothermal ablation is an emergent therapeutic technology that is used for the targeted removal of specific tissues, including cancer [24–26]. Recently, there has been an increase in the use of nanoparticles to mediate photothermal ablation due to the potential for reducing collateral damage and for enabling nanoscale ablation capabilities [27–29]. When a metal nanoparticle is resonantly stimulated at the surface plasmon's wavelength, the nanoparticle absorbs the photon energy and becomes a heat generator. Some of the thermal energy goes into heating the media surrounding the nanoparticles [25]. Depending on the volume affected, this nonradioactive photothermal transfer of energy can be used to locally denaturize proteins, ablate organelles, destroy single cell or result in the removal of the surrounding tissue [30, 31]. The efficiency of this process depends both, on the efficiency of stimulating the surface plasmon mode, the efficiency of the light-to-heat conversion and the thermodynamic processes that control the heat transfer between the nanoparticle and the surrounding media [32].

By selecting the appropriate wavelength and intensity of the same femtosecond pulsed laser used for imaging, we were able to destroy single cells containing nanostars while neighboring cells remained intact by stimulating the one-photon surface plasmon mode of the nanoparticle. Furthermore, fine spatiotemporal excitation resulted in the ablation of intracellular organelles, in particular the nucleus, allowing for targeted removal of sub-cellular structures allowing us to study **nanoparticle-assisted photothermal ablation of neuronal cells and organelles** [32]. This technique could be further refined by introducing biological functionalization of the nanoparticles' surface in order to target specific intracellular structures [33–35].

In this chapter we detail the procedure to synthesize star shaped gold nanoparticles with a strong surface plasmon resonance in the NIR. We provide a detailed explanation of the procedure to prepare mouse brain slices for imaging by the addition of fluorescent markers and nanoparticles. We also describe the procedure to stimulate the nanoparticles for photothermal ablation studies and, independently, the experimental procedure followed to perform electrophysiological recordings in the presence of the nanoparticles. We conclude the chapter with an explanation on the interpretation of the experimental data collected and potential pitfalls and caveats that may appear when following the procedures described herein.

2 Materials

The care and use of animals followed standard procedures approved by the UTSA Institutional Animal Care and Use Committee (IACUC).

2.1 Nanoparticle Synthesis

- Deionized (DI) water.
- Silver nitrate ($AgNO_3$, 204390, Sigma).
- Sodium citrate tribasic dehydrate ($Na_3C_6H_5O_7$, S4641, Aldrich).
- Sodium borohydride ($NaBH_4$, 213462, Aldrich).
- Ascorbic acid ($C_6H_8O_6$, 255564, Sigma-Aldrich).
- Hexadecyltrimethylammonium bromide (CTAB—$C_{19}H_{42}BrN$, H6269, Sigma).
- Gold chloride trihydrate ($HAuCl_4$, 520918, Aldrich).
- Pipettes.
- Beaker.
- Ice.
- 20 mL scintillation vials.
- 15 mL conical centrifuge tubes.
- Refrigerator.
- Timer.
- Stirring bar.
- Hot-plate magnetic stirrer.
- Sonicator.
- Centrifuge.
- Aluminum foil (optional).

2.2 Particle Characterization

- Transmission electron microscope (JEOL 2010-F, JEOL).
- Transmission electron microscope used in scanning electron microscope mode and for energy-dispersive X-ray spectroscopy (Hitachi S-5500, Hitachi).
- UV/Vis/NIR Spectrophotometer (Olis Cary-14 Spectrophotometer, Olis).

2.3 Brain Slice Preparation and Labeling

- C57/BL6NJ mice 12–17 days old for photothermal ablation (PTA)/14–21 days old for nonlethal effects (NLE).
- Artificial cerebrospinal fluid (aCSF) composed of
  - 125 mM of sodium chloride (NaCl, Fisher Scientific, USA).
  - 2.5 mM of potassium chloride (KCl, Fisher Scientific).
  - 2 mM of calcium chloride ($CaCl_2$, Fisher Scientific).
  - 1.3 mM of magnesium chloride ($MgCl_2$, Fisher Scientific).
  - 1.25 mM of sodium phosphate monobasic monohydrate (NaH_2PO_4, Fisher Scientific).

- 26 mM of sodium bicarbonate ($NaHCO_3$, Fisher Scientific).

- 20 mM of D-glucose (Fisher Scientific).

- Brain slices were fixed in 0.1 M Phosphate buffered solution at 10% formalin (4% formaldehyde (CH_2O,)).

- Sodium hydroxide (NaOH).

- Alcohol.

- Distilled water.

- Potassium permanganate solution ($KMnO_4$).

- 4′,6-diamidino-2-phenylindole (DAPI, Vector Laboratories, Burlingame, CA)

- Cell tracker green (CMFDA, Molecular Probes, Thermo Fisher Scientific) dissolved in Dimethyl sulfoxide (DMSO, Sigma-Aldrich) at a concentration of 1 mM.

- Microtome.

- Circulating bath chamber.

- Coverslip or microscope slide.

- Incubator.

- Glass capillary tubes.

- Electrode maker.

2.4 Electro-physiology and Analysis

- Alembic VE2 amplifier (Alembic Instruments; Montreal, Canada).

- Axon pclamp software (Molecular Devices; Sunnyvale, CA).

- Micro syringe pump (Harvard Apparatus, Cambridge, MA).

- Matlab.

2.5 Microscopy, Imaging, and Analysis

- Power meter.

- Two-photon laser scanning microscope (Prairie Technologies, Madison, WI, USA) with a 20× 0.9 N.A. water immersion objective (Olympus) coupled to a tunable (700–950 nm) femtosecond (pulse duration <150 fs) Ti:sapphire laser (Chameleon, Coherent, Santa Clara, CA, USA) with a 90 MHz repetition rate.

- 3D volumetric reconstructions (Imaris software, Bitplane, Zurich, Switzerland).

- Matlab.

3   Methods

3.1   Nanoparticle Synthesis

It is well known that the production of gold nanoparticles in solution is sensitive to the timing of the procedures as well as to the concentration of the reactants. Small variations of a specific synthesis protocol can result in nanoparticles of different size, shape, and concentration. Therefore, care should be taken to follow the protocols described below carefully.

We developed a method to chemically synthesize star shaped gold nanoparticles in suspension, also known as nanostars. This method can be broken up into two parts: (1) first we synthesize silver nanoparticles and (2) subsequently use them to grow the gold nanostars. The silver nanoparticles function as nucleation sites for the gold nanoparticles and are therefore referred to as silver seeds. Although the silver seeds are the nucleating agents for the anisotropic growth of the gold nanoparticles, the nanostars do not contain any silver atoms.

3.1.1   Preparation of Silver Seeds

Stock Solutions

1. Prepare 20 mL of a silver nitrate ($AgNO_3$) solution at 0.25 mM by mixing silver nitrate with deionized (DI) water. Keep the solution in the dark.

2. Prepare a 5 mM solution of sodium citrate tribasic ($Na_3C_6H_5O_7$) by adding 14.7 mg of sodium citrate tribasic to 10 mL of DI water and shaking the vial until the powder is fully dissolved.

3. Prepare a 40 mM solution of sodium borohydride ($NaBH_4$) by adding 15.1 mg of sodium borohydride to 10 mL of DI water. Close the vial immediately. Shake the solution gently by hand and place the vial in a beaker with ice.

4. Keep the beaker in the refrigerator for 15 min.

Silver Seed Solution

1. Place 10 mL of the silver nitrate stock solution in a 20 mL scintillation vial.

2. Place a stirring bar in the vial and start stirring.

3. Add 0.25 mL of the sodium citrate tribasic solution.

4. Remove the sodium borohydride solution from the refrigerator once the 15 min are over, and add 0.4 mL to the silver nitrate and sodium citrate tribasic mixture in a single, quick stroke. The solution will turn yellow (Fig. 1a).

5. Stir the solution for 5 min.

6. Stop stirring and remove the stirring bar. Do **not** close the vial.

7. Place the vial in the dark at room temperature. You can use aluminum foil to cover them up.

8. Wait for at least 2 h before using the seeds and at most 1 week.

3.1.2　Gold Nanostar Growth Solution

1. Turn on the hot-plate magnetic stirrer and set the temperature to 30 °C.

2. Prepare 10 mL of a concentrated solution of gold chloride ($HAuCl_4$) by mixing gold chloride with deionized (DI) water. Calculate the molarity of the solution. Keep the solution in the dark.

3. Prepare an 80 mM solution of ascorbic acid ($C_6H_8O_6$) by adding 140 mg to 10 mL of DI water.

4. Prepare a 50 mM solution of cetyltrimethylammonium bromide (CTAB—$C_{19}H_{42}BrN$) by adding 364 mg to 20 mL of DI water. Add a stirring bar to the vial and place on the warm plate. Start stirring.

5. After the CTAB powder is completely dissolved and the solution becomes transparent, turn off the heater. Keep stirring.

6. Add enough silver nitrate solution to obtain a solution with a final molarity of 0.049 mM.

7. After 1 min, add enough gold chloride solution to obtain a solution with a final molarity of 0.25 mM.

8. After one additional minute, add 0.1 mL of the ascorbic acid solution. The solution will turn colorless.

9. After 20 s add 0.05 mL of the silver seed solution and continue stirring for 15 min. The suspension will turn blue and then brown (Fig. 1b, c).

10. Stop stirring. Remove the stirring bar and keep the suspension at room temperature for 24 h.

If the gold nanostars are kept in this CTAB solution in the dark, they will retain their shape for at least 1 month after synthesis.

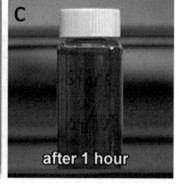

Fig. 1 Optical images of scintillation vials containing: (**a**) silver seeds, (**b**) gold nanostar growth solution immediately after application of the silver seed solutions and (**c**) after nanostar formation. *Adapted from ref.* 13

3.1.3 Gold Nanostar Extraction and Rinsing

In order to utilize the nanostars in biological experiments, it is necessary to extract them and thoroughly rinse them to remove any residual CTAB or other remaining chemical components as explained below. Note that CTAB may crystallize at room temperature. In that case you will need to immerse the suspension in hot tap water or heat it up to 30 °C on a hot plate until the crystals dissolve before collecting and rinsing the nanoparticles as described below.

1. Take 10 mL of the gold nanostar growth solution and place it in a centrifuge tube.

2. Sonicate the suspension for 2 min.

3. Centrifuge the suspension for 5 min at 730 relative centrifugal force (rcf). The nanostars will accumulate on the walls of the tube.

4. Remove the liquid from the central portion of the tube with a pipette.

5. Add DI water to the tube and sonicate for 2 min.

6. Centrifuge the suspension for 3 min at 460 rcf.

7. Remove the liquid from the central portion of the tube with a pipette.

8. Add DI water to the tube and sonicate for 2 min.

9. Centrifuge the suspension for 3 min at 380 rcf.

10. Remove the liquid from the central portion of the tube with a pipette.

11. Add DI water to the tube to obtain a 3 mL suspension.

3.1.4 Gold Nanostar Characterization

Electron Microscopy

1. Place a drop of the gold nanostar solution onto a commercial copper TEM grid.

2. Allow it to dry in air.

3. Utilize transmission electron microscope and scanning electron microscope to image the nanoparticles individually or on larger scale to obtain size and shape distribution. Representative images are shown in Fig. 2.

4. To probe the chemical composition of the nanostars, utilize an electron microscope equipped with energy dispersive spectroscopy (EDS). Typical results are shown in Fig. 3.

UV/Vis/NIR Spectra

1. Take 10 mL of the gold nanostar solution and place it in a 20 mL scintillation vial.

2. Place the scintillation vial in the sample chamber of the UV/Vis/NIR spectrometer and take a spectrum (Fig. 4).

3. Use the spectra of DI water as the baseline.

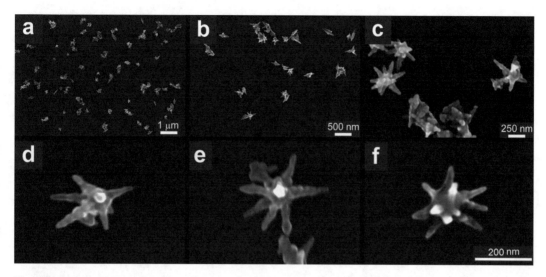

Fig. 2 Scanning electron microscope images of gold nanostars (**a–c**) and individual nanostars (**d–f**). *Modified from ref.* 13

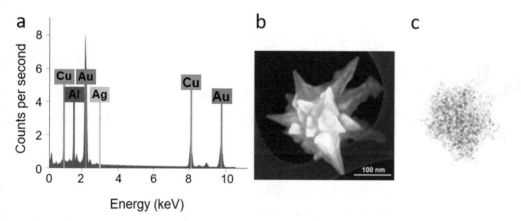

Fig. 3 Energy dispersive X-ray spectrometry. (**a**) Spectra of gold nanostars. (**b**) Scanning electron microscope image of gold nanostar and (**c**) distribution of gold inside nanostar shown in **b**)

3.2 Brain Slice Preparation

3.2.1 Acute Hippocampal Slice Preparation

1. Euthanize 12–17-day-old mice for PTA) / 14–21-day-old mice for NLE.

2. Quickly remove the brain and section into 200 μm thick slices.

3. Incubate for 30 min for PTA (35 min for NLE) in artificial cerebrospinal fluid (aCSF) at 36 °C for PTA (37 °C for NLE).

4. For NLE, transfer slices to chamber, bathe in room temperature circulating oxygenated aCSF at about 2 mL/min throughout the experiments [36, 37].

3.2.2 Fixed Tissue Preparation

1. Transfer slices to a chamber containing 1 mL of aCSF and 20 mL of 0.02 mM gold nanostars for 1–3 h.

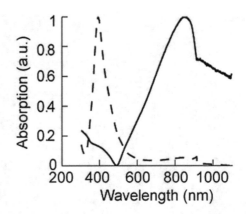

Fig. 4 Normalized absorption spectra of silver seeds (dashed line) and gold nanostars (solid line) [13]

2. Wash slices two times, that is:

  (a) Immerse slices in aCSF for 10 min.

  (b) Replace aCSF.

  (c) Immerse slices in aCSF for another 10 min.

3. Immerse slices in 0.1 M phosphate buffered at 10% formalin for 30 min.

4. Wash slices in aCSF three times as indicated above in step 2).

5. Transfer to a solution containing 1% sodium hydroxide in 80% alcohol for 5 min.

6. Transfer to a solution containing 1% sodium hydroxide in 70% alcohol for 2 min.

7. Transfer to distilled water for 2 min.

8. Postfix the slices by immersing them in a 0.06% potassium permanganate solution for 10 min.

9. Wash slices in distilled water for 2 min as indicated in step 2 above.

10. Transfer slices to a coverslip and mount them with a drop of the nuclear stain DAPI for imaging.

3.2.3 Live Tissue Preparation

1. Dissolve the cytosolic marker cell tracker green CMFDA in DMSO at a concentration of 1 mM.

2. Prepare a 2 μM solution of cell tracker in aCSF.

3. Incubate slices in the 2 μM solution for 15 min at 37 °C.

4. Wash slices in aCSF for 10 min, two times, that is

  (a) Immerse slices in aCSF for 10 min.

  (b) Replace aCSF.

  (c) Immerse slices in aCSF for another 10 min.

5. Incubate slices with 1 mL of aCSF and 20 mM gold nanostars for 2 h at room temperature.

6. Wash slices in aCSF for 10 min, two times as indicated in step 4) above.

7. Image while in a room temperature circulating bath of aCSF at about 2 mL/min.

3.3 Imaging and Nanoparticle Stimulation

A Prairie Technologies two-photon laser scanning microscope with a 20× 0.9 N.A. water immersion objective was used to collect 2D z-stacks of fixed slices and time lapse images of fixed and live brain slices. The laser is a Chameleon Ti:sapphire tunable laser that, depending on the power output, was used only to generate (1) the two-photon fluorescence of DAPI or cell-tracker and (2) the luminescence of the gold nanostars, or (3) also to stimulate the gold nanoparticles. Before starting these experiments, care should be taken to calibrate the relative laser power indicated in the software that controls the microscope to the actual power output at the imaging plane. This can be done as follows.

3.3.1 Calibrate Laser Power Output

1. Placing a power meter close to the imaging plane.

2. Measure the power.

3. By tuning the relative power indicated in the software, set the measured power to about 0.6 mW (1.6 mW)—this is the power used in our system for imaging fixed (live) brain slices.

4. Note the relative power reading corresponding to 0.6 mW (1.6 mW).

5. Increase the relative power so that the power meter reads 1.44 mW (3.6 mW)—this is the power used to simultaneously image and stimulate the nanoparticles in fixed (live) brain slices with our system.

6. Note the relative power corresponding to 1.44 mW (3.6 mW).

7. This number can be used to correlate a relative power reading to the actual power delivered at the focal point.

The powers quoted are the ones used in our experiments and should be only used as a guide. In our experience, there is some variability in the laser power needed to image and stimulate the surface plasmon mode. One of the reasons for this is the fact that we are using 200 μm thick brain slices which contain the cells, nuclei, and nanoparticles at varying depths. The deeper we are imaging in a slice, there is more light scattered and the cells experience different tensions and thermal dissipation due to the presence of the surrounding cells. In addition, the smaller excitation volume of a two-photon microscope as compared to that of a confocal microscope influences the power needed to stimulate nanoparticles that are not precisely at the focal point of the microscope. We also observed that for the live cell experiments, we had to use higher laser powers both for imaging and for stimulation compared to those used during fixed tissue experiments. This is possibly due to the cooling effect of the constantly circulating bath and the small optical distortions that it can introduce as well as the effect of the fluorescent markers used on the optical properties of the tissue.

3.3.2 Collection of z-Stacks for 3D Localization of Nanoparticles

1. For imaging the fixed (live) brain slices, tune the laser to 720 nm (750–760 nm) and set it to 0.6 mW (1.6 mW).

2. Separate the fluorescent photons into two spectral windows: one for DAPI (cell tracker) spanning 435–485 nm, and one for the nanoparticle's luminescence spanning 584–630 nm.

3. Using the scanning mode of the microscope, collect 2D image stacks separated vertically between 0.1 and 0.2 μm to a depth of 100 μm. Use a collection rate of 0.02–0.45 frames/s with dwell times of 2–30 μs.

3.3.3 Volumetric Reconstruction of the Slices for Nanoparticle Localization

The following procedure is used to find the location of the nanoparticles within the cells. Given the time it takes to implement it, it is recommended to perform this analysis off-line.

1. Merge the images collected from the z-stacks for each channel in order to generate a 3D reconstruction of the first 100 μm of the slice. The volumetric reconstruction from the channel corresponding to DAPI localizes the nuclei of the cells while the other one localizes the nanoparticles. In the case of cell tracker green, the images show the fluorescent signal from the cytosol with a dark nucleus (Fig. 5a, b).

2. Merge the volumetric reconstructions from both channels.

3. Set the transparency to 50% in order to visualize the location of the nanoparticles relative to the nuclei (Fig. 5c). This allows you to locate and count the number of nanoparticles within the first 100 μm of the slice.

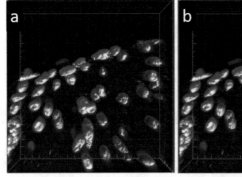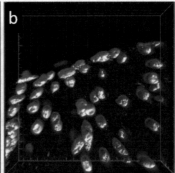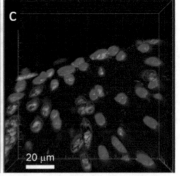

Fig. 5 Volumetric reconstruction of the molecular layer of a cerebellar slice incubated with gold nanostars. (**a**) Nuclei of molecular layer neurons. (**b**) Clusters of nanoparticles identified using the red channel. (**c**) Same data as in **b**) with the transparency of the reconstructed surfaces set to 50% showing the presence of nanoparticles inside the nuclei

3.3.4 Determine the Minimal Laser Energy Necessary to Ablate the Cells

1. Focus the microscope to an imaging plane in which both nanoparticles and cells are visible.

2. Collect an image using the scanning mode of the microscope.

3. Increase the laser's power by 2 mW and collect ten images. Use a collection rate of 0.02–0.45 frames/s with dwell times of 2–30 µs.

4. Check for any changes in the shape or fluorescence of the cells.

5. Repeat steps 3 and 4 until you see an effect.

6. Move to a different part of the same slice at a similar depth to collect a time-lapse series of images during nanoparticle stimulation.

3.3.5 Collection of Time Lapse Images During Nanoparticle Stimulation

1. Focus the microscope to an imaging plane in which both, nanoparticles and nuclei are visible.

2. Collect an image in both channels using the scanning mode of the microscope.

3. Merge the information contained in both channels to localize the nanoparticles within the cells in that plane.

4. Acquire up to ten images at this low laser power. Use a collection rate of 0.02–0.45 frames/s with dwell times of 2–30 µs.

5. Increase the laser's power based on your previous findings (Sect. 3.3.4) in order to stimulate the nanoparticles' surface plasmon.

6. Continue imaging for up to 500 s at this high power.

It is worth noting that throughout these experiments, the scanning mode of the microscope is being used, therefore every pixel in the image receives the same laser energy.

Interpreting the Experiments on Fixed Slices

The fixed brain slices are stained with the nuclear stain DAPI therefore, when imaged, the cell nuclei fluoresce. When you select cells that contain nanoparticles inside the nucleus, increasing the laser power will result in either (1) a loss of fluorescence within an approximately circular area close to where the nanoparticles are located, see Fig. 6a–e), or (2) a continuous expansion and deformation of nanoparticle containing nuclei, see Fig. 6f–i). Both of these observations can be explained by the considering that the electromagnetic radiation is absorbed by the nanoparticle during the exposure time. It is then dissipated into heat during that exposure and during the time it takes for the laser to come back to that location, i.e., time between frames. If your observations follow (1) above, then there was no cumulative effect possibly because, after stimulation, the nanoparticles moved out of the focal plane of the microscope.

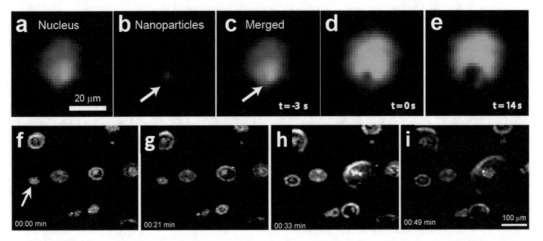

Fig. 6 Temporal sequence showing the photothermal ablation of nanoparticle containing nuclei. (**a**) DAPI stained nucleus. (**b**) Luminescence showing a cluster of nanoparticles (arrow). (**c**) Merged image from **a**) and **b**) showing location of nanoparticles within the nuclei (**d–e**). Increasing the laser power from 0.88 to 1.55 mW produces a loss of fluorescence of a section of the nucleus centered on the nanoparticles [32]. (**f**) Nanoparticle containing nucleus before ablation (arrow). (**g–i**) Cumulative effect of a continued pulsed laser irradiation at 2.4 W/cm^2. Stimulation wavelength 720 nm

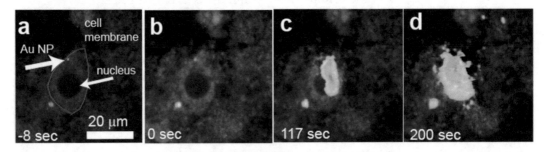

Fig. 7 Temporal sequence of single cell ablation in live cells. Purkinje cell containing nanoparticles imaged **a**) before and **b–d**) after increasing the laser power from 1.37 to 4.06 mW. The line in (**a**) locates the cell boundary [32]

If instead they follow (2) above, then there was a cumulative effect due to the continued imaging at the high (stimulating) laser power which resulted in repeated stimulation of the nanoparticles, heat transfer and a resulting deformation.

Interpreting the Experiments on Live Slice

The live brain slices were stained with a cytosolic marker, and therefore the images collected show the fluorescent signal of the cytosol and a dark nucleus. When you select cells in which the nanoparticles are located close to the nucleus, increasing the laser power will result in a region of increased fluorescence originating at the nanoparticle's location that grows as time goes on, see Fig. 7. We interpret this increased fluorescence as the destruction of the cytosol via thermally induced protein denaturation due to nanoparticle stimulation. If irradiation continues, the region continues to expand and can start to degrade the cellular membrane.

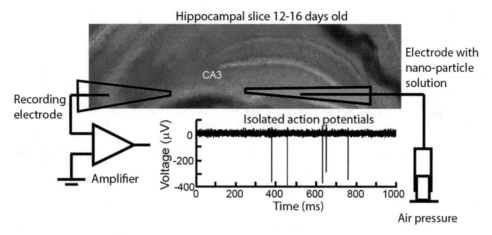

Fig. 8 Experimental setup used to record and deliver star shaped gold nanoparticles to the CA3 region of mouse hippocampal slices. Transmitted light image obtained using an Olympus BX61WI microscope with a 20× 0.95 N.A. objective [23]

3.4 Electro-physiology

Firing rate activity recordings were obtained using an Alembic VE2 amplifier together with Axon pclamp software. The signal was filtered between 1 and 2 kHz to isolate action potential waveforms and sampled at 10 kHz. Action potentials were recorded and isolated using a combination of voltage thresholds. The data was then imported into Matlab to be further analyzed.

1. Fabricate recording electrodes from glass capillary tubes (1–3 MΩ).

2. Bring a recording electrode filled with 0.2 μL of 5 M NaCl in close proximity to the CA3 area of the slice. This section contains the cell bodies of excitatory hippocampal pyramidal cells.

3. Place a second electrode containing regular aCSF connected to a micro syringe pump close to the first electrode using a micromanipulator (Fig. 8).

4. Record the firing rate activity for 20–50 min.

5. Deliver approximately 1 mL of gold nanostar solution by pressure injection during a 5 min window with continuous perfusion. This allows the nanoparticles that did not attach to neurons to be washed out.

6. Record the firing rate activity for 20–50 min.

3.4.1 Analysis of Average Firing Rate

1. Average the firing rate for 20–50 min before and after application of the gold nanoparticle solution.

2. Compare the firing rates.

The top (bottom) panel in Fig. 9 shows the firing rate activity of isolated hippocampal CA3 neurons before (after) the application of gold nanoparticles. These results indicate that acute nanoparticle application on neurons results in an increase in firing rate.

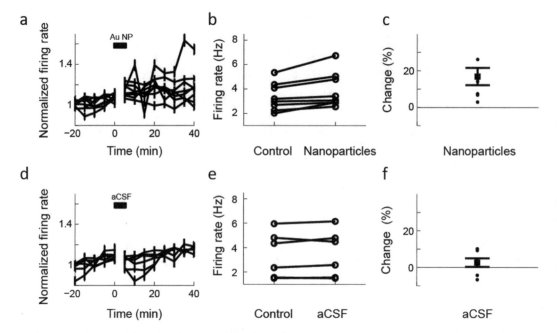

Fig. 9 Transient application of gold nanoparticles increases hippocampal neuronal activity. (**a**) Firing rate average from extracellularly recorded CA3 hippocampal neurons before and after gold nanoparticle (Au NP) application. The activity was averaged every 5 min. Bars are S.E.M. Firing rates were normalized to values before application. (**b**) Absolute firing rate of all the experiments in (**a**) before and after nanoparticle application. Firing rate after application was averaged from $t = 20$ to $t = 50$ min. (**c**) Percentage change of firing rate application from **b**. Error bars are for the S.E.M. (**d–f**) Identical analysis as in (**a–c**) when the application pipette only contained artificial cerebrospinal fluid (aCSF) and no nanoparticles [23]

3.4.2 Analysis
of the Action Potentials

1. Compare the spike height and spike width of the average action potential shape before and after nanoparticle application. As shown in Fig. 10a), we did not find any changes in these quantities.

2. Determined the time of the minimum voltage deflection in each action potential, which is the time of maximum potassium current activation.

3. Starting from the minimum voltage (maximum potassium activation) integrate the area of the voltage trace for 0.2 ms, which corresponded to about 30% of the repolarization period (shaded area Fig. 10b) before and after nanoparticle application.

4. Quantify these values in all the experiments before and after the application of nanoparticles.

Figure 10c) shows the results obtained from our experiments. They indicate that there is a significant decrease of 5.0% ± 1.8 S.E.M ($p < 0.05$) in the current associated with potassium channels. Thus, our data suggests that our gold nanoparticles preferentially affect potassium channels.

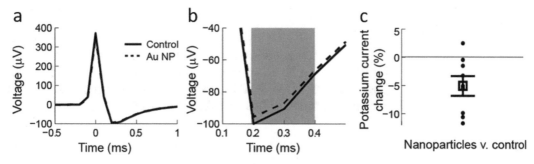

Fig. 10 Transient application of gold nanoparticles reduces the potassium associated current in CA3 hippocampal neurons. (**a**) Average action potential before (control) and after application of gold nanoparticles (Au NP). The overlay at this scale makes the traces indistinguishable. (**b**) The hyperpolarization region of the action potential (from A) is associated with potassium currents. (**c**) The percent difference in the shaded area integrated from (**b**) before and after application of gold nanoparticles for all the experiments (5.0% ± 1.8 S.E.M., $n = 8$ experiments). *Adapted from* [23]

4 Notes

4.1 On the Preparation of the Seed Solution: Sect. "Silver Seed Solution"

1. The sodiuorohydride solution should be prepared fresh every time and used within 1 h of preparation.

2. The sodium borohydride solution **must** be kept cold—step 4).

3. Make sure you add the sodium borohydride solution to the silver nitrate and sodium citrate tribasic mixture in a **single, quick** stroke—step 4).

4. After adding the sodium borohydride solution and removing the stirring bar, it is very important **not to close** the vial in order to allow hydrogen to escape—step 6).

5. If the solutions do not have the correct color, do not use. Start over.

4.2 On the Preparation of the Gold Nanostar Growth Solution: Sect. 3.1.2

1. The rate at which the compounds are added is critical for producing nanostars—steps 7 thru 9).

2. If the solution does not have the correct color, do not use. Start over.

4.3 On Imaging and Nanoparticle Stimulation: Sect. 3.1.3

1. In our experience, most of the nanoparticles that we can image with the two-photon microscope are clusters of nanoparticles. In order to estimate their size we fitted a Gaussian to the intensity profile obtained from the nanoparticle luminescence and centered it on the local intensity peak.

Acknowledgments

This project was supported by the NIH/NIGMS MARC U*STAR GM07717, the National Institute on Minority Health and Health Disparities RCMI G12MD007591 from the National Institutes of Health and the NSF PREM DMR 0934218.

References

1. Jain PK, Lee KS, El-Sayed IH, El-Sayed MA (2006) Calculated absorption and scattering properties of gold nanoparticles of different size, shape, and composition: applications in biological imaging and biomedicine. J Phys Chem B 110((14)):7238–7248. https://doi.org/10.1021/Jp0571700

2. Tsai CLC, Chen J-C, Wang WJ (2001) Near-infrared absorption property of biological soft tissue constituents. J Med Biol Eng 21(1):7–14

3. Bardhan R, Chen W, Perez-Torres C, Bartels M, Huschka RM, Zhao LL, Morosan E, Pautler RG, Joshi A, Halas NJ (2009) Nanoshells with targeted simultaneous enhancement of magnetic and optical imaging and Photothermal therapeutic response. Adv Funct Mater 19(24):3901–3909

4. Link S, El-Sayed MA (2000) Shape and size dependence of radiative, non-radiative and photothermal properties of gold nanocrystals. Int Rev Phys Chem 19(3):409–453

5. Link S, El-Sayed MA (1999) Size and temperature dependence of the plasmon absorption of colloidal gold nanoparticles. J Phys Chem B 103(21):4212–4217

6. Prescott SW, Mulvaney P (2006) Gold nanorod extinction spectra. J Appl Phys 99((12)):123504. https://doi.org/10.1063/1.2203212

7. Chen JY, Wang DL, Xi JF, Au L, Siekkinen A, Warsen A, Li ZY, Zhang H, Xia YN, Li XD (2007) Immuno gold nanocages with tailored optical properties for targeted photothermal destruction of cancer cells. Nano Letters 7((5)):1318–1322. https://doi.org/10.1021/Nl070345g

8. Kumar PS, Pastoriza-Santos I, Rodriguez-Gonzalez B, Garcia de Abajo FJ, Liz-Marzan LM (2008) High-yield synthesis and optical response of gold nanostars. Nanotechnology 19((1)):015606. https://doi.org/10.1088/0957-4484/19/01/015606

9. Nehl CL, Liao HW, Hafner JH (2006) Optical properties of star-shaped gold nanoparticles. Nano Lett 6((4)):683–688. https://doi.org/10.1021/Nl052409y

10. Oldenburg SJ, Averitt RD, Westcott SL, Halas NJ (1998) Nanoengineering of optical resonances. Chem Phys Lett 288(2–4):243–247

11. Prodan E, Radloff C, Halas NJ, Nordlander P (2003) A hybridization model for the plasmon response of complex nanostructures. Science 302(5644):419–422

12. Loo C, Lin A, Hirsch L, Lee MH, Barton J, Halas NJ, West J, Drezek R (2004) Nanoshell-enabled photonics-based imaging and therapy of cancer. Technol Cancer Res Treat 3(1):33–40

13. Kereselidze Z, Romero VH, Peralta XG, Santamaria F (2012) Gold nanostar synthesis with a silver seed mediated growth method. J Vis Exp 59:3570. https://doi.org/10.3791/3570

14. Zhang K, Fang H, Chen Z, Taylor JS, Wooley KL (2008) Shape effects of nanoparticles conjugated with cell-penetrating peptides (HIV Tat PTD) on CHO cell uptake. Bioconjug Chem 19(9):1880–1887. https://doi.org/10.1021/bc800160b

15. Zhang SL, Li J, Lykotrafitis G, Bao G, Suresh S (2009) Size-dependent endocytosis of nanoparticles. Adv Mater 21(4):419–424. https://doi.org/10.1002/adma.200801393

16. Jung S, Bang M, Kim BS, Lee S, Kotov NA, Kim B, Jeon D (2014) Intracellular gold nanoparticles increase neuronal excitability and aggravate seizure activity in the mouse brain. PLoS One 9(3):e91360. https://doi.org/10.1371/journal.pone.0091360

17. James FH, Daniel NS, Henry MS (2004) The use of gold nanoparticles to enhance radiotherapy in mice. Phys Med Biol 49(18):N309

18. Mendoza KC, VD ML, Kim S, Griffin JD (2010) In vitro application of gold nanoprobes in live neurons for phenotypical classification, connectivity assessment, and electrophysiological recording. Brain Res 1325:19–27

19. Connor EE, Mwamuka J, Gole A, Murphy CJ, Wyatt MD (2005) Gold nanoparticles are taken up by human cells but do not cause acute cytotoxicity. Small 1(3):325–327. https://doi.org/10.1002/smll.200400093

20. Iversen TG, Skotland T, Sandvig K (2011) Endocytosis and intracellular transport of nanoparticles: present knowledge and need for future studies. Nano Today 6(2):176–185. https://doi.org/10.1016/j.nantod.2011.02.003

21. Alkilany A, Murphy C (2010) Toxicity and cellular uptake of gold nanoparticles: what we have learned so far? J Nanopart Res 12(7): 2313–2333.https://doi.org/10.1007/s11051-010-9911-8

22. Yang Z, Liu ZW, Allaker RP, Reip P, Oxford J, Ahmad Z, Ren G (2010) A review of nanoparticle functionality and toxicity on the central nervous system. J R Soc Interface 7(Suppl 4):S411–S422. https://doi.org/10.1098/rsif.2010.0158.focus

23. Salinas K, Kereselidze Z, DeLuna F, Peralta XG, Santamaria F (2014) Transient extracellular application of gold nanostars increases hippocampal neuronal activity. J Nanobiotechnology 12:31

24. Hu M, Chen J, Li Z-Y, Au L, Hartland GV, Li X, Marquez M, Xia Y (2006) Gold nanostructures: engineering their plasmonic properties for biomedical applications. Chem Soc Rev 35(11):1084–1094

25. El-Sayed MA (2001) Some interesting properties of metals confined in time and nanometer space of different shapes. Acc Chem Res 34(4):257–264. https://doi.org/10.1021/Ar960016n

26. Huang XH, Jain PK, El-Sayed IH, El-Sayed MA (2007) Gold nanoparticles: interesting optical properties and recent applications in cancer diagnostic and therapy. Nanomedicine (Lond) 2(5):681–693. https://doi.org/10.2217/17435889.2.5.681

27. Lu W, Xiong C, Zhang G, Huang Q, Zhang R, Zhang JZ, Li C (2009) Targeted photothermal ablation of murine melanomas with melanocyte-stimulating hormone analog-conjugated hollow gold nanospheres. Clin Cancer Res 15(3):876–886. [pii]. https://doi.org/10.1158/1078-0432.CCR-08-1480

28. Huang YF, Sefah K, Bamrungsap S, Chang HT, Tan W (2008) Selective photothermal therapy for mixed cancer cells using aptamer-conjugated nanorods. Langmuir 24(20):11860–11865. https://doi.org/10.1021/la801969c

29. Tam F, Chen AL, Kundu J, Wang H, Halas NJ (2007) Mesoscopic nanoshells: geometry-dependent plasmon resonances beyond the quasistatic limit. J Chem Phys 127(20):6. https://doi.org/10.1063/1.2796169

30. Hirsch LR, Stafford RJ, Bankson JA, Sershen SR, Rivera B, Price RE, Hazle JD, Halas NJ, West JL (2003) Nanoshell-mediated near-infrared thermal therapy of tumors under magnetic resonance guidance. Proc Natl Acad Sci U S A 100(23):13549–13554. https://doi.org/10.1073/pnas.2232479100

31. Huang X, Qian W, El-Sayed IH, El-Sayed MA (2007) The potential use of the enhanced non-linear properties of gold nanospheres in photothermal cancer therapy. Lasers Surg Med 39(9):747–753. https://doi.org/10.1002/Lsm.20577

32. Romero VH, Kereselidze Z, Egido W, Michaelides EA, Santamaria F, Peralta XG (2014) Nanoparticle assisted photothermal deformation of individual neuronal organelles and cells. Biomed Opt Express 5(11):4002–4012. https://doi.org/10.1364/Boe.5.004002

33. Wang AZ, Gu F, Zhang L, Chan JM, Radovic-Moreno A, Shaikh MR, Farokhzad OC (2008) Biofunctionalized targeted nanoparticles for therapeutic applications. Expert Opin Biol Ther 8(8):1063–1070. https://doi.org/10.1517/14712598.8.8.1063

34. Shenoy D, Fu W, Li J, Crasto C, Jones G, DiMarzio C, Sridhar S, Amiji M (2006) Surface functionalization of gold nanoparticles using hetero-bifunctional poly(ethylene glycol) spacer for intracellular tracking and delivery. Int J Nanomedicine 1(1):51–57

35. Albanese A, Tang PS, Chan WCW (2012) The effect of nanoparticle size, shape, and surface chemistry on biological systems. Annu Rev Biomed Eng 14(16):1. https://doi.org/10.1146/annurev-bioeng-071811-150124

36. Santamaria F, Wils S, De Schutter E, Augustine GJ (2011) The diffusional properties of dendrites depend on the density of dendritic spines. Eur J Neurosci 34(4):561–568. https://doi.org/10.1111/j.1460-9568.2011.07785.x

37. Santamaria F, Wils S, De Schutter E, Augustine GJ (2006) Anomalous diffusion in Purkinje cell dendrites caused by spines. Neuron 52(4):635–648. https://doi.org/10.1016/j.neuron.2006.10.025

Chapter 7

Regulating Growth Cone Motility and Axon Growth by Manipulating Targeted Superparamagnetic Nanoparticles

Tanchen Ren, Jeffrey L. Goldberg, and Michael B. Steketee

Abstract

Central nervous system (CNS) neurons fail to regenerate after injury or disease due, in part, to a reduced intrinsic axon growth ability, which is regulated at the growth cone. Recently, we showed that growth cone motility can be regulated by applying a magnetic field to superparamagnetic iron oxide nanoparticles (SPIONs) targeted either intracellularly to signaling endosomes or extracellularly to cell surface receptors. By applying mechanical forces to extracellular SPIONs, filopodia can be elongated and the rate and the direction controlled. Here, we describe the methods for each of these approaches with additional notes on important caveats and experimental design considerations. These methods offer new approaches to studying growth cone motility and axon growth biology, expanding our knowledge and thus our ability to develop new therapies to promote axon regeneration after nervous system trauma or disease.

Key words SPION, Magnetic nanoparticles, Signaling endosomes, Growth cone, Filopodia, Axon regeneration

1 Introduction

In the mammalian central nervous system (CNS), injured axons fail to regenerate, leading to permanent functional impairment as observed in glaucoma, stroke, and traumatic brain injury. Injured CNS neuron axons initially display transitory axonal sprouting [1], indicating some intrinsic capacity for regeneration exists. However, this initial sprouting is often unsuccessful in restoring neuronal function and many CNS neurons ultimately undergo apoptotic cell death [2]. The inability of CNS neurons to regenerate axons back to their targets is attributed to numerous factors that suppress axon regeneration, including poor intrinsic axon growth ability [3], dysfunctional organelle dynamics [4], lost neurotrophic support [5], inhibitory molecules expressed by glia [6] or released during cellular injury [7], and a tissue destructive inflammatory immune

Fidel Santamaria and Xomalin G. Peralta (eds.), *Use of Nanoparticles in Neuroscience*, Neuromethods, vol. 135,
https://doi.org/10.1007/978-1-4939-7584-6_7, © Springer Science+Business Media, LLC 2018

response [8] that leads to permanent scarring. In recent optic nerve trauma studies, combinatorial approaches, targeting one or more axon growth inhibiting factors by molecular and/or genetic manipulations, can improve neuron survival and promote long distance axon regeneration in mouse, indicating adult mammalian CNS axons can regenerate through the host optic nerve [9] and in some cases reinnervate the visual cortex in the mouse brain [10]. However, the number of neurons that regenerated successfully was low with some, but little functional recovery. Additionally, studies using peripheral nerve grafts and bioengineered neural bridges have shown transected retinal ganglion cells (RGC) axons can also regenerate long distances through nonhost tissues and materials. In some cases, transected RGC axons regenerated sufficiently to reinnervate the brain. However, these studies also reported that a low percentage of axons regenerated successfully [11].

A more direct and translational approach may be to manipulate axon growth by noncontact means for instance by applying a magnetic field to growth cone targeted magnetic nanoparticles. Targeting numerous growth inhibiting factors in a combinatorial fashion may lead to complicated off-target effects and foreign tissue grafts or bioengineered neural bridges may not be compatible long-term and are difficult to envision routing properly to the numerous visual centers in the brain. Thus, a more direct and less invasive approach may be to stimulate axon growth mechanically by manipulating superparamagnetic nanoparticles (SPIONs) targeted to specific subcellular domains or extracellular receptors. Additionally, the ability to manipulate specific vesicle, organelle, or receptor complexes by noncontact means provides a powerful tool for complimenting molecular and genetic approaches to understanding the underlying mechanisms regulating growth cone motility and axon growth.

SPIONs have desirable physical properties for manipulating organelles or receptor complexes to study the resulting biological on growth cone motility and axon growth. SPIONs have a size (1–100 nm) within the size range of macromolecules and organelles and can be injected by minimally invasive methods. SPIONs can be targeted by coupling to moieties that bind specific macromolecules or organelles (e.g., antibodies), or to functional proteins [12]. Functionalizing SPIONs with ligands to different extracellular receptors, SPION can induce receptor clustering, activation, and direct intracellular signaling pathways and cellular functions [13], cytokine secretion [14], and cell apoptosis [15], among others. Macromolecules or organelles bound to SPIONs can then be manipulated in spatiotemporally using magnetic fields [16]. Recent studies have also shown that new neurites can be initiated and directed to form synapses, essentially rewiring neuronal networks [17]. Our lab has manipulated neuronal signaling endosome transport [18] and filopodial elongation [19] in order to study

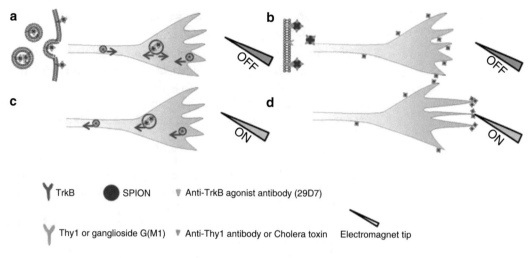

Fig. 1 Superparamagnetic iron oxide nanoparticles (SPIONs) as tools to study and to promote axon growth. (**a**) SPIONs can bind with TrkB on cell membrane to facilitate cellular uptake and form endosomes. Endosomes can traffic bidirectionally in nascent RGC neurites. (**b**) Anti-Thy1 SPIONs Thy1 on cell membrane and stitch at the axon membrane. (**c**) Electromagnet can stop the retrograde transport of TrkB endosomes and halt neurite growth. (**d**) Electromagnet can induce clustering of Thy1 receptors at the end of growth cone and rapid filopodia elongation

their effects on axon growth (Fig. 1). Below we discuss the methods underlying our approaches to targeting SPIONs to specific growth cone subdomains as well as provide additional notes and tips on how to successfully implement these methods.

2 Materials and Methods

All solutions were prepared using analytical grade reagents and nanopure water with an electrical conductivity of at least 18 $M\Omega$ cm at 25 °C. All procedures were done in a class II biosafety cabinet (NuAire) or approved chemical safety cabinet. All reagents and nanoparticles were stored at 4 °C, unless noted otherwise.

2.1 Antibodies and Growth Cone Targeting Molecules

To target SPIONs to RGC growth cones in vitro, we used antibodies against the tropomyosin related kinase B (TrkB) receptor and Thy-1 (CD90) and cholera toxin subunit B. Anti-TrkB agonist antibody (29D7) conjugated to Alexa 594 was purchased from Wyeth (Dallas, TX). 29D7 activates TrkB receptors and enhances RGC survival and neurite growth in vitro and in vivo [20]. Anti-Thy1 was purchased from Millipore (Billerica, MA). Thy1 is highly expressed in mature RGC axons within the outer leaflet of lipid rafts and crosslinking anti-Thy1 antibodies can promote neurite outgrowth [21]. Cholera toxin (Sigma-Aldrich, St Louis, MO), can specifically bind to ganglioside G(M1) on RGC membranes [22] as well as other glycosylated surface proteins [23].

2.1.1 Antibody Notes

1. Using antibodies free of the preservative sodium azide is advisable to minimize cellular toxicity.

2. Antibody clone and epitope selectivity is critical to optimal targeting specificity. After antibodies or other targeting molecules are conjugated to SPIONS, targeting specificity and kinetics should be reconfirmed with cell lines expressing the target of choice or by ELISA since conjugation can alter epitope recognition and specificity.

3. Consideration should also be taken on whether the targeting antibody should be conjugated directly to the SPIONs or indirectly through a secondary antibody conjugated to the SPION. Indirect conjugation is less likely to lead to loss of epitope recognition but may lead to decreased specificity and increased background.

2.2 Functionalizing SPIONs for Signaling Endosome Targeting

To target SPIONs to endosomes, 50 nm superparamagnetic MACS MicroBeads, conjugated with a rat anti-mouse IgG antibody (IgG-SPIONs, Miltenyi Biotec), were functionalized with an agonist anti-TrkB receptor antibody as described [18]. First, 12.5 μL of IgG-SPIONs were dispersed in 200 μL of PBS, 0.02% BSA in a 2.0 mL microcentrifuge tube and mixed by pipetting. Second, 10 μL of Alexa 594 conjugated mouse 29D7 (10 μg/mL in PBS) was added, mixed by pipetting, and incubated for 10 min at 37 °C. The resulting functionalized 29D7-SPIONs were then collected by centrifugation in a table top microfuge ($2000 \times g$, 1 min), washed twice with PBS, collected by centrifugation, and finally resuspended in Hibernate E (Brain bits) SATO media (H-SATO; [18]. Control SPIONs (cMNPs) were functionalized with goat anti-rat IgG Alexa 594 (Invitrogen) and prepared as above.

2.3 SPION Signaling Endosome Functionalization Notes

1. Diligently follow all laboratory safety and waste disposal regulations. Nanoparticles are often not removed by standard wastewater purification techniques and therefore should not enter the public wastewater system, which can lead to undesired environmental impacts.

2. For some antibodies or targeting molecules, centrifugation may lead to irreversible clumping. Alternatively, placing a magnet on the side of the microcentrifuge tube can collect functionalized SPIONs. Then, the PBS can be carefully pipetted off. Please note, after magnetizing SPIONs, the SPIONs require time to return to a superparamagnetic state based on the Néel relaxation time [24]. Practically, we found that waiting for 5–10 min and then pipetting will redisperse the SPIONs.

3. The clumping of functionalized SPIONs may also indicate degraded or contaminated SPIONs or antibodies.

4. When choosing antibodies or proteins to target subcellular domains, like signaling endosomes, running stringent controls and detailed analyses of organelle and cellular behaviors is critical to determining if the SPIONs have reached the proper domain without detectably altering normal cellular functions. For example, after targeting SPIONs to TrkB signaling endosomes, we analyzed growth cone filopodial and lamellar dynamics, axon growth rate, TrkB receptor phosphorylation, TrkB signaling cascade activation, and SPION signaling endosome transport directions and rates, among other factors.

5. SPIONs functionalized indirectly with antibodies, like adding a primary antibody to secondary antibody conjugated SPION, should always be made fresh and used immediately for consistent and reproducible results.

2.4 Functionalizing SPIONs for RGC Outer Membrane Targeting

To target SPIONs to outer RGC growth cone membranes, anti-Thy1 antibody (Thy1, 0.2 mg; Millipore, Billerica, MA) or cholera toxin subunit b (Ctxb, 0.1 mg; Sigma-Aldrich, St. Louis, MO) was conjugated to 1 mg of 40 nm carboxyl SPIONs using the carbodiimide conjugation method (Ocean NanoTech, Springdale, AR). SPION coupling was verified by agarose electrophoresis by Ocean NanoTech. Two size SPIONs, with iron core diameters of either 10 or 40 nm, were used for these experiments to provide more control over the forces applied to the growth cones membranes since the 40 nm SPIONs generate forces about tenfold greater than the 10 nm SPIONs as measured 10–100 μm from the electromagnet tip, described below. Once received, the Thy-1 and CtxB conjugated SPIONs were centrifuged at $2000 \times g$ in a microcentrifuge tube for 1 min, washed twice with PBS, 0.2% BSA, and then resuspended in 1.0 mL of PBS to characterize particle size and concentration using a light scattering device (Wyatt Technology Corporation, Santa Barbara, CA). For cell culture experiments, the Thy-1 and Ctxb SPIONs were resuspended in H-SATO as described above.

2.5 Functionalization Notes

1. Washing the functionalized SPIONs is necessary to remove sodium azide, often used as a preservative by nanoparticle manufacturers, and other potentially toxic chemical reagents that may be leftover from the particle manufacturing and conjugation processes that can lead to cellular toxicity.

2. Despite the manufacturer's documentation on SPION size and concentration, independently characterizing particle sizes and concentrations are important to independently verify the documented sizes and concentrations and to detect clumping induced during processing and shipping, which can be a problem with some particles, manufacturers, and conjugations.

Clumped particles should be returned to the manufacturer or remade if manufactured in-house.

3. Nanoparticles should not be filtered after conjugation and/or functionalization.

4. Functionalized SPIONs are generally stored at 4 °C and should be used within 1–2 weeks based on our experience; functionalized SPIONs tend to degrade leading to clumping, poor binding and internalization, and increased background. Additionally, we do not recommend aliquoting or freezing conjugated nanoparticles, which can also lead to rapid degradation, increased background, and nonspecific binding.

2.6 Retinal Ganglion Cell Culture

RGCs were purified from embryonic or postnatal male and female Sprague–Dawley rat pups (Harlan Laboratories) to >99% purity by immunopanning as described [25]. Briefly, embryonic (E) day 20 to postnatal (P) day 8 Sprague–Dawley rats were decapitated, the eyelids removed, and the retinas enucleated using a 1.8–3 mm iris spatula (Fine Science Tools). To remove nonretinal tissue, the retinas were put in a petri dish containing room temperature Dulbecco's PBS with phenol red (Gibco). Cleaned retinas were treated enzymatically with 165 U papain (Worthington, Freehold, NJ) at 37 °C for 30 min before homogenization into a single cell suspension. RGCs were purified from the cell suspension by first immunopanning using negative selection anti-macrophage antibody (AIA51240; Accurate chemical, Westbury, NY) and then second by positive selection using an anti-Thy1 antibody (clone T11D7). Purified RGCs were cultured on poly-D-lysine (PDL, 70 kDa, 10 μg/mL for 30 min, at r.t.; Sigma) coated and then laminin-coated (2 μg/mL in NB-SATO for 3–12 h at 37 °C and 10% CO_2) glass bottom dishes (P35G-1.0-2-C, MatTeK) in H-SATO containing Hibernate-E (BrainBits) with 5 mg/mL insulin, 1 mM sodium pyruvate, 1 mM L-glutamine, 40 ng/mL triiodothyronine, 5 mg/mL N-acetyl cysteine, B27, with or without 50 ng/mL BDNF (Peprotech, Rocky Hill, NY), 10 μg/mL ciliary neurotrophic factor (Peprotech, Rocky Hill, NY), and 5 μM forskolin. For electron microscopy, RGCs were cultured on formvar/carbon film nickel or gold grids (100 mesh; Electron Microscopy Sciences, Hatfield, PA) precoated with PDL and laminin as above. Generally, RGC were cultured at a density of ~3 cells per mm² (~1000 cells per 20 mm MatTek well).

2.7 Microscopy Notes

1. Do not use copper electron microscope grids for neuronal culture due to cytotoxicity.

2. Phenol red may be omitted from all of the above solutions as long as the remaining solutions are not going to be stored for later use. Though phenol red is a common and useful pH indicator, phenol red is not an optimal choice for high intensity

epifluorescent imaging, as required to image fluorescent nanoparticles, since phenol red is phototoxic to many cell types [26]. Moreover, common contaminants in phenol red reagents can act as an estrogen receptor agonist [27] and a prostaglandin receptor antagonist [28].

2.8 Loading RGCs with TrkB Receptor Functionalized SPIONs

RGCs were loaded with SPIONs prior to culturing by incubating ~5 × 10⁵ cells/mL with for 12–16 h in suspension cultures under constant agitation at 37 °C. After incubation, SPION-loaded RGCs were centrifuged at $80 \times g$ at r.t. for 15 min and then resuspended in fresh H-SATO and cultured on laminin as above. RGCs were maintained 37 °C and 10% CO_2 on a Zeiss with a heated stage (Zeiss) and recorded with a Zeiss camera at 12 frames/min with differential interference contrast (DIC) optics (Plan Apo 63 Å~/1.40 DM objective; Zeiss). Functionalized SPIONs were then slowly added by pipetting to RGCs cultured in glass bottom wells in 35 mm MatTek (dishes and allowed to incubate for X minutes. Noninternalized or unbound particles were then removed by carefully perfusing 2.0 mL of NB-SATO medium 3×.

2.9 Notes

1. Optimal cell loading concentrations and time points for visualization of the functionalized nanoparticles will need to be determined for each cell type and culture condition.

2. Toxicity analyses should be performed using an MTT or similar live/dead assay.

3. When diluting the IgG conjugated SPIONs, including BSA is important to prevent clumping and to reduce nonspecific binding sites, which can lead to increased background.

4. Osmolality and pH should always be checked before and after adding SPIONs to cell cultures for optimal receptor expression, SPION binding kinetics, and overall cellular health.

3   Electromagnet Construction and Set-Up

An electromagnet was constructed as described (Fig. 2) [29]. To make the electromagnet core, a1018 grade cold-rolled steel rod (Home Depot, Atlanta, GA) was machined to the following dimensions: rear diameter: 8 mm by 17 cm length. One end of the core was machined down to a point, permitting closer positioning to the neurons, a more focal magnetic field, and a higher magnetic gradient near the core's tip. The core was fit tightly into a 120/60 HT coil solenoid (216–758-1D, ASCO power technologies, Florham Park, NY), which was connected to a variable autotransformer voltage source (variable autotransformer 0–100; Staco, Dayton, OH) and mounted on a micromanipulator (Narishige, Setagaya-Ku, Japan). The electromagnet force was measured using

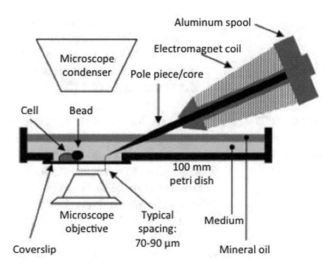

Fig. 2 Electromagnet set-up [29]

100 nm nanoparticles (Chemicell, Berlin, Germany) in glycerol (4080 centipoise or kg/(m*s); Sigma). To measure the magnetic force, fluorescent time-lapse recordings of nanoparticles were recorded at electromagnet autotransformer settings of 15, 30, and 60. Stoke's law ($F = 6 \cdot \pi \cdot \eta \cdot R \cdot v$) was used then used to calculate the force at distances of ~5–100 μm from the electromagnet tip, where η represent viscosity, R represent radius and v represent follow velocity. A logarithmic relationship was observed between distance and force ($R^2 \geq 0.967$). Assuming that the iron oxide material is the same within all SPIONs, the forces applied to 40 nm and 10 nm iron core SPIONs at different distances were calculated using the force values obtained for the fluorescent SPIONs and the formula ($F = V(M \cdot \nabla)B$) where V is the volume of the iron core, M is volumetric magnetization, and B is magnetic field. For some experiments, the tip of the magnet was dipped into a 50 μg/mL 70 kDa PDL solution in water for 30 min and then into a 2 μg/mL laminin (Sigma) in Neurobasal for 4 h.

3.1 Electromagnet Notes

1. Cleaning all machine oil residues from the core is critical to preventing cytotoxicity.

2. When filing the core end down to a point, care should be taken to not machine too fine of a tip, i.e., below 2 mm, since applying high current can lead to more rapid tip degradation.

3. The current settings should not exceed the settings recommended above. Increased current can lead to an increased rate of core degradation and heat.

4. Prior to conducting experiments with live cells, the temperature of the media around the core tip should be measured to make sure the tip is not heating the media above optimum cellular

temperatures, which can lead to improper pH control in CO_2 buffered medias as well as disruption of cellular functions. When recording mammalian neuronal growth cones, changes in temperature of 1–2 °C can lead to significant changes in growth cone motility and axon growth. If increased heating around the tip is an issue, slowly perfusing media through the culture dish may lead to more accurate temperature control around the tip.

5. The core tip must be coated with acetone free nail polish to prevent microscopic iron filings from being released from the tip. We recommend coating the tip three times, allowing for drying in between coats. Recoating with nail polish is required periodically. When new coats are required, iron filings will be visible under the microscope. They tend to gradually build up around the tip on the glass bottom coverslip when the electromagnet is powered on.

6. Micromanipulator setup should be performed prior to conducting live cell experiments. Adjusting the core tip without contacting the cells or breaking the coverslip may require some practice. During the initial set-up, the objective turret should be turned to an open slot to avoid damaging an objective.

7. The microscope should be set up on an antivibration table and the micromanipulator affixed to the table, not the microscope, to prevent interference between the microscope condenser and the magnet while moving the stage.

3.2 Light and Electron Microscopy

A Zeiss DIC/Axiovert inverted fluorescence time-lapse microscope with a heated stage (Zeiss) and incubation chamber (Zeiss) was used to conduct and to record RGC and SPION experiments. During time-lapse recordings, the H-SATO volume was increased to 2.7 mL and overlaid with 2.0 mL of mineral oil (Sigma) to minimize evaporation. Cultures were maintained at 37 °C and recorded at 12 frames/min with differential interference contrast (DIC) optics (Plan Apo 63×/1.40 DM objective; Zeiss) under the control of Axio Vision software. To fix recorded growth cones, 2.0 mL of 37 °C, 8% paraformaldehyde in Krebs/Sucrose buffer is layered on top of the mineral oil. The fixative gently drops through the oil, rapidly fixing the growth cones during the recording. The focus will need to be adjusted slightly.

3.3 Light Microscopy Fixation Notes

1. The high sucrose Krebs/sucrose buffer is excellent at preserving delicate growth cone processes like filopodia, which can be verified by fixing during time-lapse recording [30].

3.4 Extracellular SPION Targeting Analysis

Transmission electron microscopy was used in some experiments to analyze intracellular and extracellular SPION localizations (Fig. 3). RGCs were cultured overnight on gold or nickel mesh

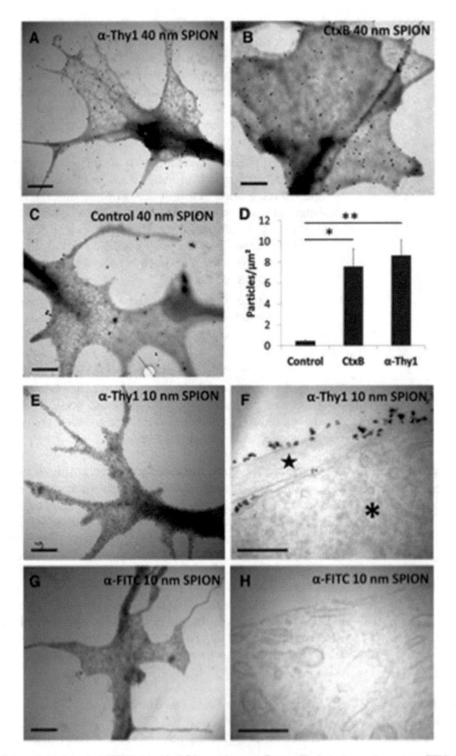

Fig. 3 Extracellular targeted SPIONs bind to RGC membranes. Transmitted electron microscope (TEM) images of RGC growth cones treated with (**a**). 40 nm anti-Thy1 SPIONs and (**b**) 40 nm anti-CtxB SPIONs for 30 min show abundant SPION binding to filopodia and other parts of the growth cone. (**c**) In contrast, few nonfunctionalized, control SPIONs were bound to RGC growth cones. (**d**) Quantification of control and targeted SPIONs

grids (~3 RGCs/mm²) and then treated with either anti-Thy1 (10 µL, 40 µg/mL) 40 nm SPIONs (Miltenyi Biotec, Bergisch-Gladbach, Germany) or 10 nm ctxb SPIONs (Ocean NanoTech) or control SPIONs for 15 min and then fixed with 2% glutaraldehyde (Electron Microscopy Sciences, Hatfield, PA) in PHEM-N buffer (600 mM PIPES, 25 mM HEPES, pH 6.9, 10 mM EGTA, 20 mM MgCl2, and 7.4 mM NaCl; [31]) for 2 days at 4 °C, stained with 1% osmium tetroxide (Electron Microscopy Sciences) in 0.15 M cacodylate for 1 h and then dehydrated incrementally in EtOH (20, 30, 40 and 60% for 1 min each, 70% for 1.5 min, 80% for 1.5 min, 95% for 3 min, and 100% for 5 min), and then visualized by TEM (Philips CM10; FEI, Hillsboro, OR). To determine the percentage of extracellular versus intracellular SPIONs, SPION localization was quantified by TEM in ultrathin sections. RGCs were cultured on glass coverslips and treated with SPIONs as above. The RGCs were fixed with ½ Karnovsky's fixative (2.5% glutaraldehyde, 2.0% paraformaldehyde, 0.1 M cacodylate buffer; Electron Microscopy Sciences) for 2 days at least 4 h, dehydrated with EtOH, as described above, and embedded in epon (Embed, Electron Microscopy Sciences) at 64 °C. The glass coverslip was then removed using hydrofluoric acid and the block was sectioned on a Leica UCT ultramicrotome. The sections (50–60 nm) were picked up on formvar carbon-coated 150 mesh copper grids (Electron Microscopy Sciences) and visualized using the TEM as described above.

3.5 Electron Microscopy Fixation Notes

1. For TEM fixation, most of the media is aspirated from the MatTek dish, leaving about 150 µL of media in the glass bottom well only. Then, approximately 2.0 mL of room temperature ½ Karnovsky's fixative is rapidly but gently perfused into the dish by pipetting into the side of the dish, not directly across the well. The fixative is then removed from the dish, except for the glass bottom well, and then another 2.0 mL is perfused into the dish.

2. The neurons should be maintained at 37 °C until just before fixation. Keeping the 35 mm MatTek dishes within a covered 150 mm petri dish or other sterile plastic container within the incubator permits rapid transport of 2–3 dishes to the chemical fume hood for fixation with minimal temperature loss. Karnovsky's fixative should not be used outside of a chemical fume hood.

Fig. 3 (continued) (mean ± SEM). *$P < 0.05$, **$P < 0.01$. (**e**) TEM images of RGC growth cones incubated with 10 nm anti-Thy1 SPIONs show abundant attachment. (**f**) TEM images of RGC ultrathin sections show that anti-Thy1 SPIONs remain on the outside membrane of the cell body (*) and neurite (star). (**g**) In contrast, anti-FITC SPIONs were not found on RGC growth cones or in (**h**). RGC ultrathin sections. Scale bar is 1 µm a, b, c, e, and g, and 200 nm in f and h [19]

3. The fixative osmolarity (650–1000 mOsM) and pH (7.4) should both be optimized to preserve the morphology of fine processes. For preserving growth cone lamellar and filopodial morphology, we have had good luck using higher osmolarity fixatives (1000 mOsM), increased by adding sucrose [31].

3.6 SPION Targeting to Signaling Endosomes

To detect SPION endocytosis, primary RGCs were allowed to extend neurites for 12–24 h before perfusing 29D7 functionalized SPIONs into the culture medium for 15 min. SPIONs were detected as puncta in all growth cones, neurites, and somas with SPION detection more prominent in growth cone central domain, indicating selective endocytosis or transport. 29D7 functionalized SPIONs increased phosphoTrkB (p-TrkB) that colocalized with 92% of the SPION puncta whereas control SPIONs were virtually undetectable in RGC growth cones, neurites, or somas, and neither increased p-TrkB nor colocalized with p-TrkB puncta (Fig. 4).

3.7 Notes

1. Before adding the SPIONs, check that they are unclumped and free of contaminants by placing 1 µL on a coverslip and checking the SPION fluorescent signal. Clumped SPIONs are generally obvious as are contaminants, which generally exhibit obvious tumbling behaviors.

2. To fix recorded growth cones, pipette 1 mL of 8% paraformaldehyde in Krebs Sucrose buffer onto the top of the oil while recording as described [32]. This method permits rapid fixation and excellent preservation of growth cone morphology while recording. The focus will need to be adjusted slightly. Never fixative while the electromagnet pole is in the dish.

3. To easily locate the recorded growth cones, which can be very difficult after fixation and permeabilization, mark the front of the dish and note the X and Y coordinates on the stage.

4 SPION Endosomal Transport and Trafficking

To determine whether SPION loaded signaling endosomes were transported and trafficked into nascent neurites, purified RGCs were preloaded with functionalized or control SPIONs by incubating ~5×10^5 cells/mL with 40 µg/mL 29D7 SPIONs or control SPIONs for 12–16 h in BDNF(−) suspension cultures under constant agitation at 37 °C. After incubation, SPION-loaded RGCs were resuspended in fresh H-SATO and cultured on PDL/laminin-coated coverslips. Under preloading conditions, functionalized and control SPIONs were detected in 95% and 7% of RGC somas, and 80% and 0% of RGC neurites and growth cones, respectively. SPION loaded signaling endosomes moved bidirectionally in

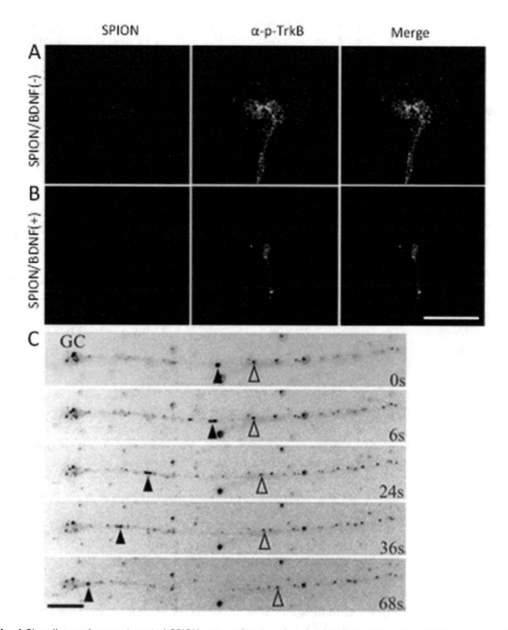

Fig. 4 Signaling endosome targeted SPIONs are endocytosed and colocalize with activated TrkB receptors in primary neurons. (**a**) Anti-TrkB antibody SPIONs but not cMNPs (not shown) were detectable in RGC neurites and growth cones as puncta of varying sizes that colocalized with antibodies against phospho-TrkB (α-p-TrkB), in the absence of BDNF. (**b**) In RGCs cultured with BDNF, neither binding nor endocytosis of SPIONs was detected. (Scale bar: 10 μm.) (**c**) Time-lapse microscopy shows that SPION signaling endosomes are trafficked bidirectionally in nascent RGC neurites. Discrete SPION puncta were detected in RGC neurites and growth cones (GC). Within neurites, SPION signaling endosomes were transported both anterogradely (filled arrowhead) and retrogradely (open arrowhead) between the soma (right) and growth cone. Time in seconds (s) is indicated. (Scale bar: 10 μm.) [18]

nascent neurites between soma and growth cones independently (Fig. 4) at rates that varied from 0 to 12 μm/s, comparable with fast axonal transport [33].

4.1 Endosomal Targeting Notes

1. Confirming that axonal growth rates and preloaded signaling endosomes are transported as expected is important for verifying that SPION loading did not disrupt axonal transport.

2. During fluorescent imaging, minimizing the exposure of growth cones to strong fluorescent light is important since prolonged exposure can lead to phototoxicity, reduced growth cone motility, and axonal retraction. Therefore, we recommend optimizing the imaging system to collect as much light as possible and as quickly as possible by using high quality objectives and cameras [26].

4.2 Modulating Neurite Growth by Altering SPION Mediated Signaling Endosomes

To characterize SPION loaded signaling endosome responses to a focal magnetic force, we analyzed at least 10 RGC neurites from both SPION-loaded and non-SPION-loaded RGCs (Fig. 5). The number of arbitrary fluorescence light units (FLU) was measured in Axiovision at 0 and 20 min. To monitor effects on the neurites, cells were recorded with DIC optics in between fluorescent exposures at 12–15 frames per min. To quantify the SPION response to a focal magnetic force, SPION-loaded neurites were exposed to a constant 15-pN force and three SPION-loaded neurites exposed to fluorescent light only were analyzed. FLU units were measured in Axiovision. Exposure times were matched in each experiment. Magnetic forces were applied to SPION-loaded RGCs via an electromagnetic needle. Force was controlled by varying the distance between the needle tip and the RGC. Growth cone protrusions were analyzed as described [32].

At approximately 15 pN, we observed net SPION signaling endosome transport halted anterogradely, and SPION signaling endosomes moved retrogradely out of distal neurites into proximal neurites and somas. After treating at 15 pN for 1–5 min, neurite growth was halted without retraction. Over a 20-min exposure, the mean fluorescent light units decreased $90.1 \pm 7\%$ in neurites and increased $41.5 \pm 4.8\%$ in somas ($n = 3$). This net retrograde transport was not accompanied by process retraction or evacuation. Other vesicles, mitochondria, and vacuoles continued to move bidirectionally. Thus, a focal 15 pN magnetic force biased SPION signaling endosome transport away from growth cones. At forces >15 pN, SPIONs changed from punctate to diffuse in neurites and growth cones, indicating we were unable to pull MNP signaling endosomes toward the magnet without disrupting their integrity ($n = 5$). Thus, we were unable to pull signaling endosomes toward the magnet without disrupting endosome integrity similar to results in nonneuronal cells [34]. The growth cone's central domain

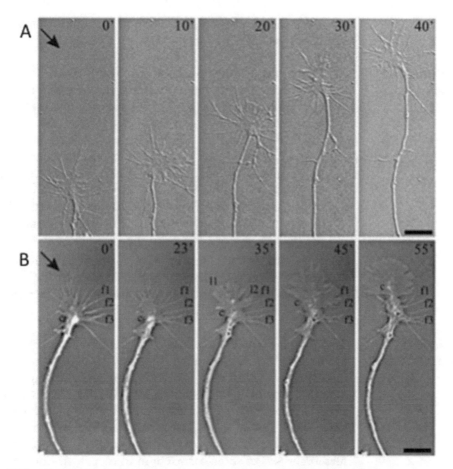

Fig. 5 In RGCs with SPION-loaded signaling endosomes, a focal magnetic force alters growth cone motility and halts neurite growth. (**a**) In control RGCs, a constant 15-pN force failed to alter either growth cone motility or neurite growth rate. (**b**) In SPION-loaded RGCs, a 15-pN force applied for 3 min was sufficient to immobilize both lamellar and filopodial protrusions and halt neurite growth. All active lamella (l) and filopodia (f) immobilized for 20 min after removing the magnet (compare 0′ and 23′). By 35 min, both lamellar (l1 and l2) and filopodial protrusions reinitiated at the leading edge in concert with resumed axon growth (compare 23′ and 55′). Time in minutes (′) is indicated. Electromagnet tip indicated by black arrows. (Scale bars: 5 μm.) [18]

stopped advancing, and both lamellar and filopodial protrusions immobilized in the peripheral domain. Approximately 15–30 min after removing the magnet, protrusion and central domain advance resumed, typical of neurite growth. This delayed recovery indicates the force did not simply hold the growth cone but likely altered signaling presumably because of altered SPION signaling endosome localization. Interestingly, even as the central domain advanced after recovery, the previous central domain failed to consolidate into neurite and previously immobilized lamella and filopodia failed to remobilize (Fig. 5b). Thus, in SPION-loaded RGCs, a focal magnetic force immobilizes peripheral domain protrusions concomitant with reversibly inhibiting peripheral protrusive activity and neurite growth.

4.3 Electromagnet Set-Up Notes

1. Electromagnet tip positioning is critical to avoid contacting the delicate growth cones, breaking the coverslip, and damaging the microscope objectives. After setting up the electromagnet on the micromanipulator, a test culture dish with media should be put on the scope and the electromagnet needle positioned and tested before conducting cell experiments. Placing the magnetic tip just above the glass coverslip at the edge of the visual field using a 63× objective works well. To generate sufficient forces using this electromagnet here and below, the tip must be positioned close to the neurons ~10–50 μm from the cell of interest.

2. In cellular cultures, moving the cell toward the magnetic tip, positioned just within the visual field is easier for controlling distance than moving the tip toward the cell even with a high quality micromanipulator.

4.4 Elongating Filopodia with SPIONs

To determine whether applying a magnetic force to SPIONs binding on RGC membrane alters neurite elongation, growth cone movement was observed with either 40 nm anti-Thy1 or CtxB SPIONs, focal magnetic fields was varied in the range of 3–50 pN/particle. Under magnetic field, membrane and cytoplasmic accumulation at filopodial tips, consistent with SPION movement and accumulation (Fig. 6). This indicates that SPION linkage to the membrane was of sufficient strength to remain attached to filopodia membranes in the presence of applied mechanical tension. Immediately following this membrane accumulation at the filopodial tips, the tips lifted off the substrate and elongated in RGCs treated with either 40 nm anti-Thy1 or 40 nm CtxB SPIONs but not after exposure to nonfunctionalized. When we used the 10 nm anti-Thy1 SPIONs, which generate smaller magnetic forces, filopodia elongation was also observed in all cases and was indistinguishable from the 40 nm cases demonstrating that filopodia elongation can be achieved with SPIONs that generate forces of 3–6 pNs.

After elongation was initiated, the rate of ongoing elongation could be regulated by reducing or increasing the voltage to the electro magnet, or by changing the distance between the electromagnet tip and the RGC filopodia-bound SPION. Since the electromagnet pole entered the dish at an angle, the magnetic field was not completely parallel to the dish surface and therefore, filopodia lifted off the substrate into a slightly different focal plane than the growth cone. It is essential to note that the electromagnet tip had to be well sharpened and placed ahead of the growth cone at approximately a 45° angle while physically contacting the bottom

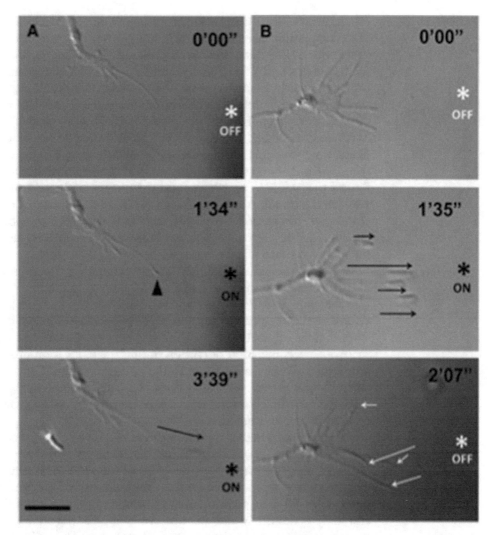

Fig. 6 In RGCs treated with extracellular targeted SPIONs, a focal magnetic force elongates filopodia. (**a**). A filopodium (arrowhead) from an RGC growth cone treated with anti-Thy1 SPIONs elongated (black arrow) toward the tip of the applied electromagnet (asterisk). (**b**). Similarly, in RGCs treated with CtxB SPIONs, several filopodia elongated (black arrows) in response to an applied magnetic field and then retracted (white arrows) when the electromagnet was turned off (OFF). Scale bar 10 μm in **a** and **b** [19]

of the culture dish in order to induce filopodial elongation. This technique allows for sufficient SPION accumulation at the filopodial tip due to close proximity of the magnetic field to the growth cone. Notably, this approach did not break a single filopodium. Thus, in response to applied mechanical tension, filopodia reconfigure cellular membrane to elongate at high rates and without breakage.

To determine if elongating filopodia by applying magnetic tension stimulates actin filament polymerization, RGCs were electroporated immediately after immunopanning with the LifeAct vector

(Ibidi, Martinsried, Germany) [35] using an Amaxa Nucleofector II electroporator using the neuron small cell number program one as described [36]. One hundred thousand RGCs were transfected with 1 nmol (1 µL) of LifeAct vector in 27 µL of electroporation buffer (25 mM sodium pyruvate, 0.3% isobutyric acid, 50 µg/mL bovine serum albumin, 1 mM magnesium sulfate, and 100 mM dipotassium phosphate), Electroporated RGCs were seeded and incubated overnight. The fluorescent LifeAct peptide can bind to filamentous but not monomeric actin. Fluorescent images were recorded as described above using a 63× Plan APO objective at 12 frames per minute. To obtain optimal images of F-actin inside elongated filopodia, the filopodia were allowed to contact the electromagnet tip coated with PDL and laminin. After filopodial attachment the electromagnet was turned off to prevent movement artifacts seen when the magnet is on.

To verify whether SPION-induced elongating filopodia are exerting retraction forces, the magnetic force was removed to analyze the elongated filopodial response. We observed that just after removing the magnetic force, elongated filopodia were pulled back toward their original position (Fig. 6). This result confirms that there is an accumulation of tension during the elongation process that induces filopodia to retract to their original position once the external force is removed. Thus, these elongated filopodia are generating mechanical tension that could be transmitted to the growth cone to promote elongation.

To examine whether this filopodial tension is able to promote axon elongation when integrins are linked to the substrate, we anchored the tip of the elongated filopodia to the PDL-laminin electromagnet pole and focused on the response of the growth cone. After 30 min, growth cones remained at the same position and only sporadically did we observe filopodia engorging and adapting their shape to the new anchorage point. Thus, filopodial retraction forces from elongated filopodia are unable to induce growth cone advance when filopodia integrin receptors are linked to a laminin substrate, consistent with previous reports on the role of filopodial adhesions in growth cone advance [30].

4.5 Filopodial Elongation Notes

1. To permit F-actin visualization in elongated filopodia, the electromagnet tip was coated with PDL-laminin using a standard coverslip PDL/laminin coating protocol to bind and stabilize elongated filopodia for imaging.

5   Conclusions

The small size and highly diverse methods that can be used to deliver and to target SPIONs to intracellular or extracellular domains provides a wide range of useful tools for manipulating specific cellular activities by noncontact means. Thus, functionalized SPIONs provide useful tools for understanding cellular mechanisms that cannot be manipulated by conventional molecular and genetic techniques. We have successfully targeted SPIONS to rat RGC axons in vivo as well [18] and even detected SPIONs injected into the vitreous of the eye, after optic nerve crush, in RGC axons at the crush site (unpublished observations). Thus, advancements in applying focal magnetic fields in vivo may ultimately permit axons to be directed back to their targets after injury.

References

1. Sipperley JO, Quigley HA, Gass DM (1978) Traumatic retinopathy in primates. The explanation of commotio retinae. Arch Ophthalmol 96:2267–2273

2. Berkelaar M, Clarke DB, Wang YC, Bray GM, Aguayo AJ (1994) Axotomy results in delayed death and apoptosis of retinal ganglion cells in adult rats. J Neurosci 14:4368–4374

3. Goldberg JL, Klassen MP, Hua Y, Barres BA (2002) Amacrine-signaled loss of intrinsic axon growth ability by retinal ganglion cells. Science 296:1860–1864

4. Lathrop KL, Steketee MB (2013) Mitochondrial dynamics in retinal ganglion cell axon regeneration and growth cone guidance. J Ocul Biol 1:9

5. Mansour-Robaey S, Clarke DB, Wang YC, Bray GM, Aguayo AJ (1994) Effects of ocular injury and administration of brain-derived neurotrophic factor on survival and regrowth of axotomized retinal ganglion cells. Proc Natl Acad Sci U S A 91:1632–1636

6. McKeon RJ, Jurynec MJ, Buck CR (1999) The chondroitin sulfate proteoglycans neurocan and phosphacan are expressed by reactive astrocytes in the chronic CNS glial scar. J Neurosci 19:10778–10788

7. Tang S, Qiu J, Nikulina E, Filbin MT (2001) Soluble myelin-associated glycoprotein released from damaged white matter inhibits axonal regeneration. Mol Cell Neurosci 18:259–269

8. Horn KP, Busch SA, Hawthorne AL, van Rooijen N, Silver J (2008) Another barrier to regeneration in the CNS: activated macrophages induce extensive retraction of dystrophic axons through direct physical interactions. J Neurosci 28:9330–9341

9. Kurimoto T, Yin Y, Omura K, Gilbert HY, Kim D, Cen LP, Moko L, Kugler S, Benowitz LI (2010) Long-distance axon regeneration in the mature optic nerve: contributions of oncomodulin, cAMP, and pten gene deletion. J Neurosci 30:15654–15663

10. de Lima S, Habboub G, Benowitz LI (2012) Combinatorial therapy stimulates long-distance regeneration, target reinnervation, and partial recovery of vision after optic nerve injury in mice. Int Rev Neurobiol 106:153–172

11. Bray GM, Vidal-Sanz M, Aguayo AJ (1987) Regeneration of axons from the central nervous system of adult rats. Prog Brain Res 71:373–379

12. Pita-Thomas W (2015) Magnetic nanotechnology to study and promote axon growth. Neural Regen Res 10:1037–1039

13. Lee JH, Kim ES, Cho MH, Son M, Yeon SI, Shin JS, Cheon J (2010) Artificial control of cell signaling and growth by magnetic nanoparticles. Angew Chem 49:5698–5702

14. Mannix RJ, Kumar S, Cassiola F, Montoya-Zavala M, Feinstein E, Prentiss M, Ingber DE (2008) Nanomagnetic actuation of receptor-mediated signal transduction. Nat Nanotechnol 3:36–40

15. Cho MH, Lee EJ, Son M, Lee JH, Yoo D, Kim JW, Park SW, Shin JS, Cheon J (2012) A magnetic switch for the control of cell death signalling in in vitro and in vivo systems. Nat Mater 11:1038–1043

16. Hoffmann C, Mazari E, Lallet S, Le Borgne R, Marchi V, Gosse C, Gueroui Z (2013) Spatiotemporal control of microtubule nucleation and assembly using magnetic nanoparticles. Nat Nanotechnol 8:199–205

17. Magdesian MH, Lopez-Ayon GM, Mori M, Boudreau D, Goulet-Hanssens A, Sanz R, Miyahara Y, Barrett CJ, Fournier AE, De Koninck Y, Grutter P (2016) Rapid mechanically controlled rewiring of neuronal circuits. J Neurosci 36:979–987

18. Steketee MB, Moysidis SN, Jin XL, Weinstein JE, Pita-Thomas W, Raju HB, Iqbal S, Goldberg JL (2011) Nanoparticle-mediated signaling endosome localization regulates growth cone motility and neurite growth. Proc Natl Acad Sci U S A 108:19042–19047

19. Pita-Thomas W, Steketee MB, Moysidis SN, Thakor K, Hampton B, Goldberg JL (2015) Promoting filopodial elongation in neurons by membrane-bound magnetic nanoparticles. Nanomedicine 11:559–567

20. Hu Y, Cho S, Goldberg JL (2010) Neurotrophic effect of a novel TrkB agonist on retinal ganglion cells. Invest Ophthalmol Vis Sci 51:1747–1754

21. Beale R, Osborne NN (1982) Localization of the Thy-1 antigen to the surfaces of rat retinal ganglion cells. Neurochem Int 4:587–595

22. Hansson HA, Holmgren J, Svennerholm L (1977) Ultrastructural localization of cell membrane GM1 ganglioside by cholera toxin. Proc Natl Acad Sci U S A 74:3782–3786

23. Blank N, Schiller M, Krienke S, Wabnitz G, Ho AD, Lorenz HM (2007) Cholera toxin binds to lipid rafts but has a limited specificity for ganglioside GM1. Immunol Cell Biol 85:378–382

24. Néel L (1949) Théorie du traînage magnétique des ferromagnétiques en grains fins avec application aux terres cuites. Annales de Géophysique 5:99–136

25. Barres BA, Silverstein BE, Corey DP, Chun LLY (1988) Immunological, morphological, and electrophysiological variation among retinal ganglion-cells purified by panning. Neuron 1:791–803

26. Frigault MM, Lacoste J, Swift JL, Brown CM (2009) Live-cell microscopy - tips and tools. J Cell Sci 122:753–767

27. Grady LH, Nonneman DJ, Rottinghaus GE, Welshons WV (1991) pH-dependent cytotoxicity of contaminants of phenol red for MCF-7 breast cancer cells. Endocrinology 129:3321–3330

28. Greenberg SS, Johns A, Kleha J, Xie J, Wang Y, Bianchi J, Conley K (1994) Phenol red is a thromboxane A2/prostaglandin H2 receptor antagonist in canine lingual arteries and human platelets. J Pharmacol Exp Ther 268:1352–1361

29. Fass JN, Odde DJ (2003) Tensile force-dependent neurite elicitation via anti-beta1 integrin antibody-coated magnetic beads. Biophys J 85:623–636

30. Steketee MB, Tosney KW (2002) Three functionally distinct adhesions in filopodia: shaft adhesions control lamellar extension. J Neurosci 22:8071–8083

31. Steketee M, Balazovich K, Tosney KW (2001) Filopodial initiation and a novel filament-organizing center, the focal ring. Mol Biol Cell 12:2378–2395

32. Steketee MB, Tosney KW (1999) Contact with isolated sclerotome cells steers sensory growth cones by altering distinct elements of extension. J Neurosci 19:3495–3506

33. Viancour TA, Kreiter NA (1993) Vesicular fast axonal transport rates in young and old rat axons. Brain Res 628:209–217

34. Won J, Kim M, Yi YW, Kim YH, Jung N, Kim TK (2005) A magnetic nanoprobe technology for detecting molecular interactions in live cells. Science 309:121–125

35. Riedl J, Crevenna AH, Kessenbrock K, Yu JH, Neukirchen D, Bista M, Bradke F, Jenne D, Holak TA, Werb Z, Sixt M, Wedlich-Soldner R (2008) Lifeact: a versatile marker to visualize F-actin. Nat Methods 5:605–607

36. Corredor RG, Trakhtenberg EF, Pita-Thomas W, Jin XL, Hu Y, Goldberg JL (2012) Soluble adenylyl cyclase activity is necessary for retinal ganglion cell survival and axon growth. J Neurosci 32:7734–7744

Chapter 8

Assessment of the Effects of a Wireless Neural Stimulation Mediated by Piezoelectric Nanoparticles

Attilio Marino, Satoshi Arai, Yanyan Hou, Mario Pellegrino, Barbara Mazzolai, Virgilio Mattoli, Madoka Suzuki, and Gianni Ciofani

Abstract

Wireless neuronal stimulation, mediated by ultrasounds and piezoelectric nanoparticles, represents an unprecedented approach aimed at cell activation. Recently, we demonstrated that barium titanate nanoparticles behave as excellent nanotransducers, by eliciting specific cell response following treatment with ultrasounds. In this chapter, we describe in detail the techniques exploited to investigate the nanoparticle/cell interactions and the activation of the neuronal-like cultures in terms of sodium and calcium fluxes.

Key words Barium titanate nanoparticles, Ultrasounds, Piezoelectricity, SH-SY5Y cells, Calcium imaging, Sodium imaging

1 Introduction

Several approaches, such as optogenetics [1], deep brain stimulation [2, 3], and trans-cranial direct current/magnetic stimulation [4, 5], have been developed for a "wireless" stimulation of the neural activity. Some drawbacks of these innovative neuro-techniques include the scarce light penetration through the tissues [6], the necessity of an invasive surgical operation [7], and the low spatial resolution (in the order of cm) [8, 9].

Recently, different works demonstrated as ultrasounds (US) are an innovative tool for the trans-cranial stimulation with higher resolution (in the order of mm) [10]. Furthermore, US can be exploited in combination with piezoelectric nanomaterials, such as barium titanate nanoparticles (BTNPs) and zinc oxide nanowires, to generate direct-current output inside a biological liquid [11, 12].

Concerning cell stimulation with piezoelectric materials, mechanically deformed poly(vinylidene difluoride) (PVDF) scaffolds have been proven to be able to produce an alternating

Fidel Santamaria and Xomalin G. Peralta (eds.), *Use of Nanoparticles in Neuroscience*, Neuromethods, vol. 135, https://doi.org/10.1007/978-1-4939-7584-6_8, © Springer Science+Business Media, LLC 2018

electric field and consequently to significantly increase the neurite length of rat spinal cord neurons, with respect to the cells cultured on non-piezoelectric stimulated control scaffolds [13]. As another example, piezoelectric PVDF membranes resulted to be sensitive to acoustic vibrations, so resulting in an artificial system bio-mimicking the cochlear functions: the amplified electric signal generated by this system was used for replacing the inner ear function in deafened guinea pigs [14]. In a previous work of our group [15], PC12-derived neuron-like cells were piezoelectrically stimulated by US-activated boron nitride nanotubes (BNNTs). Particularly, it was observed a significant enhancement of the neurite outgrowth of PC12 cells stimulated by US + BNNTs compared to the cells treated with US but without the presence of the piezoelectric nanoparticles. The influx of calcium ions (Ca^{2+}) was supposed to mediate the above-mentioned increase of the neurite length, but, unfortunately, the impossibility to carry out electrophysiological investigations in the presence of US-generated vibrations limited the analysis of the involved biological mechanisms, such as the possible depolarization of the neuron membrane induced by piezoelectricity. For this reason, we recently exploited imaging techniques for detecting the Ca^{2+}/Na^+ fluxes in response to the stimulation of US + piezoelectric BTNPs [16]. The combined stimulation was able to induce tetrodotoxin (TTX) and cadmium (Cd^{2+}) sensitive high-amplitude Ca^{2+} transients, so demonstrating the involvement of voltage-gated membrane channels. Finally, the lack of cellular response when stimulating with US + non-piezoelectric BTNPs, as negative controls, strongly supported the hypothesis of a piezoelectric stimulation.

In this chapter, we summarize the methodologies exploited in our laboratories for the preparation and the characterization of piezoelectric/non-piezoelectric nanoparticles, for the treatment of SH-SY5Y-derived neurons with US and BTNPs, for the evaluation of the cytocompatibility of the adopted nanomaterials, and for the imaging of Ca^{2+}/Na^+ fluxes during stimulations. SH-SY5Y human neuroblastoma cell line is a widely used model for in vitro neurobiology investigation, and it was adopted as model of differentiation toward neurons characterized by adrenergic and cholinergic phenotypes [17].

2   Materials

2.1   Preparation/ Characterization of BTNP Dispersions

1. Gum Arabic from acacia tree (Sigma, G9752) and PBS (Sigma, D8662).

2. Barium titanate nanoparticles (BTNPs, Nanostructured & Amorphous Materials, 1144DY). For control experiments, analogous nanoparticles but with cubic crystal structure (non-

piezoelectric, Nanostructured & Amorphous Materials, 1143DY).

3. Bransonic sonicator 2510.

4. Characterization of BTNPs was performed with following instruments:

(a) Sputter (Quorum, Q150R ES); scanning electron microscope (SEM, Helios NanoLab 600i FIB/SEM, FEI).

(b) Transmission electron microscope (TEM, Zeiss 902).

(c) Confocal laser scanning microscopy (FluoView FV1000 equipped with an objective PLAPON 60XO, NA1.42, Olympus).

(d) Nano Z-Sizer 90 (Malvern Instrument).

(e) X-ray powder diffractometer (Kristalloflex 810, Siemens).

2.2 Cell Culture, Cell Differentiation, and BTNP Treatment

1. Human neuroblastoma-derived cells (SH-SY5Y, ATCC CRL-2266); Dulbecco's modified Eagle's medium/Hams F-12 medium (DMEM/F-12; Lonza, BE04-687Q); penicillin/streptomycin (Gibco, 15140-122); fetal bovine serum (FBS, Gibco, 16000-044).

2. Trypsin and ethylenediamine tetraacetic acid (1 mM, Gibco, 25300-054); dimethyl sulfoxide (DMSO; Sigma D2650).

3. 35 mm diameter μ-dishes (Ibidi, 81156).

4. DMEM (Lonza, BE12-604F); all-trans-retinoic acid (Sigma, R2625).

2.3 Metabolic and Viability Assay

1. 2-(4-iodophenyl)-3-(4-nitophenyl)-5-(2,4disulfophenyl)-2H–tetrazoilium monosodium salt provided in a premix electro-coupling solution (WST-1, BioVision, K301-2500); micro-plate reader (Victor3, PerkinElmer).

2. Propidium iodide (PI, Molecular Probes, P3566); Hoechst 33342 (Invitrogen, H1399); fluorescence microscope imaging (TE2000U, Nikon).

2.4 3D Confocal Analysis of the BTNP/ Neuron Interaction

1. CellMask green plasma membrane stain (Invitrogen, C37608); Hoechst 33342 (Invitrogen, H1399).

2. Artificial cerebrospinal fluid (aCSF): NaCl (140 mM, Sigma, S9888), KCl (5 mM, Sigma, P3911), $CaCl_2$ (2 mM, Sigma, 223,506), $MgCl_2$ (2 mM, Sigma, M9272), HEPES (10 mM, Sigma, H3375), D-glucose (10 mM, Wako, 041-00595); adjust to 7.4 using NaOH (Sigma, S2770).

3. 633 (Melles Griot), 488 (Melles Griot), and 405 (Olympus FV5-LDPSU) nm lasers.

4. For 3D rendering of confocal z-stacks, ImageJ software (http://rsbweb.nih.gov/ij/).

2.5 Calcium and Sodium Imaging During Ultrasound/ BTNP Stimulation

1. Fluo-4AM (Invitrogen, F-14201); CoroNa Green AM (Invitrogen, C36676); DMEM (Lonza, BE12-604F).

2. Calcium imaging was performed with a IX81 microscope (Olympus) equipped with an objective UPLFLN 40XO, NA 1.3, and a cooled CCD camera (Cool SNAP HQ2, Photometrics) by using 460-480HQ as an excitation filter, DM485 as a dichroic mirror, and 495-540HQ as an emission filter (all from Olympus). Sodium imaging was performed with a IX83 microscope (Olympus) equipped with an objective UPLFLN 40XO, NA 1.3, and an electron multiplying charge-coupled device camera (iXon3, Andor Technology) by using BP470-495, DM505, and BP510-550 as an excitation filter, a dichroic mirror, and an emission filter, respectively (all from Olympus).

3. Sonitron GTS Sonoporation System (ST-GTS, Nepagene) equipped with a plane wave transducer module (PW-1.0-6 mm, 6 mm diameter tip, 1 MHz).

4. Software for microscope time-lapse imaging (MetaMorph NX).

5. Cadmium chloride ($CdCl_2$, Sigma, C-2544); tetrodotoxin (TTX, Tocris, 1078); gentamicin (Sigma; G1272).

2.6 Image and Statistical Analysis

1. Time lapses analyzed with ImageJ software (http://rsbweb. nih.gov/ij/).

2. $\Delta F/F_0$ traces plotted by using Calc (Apache OpenOffice for MacOS X).

3. Statistical tests performed by using R software (R Project for Statistical Computing; https://www.r-project.org/).

3   Methods

3.1 Preparation/ Characterization of BTNP Dispersions

1. Prepare a solution of 1 mg/ml gum Arabic in PBS by weighing a 10 mg of the powder and mixing this with 10 ml of PBS in a polystyrene tube. Warm the sample at 37 °C for 1 h. *See* **Note 1**.

2. Weigh 10 mg of BTNPs and then mix with 10 ml of 1 mg/ml gum Arabic solution.

3. Sonicate the sample for 12 h by using an output power of 20 W to obtain a stable dispersion.

4. Before use in biological assays, characterize the BTNP size, dispersion, Z-potential, and crystallographic structure by using SEM, TEM, confocal fluorescence microscopy, Z-sizer, and X-ray powder diffractometer (Fig. 1).

   (a) For SEM imaging, deposit a drop of 50 μl of the nanoparticle dispersion on a silica surface. Let the drop completely

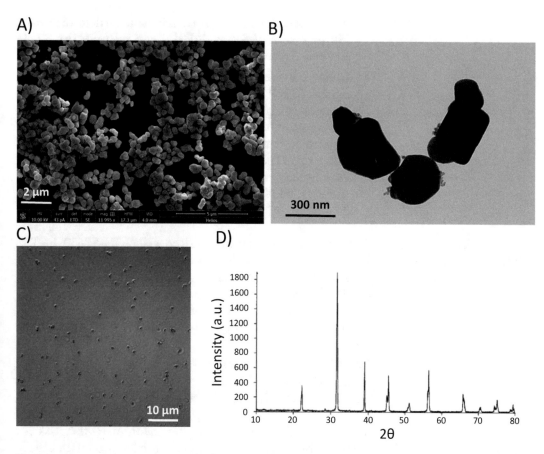

Fig. 1 Imaging and characterization of barium titanate nanoparticles (BTNPs) presenting a tetragonal crystalline structure. Imaging of BTNPs performed with SEM (**a**), TEM (**b**), and confocal laser scanning microscopy (**c**). The signal of the confocal laser image (in red) is merged with the transmitted light image (gray). XRD analysis (**d**) shows the perovskite-like crystallographic structure (two close peaks at $2\theta = 44.85°$ and $2\theta = 45.38°$), which is typical of the piezoelectric tetragonal configuration. Reproduced with permission from [16]; copyright (2015) American Chemical Society

evaporate and then gold-sputter the sample (25 s at 60 nA) before SEM analysis. We used specific SEM parameters for an optimal imaging of the BTNPs. *See* **Note 2**.

(b) For TEM imaging, deposit a 6 µl drop of solution onto a carbon-coated copper grid, remove the excess of solution with Whatman filtration paper.

(c) Concerning confocal fluorescence microscopy, put a 50 µl drop of the nanoparticle dispersion on a glass coverslip; use laser excitation at 633 nm and a collection from 645 to 745 nm for detecting the BTNP signal. We used specific laser and photomultiplier parameters for an optimal imaging of the BTNPs. *See* **Note 3**.

(d) Fill a cuvette with 1 ml of the nanoparticle dispersion (50 μg/ml in aCSF or ddH$_2$O) and measure the BTNP size and Z-potential by using the Z-sizer.

(e) Check the X-ray diffraction (XRD) patterns of the evaporated BTNP dispersion by using X-ray powder diffractometer using Cu Kα radiation (λ = 1.5406 A) at the temperature of 25 °C and at the scanning rate of 0.016°/s^{-1} with 2θ ranging in 10°–80°. Tetragonal crystal (piezoelectric) BTNPs are characterized by two peaks at 2θ = 44.85° and 2θ = 45.38°. Cubic crystal (non-piezoelectric) BTNPs are characterized by a single peak at 2θ = 45°.

3.2 Cell Culture, Cell Differentiation, and BTNP Treatment

1. Culture the neuroblastoma-derived cells (SH-SY5Y) in T75 flasks with 12 ml of proliferation medium (DMEM/F12 supplemented with 10% heat inactivated fetal bovine serum, 100 IU/ml penicillin, and 100 μg/ml streptomycin) at 37 °C in a saturated humidity atmosphere containing 5% CO$_2$. Coating is not necessary. The medium is changed every 3–4 days.

2. Split SH-SY5Y cells when 85% of confluence (*i.e.*, flask area covered by the cells) is reached. Remove the medium from the T75 flask, wash twice the cells with sterile PBS, and incubate at 37 °C for 5 min with 2 ml of 0.05% trypsin/EDTA. Add 8 ml of proliferation medium to the flask for blocking the trypsin action, transfer the solution with cells in suspension to a polystyrene tube, centrifuge the cells at 150 rcf, discharge the supernatant, and resuspend the pellet of cells in 12 ml of proliferation medium; finally seed 1 ml of this new cell suspension in a T75 flask. Add 350 μl of dimethyl sulfoxide (DMSO) to the remaining 7 ml of solution with cells and split the volume into seven criovials to be stored in liquid nitrogen.

3. For the experiments, seed cells on 35 mm diameter μ-dishes at a density of 20,000 cell/cm^2 with 1.5 ml of proliferation medium.

4. Switch the proliferation medium with the differentiation medium (DMEM supplemented with 1% heat inactivated FBS, 10 μM all-trans-retinoic acid, 100 IU/ml penicillin, and 100 μg/ml streptomycin) 24 h after the cell adhesion.

5. After 4 days of differentiation, treat the SH-SY5Y-derived neurons for 24 h with BTNPs at the final concentration of 50 μg/ml in a differentiation medium or with the vehicle of the nanoparticles (50 μg/ml gum Arabic in a differentiation medium) as a control. *See* **Note 4**.

3.3 Metabolic and Viability Assay

1. To determine the effect of BNNTs on cell metabolism and membrane integrity, WST-1 assay and propidium iodide (PI) staining were carried out on SH-SY5Y-derived neurons pre-treated with BTNPs and on control samples. *See* **Note 5**.

   (a) For WST-1 experiments, dilute 1:11 the premix solution of 2-(4-iodophenyl)-3-(4-nitrophenyl)-5-(2,4 disulfophenyl)-2H-tetrazoilium monosodium salt in the differentiation medium. Replace the differentiation medium with 1.5 ml of the obtained solution and incubate for 2 h at 37 °C. Read the absorbance at 450 nm of 100 μl of the solution *per* sample by using a micro-plate reader. Subtract the value of the absorbance of a "white" sample (an aliquot of the diluted premixed solution not incubated with cells) and finally normalize with respect to the value of the control cultures. *See* **Note 6**.

   (b) For the evaluation of the membrane integrity, treat cells with 1 μg/ml of PI and Hoechst 33342 (1 μg/ml) for 10 min in a differentiation medium at room temperature in the dark. Carry out fluorescence microscope imaging with the appropriate filters (DAPI for nuclei and TRITC for propidium iodide).

2. SHSY5Y-derived neurons incubated for 24 h with 50 μg/ml of BTNPs show no evidence of impaired membranes/metabolism with respect to the control untreated cultures.

3.4 3D Confocal Analysis of the BTNPs/Neurons Interaction

1. Stain plasma membranes and nuclei of living neurons pre-treated with BTNPs by incubating cells with 1:1000 CellMask green plasma membrane stain and Hoechst 33342 (1 μg/ml) for 10 min at 37 °C.

2. Rinse the μ-dishes with PBS and then fill the samples with 1.5 ml of artificial cerebrospinal fluid (aCSF).

3. Carry out confocal laser scanning of z-stacks of cells: BTNPs, CellMask green, and Hoechst 33342 were excited by 633, 488, and 405 nm lasers, and the emission signals were collected at 645–745, 500–555, 425–525 nm, respectively.

4. Perform 3D confocal rendering of the merged signals by using ImageJ software (http://rsbweb.nih.gov/ij/).

5. BTNPs are associated with the plasma membrane of both cell body and neurites without a significant cellular internalization (Fig. 2).

3.5 Calcium and Sodium Imaging During Ultrasounds/BTNPs Stimulation

1. Before performing Ca^{2+} or Na^+ imaging, incubate SH-SY5Y-derived neurons with 1 μM Fluo-4 AM or 1 μM CoroNa Green AM, respectively, in serum-free DMEM for 30 min at 37 °C.

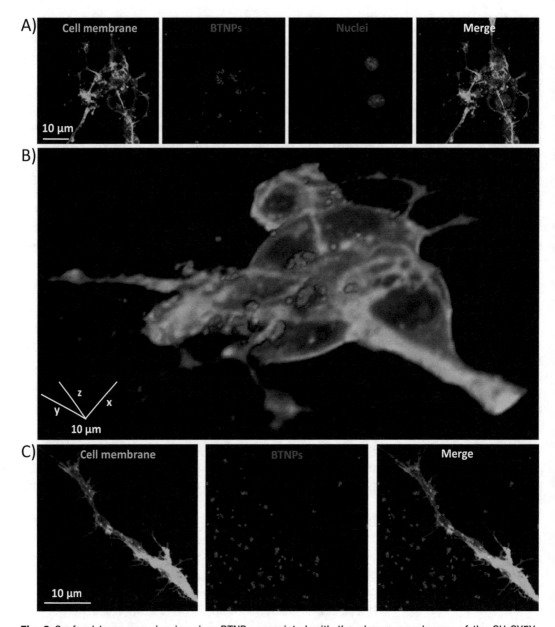

Fig. 2 Confocal laser scanning imaging: BTNPs associated with the plasma membranes of the SH-SY5Y-derived neurons. (**a**) Single *z*-stack and (**b**) 3D confocal reconstruction of the same field of cells. (**c**) BTNPs also associate with membranes of SH-SY5Y neurites (neuronal plasma membranes, BTNPs, and nuclei are respectively shown in green, in red, and in blue). Reproduced with permission from [16]; copyright (2015) American Chemical Society

2. After the reagent incubation, rinse and supply samples with aCSF. Put the samples on an inverted fluorescence microscope. Wait 20 min for the stabilization of cell conditions and for the complete de-esterification of the AM groups. *See* **Note 7**.

3. Put the ultrasound (US) probe tip over the dish, in contact with the aCSF, at a distance of 5 mm from the cells.

4. Perform fluorescence time-lapse acquisition by using the appropriate filter (FITC) and by acquiring images at a constant rate (at least 1 Hz). Stimulate the samples with US in the presence or in the absence of BTNPs. We performed US stimulations for 5 s at 0.8 W/cm². High-amplitude calcium and sodium peaks were detected when stimulating in the presence of piezoelectric BTNPs. These peaks were not observed in the case of US without nanoparticles or in the presence of non-piezoelectric BTNPs (characterized by a cubic crystal configuration). Representative calcium imaging time-lapses are reported in Fig. 3.

5. For the investigation of the channels involved in the cell excitation, add specific channel blockers to the aCSF used during the US stimulation tests. In particular, use cadmium chloride ($CdCl_2$, non-specific blocker of Ca^{2+} channels, 100 μM), tetrodotoxin (TTX, blocker of voltage-gated Na^+ channels, 100 nM), or gentamicin (blocker of mechano-sensitive cation channels, 200 μM). High-amplitude calcium peaks evoked by US + BTNPs stimulations resulted to be sensitive both to $CdCl_2$ and TTX, but not to gentamicin, suggesting that the induction of the high-amplitude Ca^{2+} transients by the US + BTNPs stimulation is mediated by both Ca^{2+} and Na^+ voltage-gated channels, but not by mechanosensitive channels. Obtained findings support the hypothesis of a piezoelectric neural stimulation able to elicit a significant Ca^{2+} influx.

3.6 Image and Statistical Analysis

1. Analyze the acquired images with ImageJ software (http://rsbweb.nih.gov/ij/). Import stack (time-lapse of fluorescence imaging) and convert it to 8-bit images.

2. Threshold images to define the regions of interest (ROIs).

3. Convert the pixel intensity of the time-lapse in $\Delta F/F_0$ by using the divide-and-subtract function of the Math process, and by defining the F_0 image as the average intensity of the first five images.

4. Perform a double smoothing and finally measure the averaged values of the pixels inside the ROIs by using the multi-measure function of the ROI manager.

5. Plot $\Delta F/F_0$ traces of single ROIs on the graphs. For a statistic comparison, consider only $\Delta F/F_0$ peaks threefold higher than

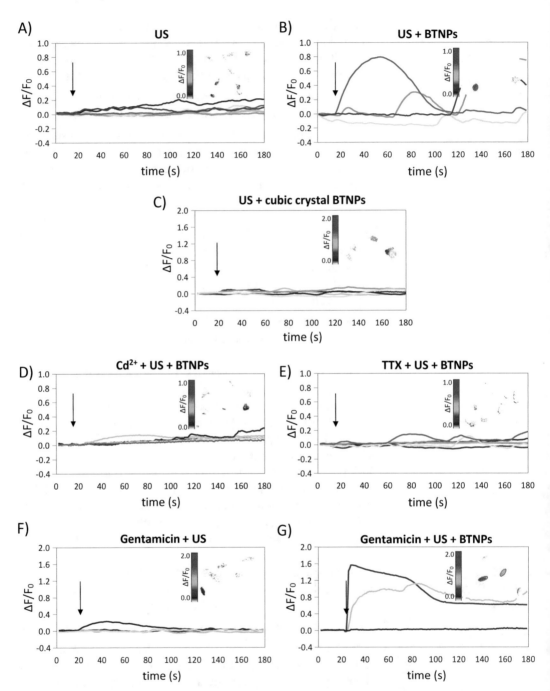

Fig. 3 Calcium time-lapses recorded during the following stimulation tests: US (**a**), US + piezoelectric BTNPs (**b**), US + non-piezoelectric BTNPs (**c**), US + piezoelectric BTNPs in the presence of the Cd^{2+} blocker (**d**), US + piezoelectric BTNPs in the presence of the TTX blocker (**e**), US in the presence of gentamicin (**f**), and US + piezoelectric BTNPs in the presence of gentamicin (**g**). The starting of the US stimulation is marked in all the time-courses by a black arrow; each colored trace corresponds to a single analyzed cell. In the inlet of each graph a representative frame of the calcium imaging is shown (at $t = 50$ s). Adapted with permission from [16]; copyright (2015) American Chemical Society

the standard deviation of the noise (calculated before the US stimulation).

6. Report and compare the average amplitude of the $\Delta F/F_0$ peaks recorded in response to the different treatments (at least the response of 20 cells for all the conditions was analyzed, triplicate experiments).

7. After testing the normality of the different $\Delta F/F_0$ peak distributions with the Shapiro test, perform the ANOVA parametric test and Tukey's HSD *post-hoc* test to compare the different distributions.

4 Notes

1. The results obtained with WST-1 assay are directly correlated to the cell number in each well. In order to have good and reliable results, the same cell number has to be seeded in each well for all the samples.

2. SEM parameters for an optimal imaging of the BTNPs: operative current = 43 pA, high voltage = 10 kV.

3. Confocal laser scanning microscope parameters for the BTNP imaging: laser power = 1%, photomultiplier high voltage = 700 V.

4. After 5 days of differentiation (4 days in differentiation medium and 1 day in differentiation medium supplemented with BTNPs or vehicle) SH-SY5Y cells express peculiar markers of differentiated neurons, such as β-3 tubulin.

5. It is possible to perform WST-1 and PI assays to also evaluate possible potential harmful effects of the ultrasound stimulation.

6. For WST-1 experiments, it is preferable to use a differentiation medium with phenol red-free DMEM.

7. During the de-esterification of the AM groups (20 min) and before the US stimulation experiments, switch on the mercury lamp to stabilize the excitation light intensity.

5 Summary

In this work we successfully performed US-driven piezoelectric neural stimulation by exploiting BTNPs, by analyzing ion fluxes at the base of the observed phenomenon. This approach can be considered a novel tool for a non-invasive wireless neural stimulation, which opens up interesting perspectives in the field of neuronal prosthetics, tissue engineering, and even in bionics applications.

Acknowledgments

The authors gratefully thank Mr. Piero Narducci (Department of Chemical Engineering, University of Pisa, Pisa, Italy) for XRD technical assistance. This research was partially supported by the Italian Ministry of Health Grant Number RF-2011-02350464 (to G.C.), by the JSPS KAKENHI Grant Number 26107717 (to M.S.), and by the JSPS Core-to-Core Program, A. Advanced Research Networks (to M.S.).

References

1. Deisseroth K (2015) Optogenetics: 10 years of microbial opsins in neuroscience. Nat Neurosci 18:1213–1225

2. Benabid AL, Chabardes S, Mitrofanis J, Pollak P (2009) Deep brain stimulation of the subthalamic nucleus for the treatment of Parkinson's disease. Lancet Neurol 8:67–81

3. Vidailhet M, Vercueil L, Houeto J-L, Krystkowiak P, Benabid A-L, Cornu P, Lagrange C, Tézenas du Montcel S, Dormont D, Grand S, Blond S, Detante O, Pillon B, Ardouin C, Agid Y, Destée A, Pollak P, French Stimulation du Pallidum Interne dans la Dystonie (SPIDY) Study Group (2005) Bilateral deep-brain stimulation of the globus pallidus in primary generalized dystonia. N Engl J Med 352:459–467

4. Schlaug G, Renga V, Nair D (2008) Transcranial direct current stimulation in stroke recovery. Arch Neurol 65:1571–1576

5. Reithler J, Peters JC, Sack AT (2011) Multimodal transcranial magnetic stimulation: using concurrent neuroimaging to reveal the neural network dynamics of noninvasive brain stimulation. Prog Neurobiol 94:149–165

6. Stanley SA, Gagner JE, Damanpour S, Yoshida M, Dordick JS, Friedman JM (2012) Radio-wave heating of iron oxide nanoparticles can regulate plasma glucose in mice. Science 336:604–608

7. Seijo FJ, Alvarez-Vega MA, Gutierrez JC, Fdez-Glez F, Lozano B (2007) Complications in subthalamic nucleus stimulation surgery for treatment of Parkinson's disease. Review of 272 procedures. Acta Neurochir 149:867–875; discussion 876

8. Wagner T, Valero-Cabre A, Pascual-Leone A (2007) Noninvasive human brain stimulation. Annu Rev Biomed Eng 9:527–565

9. Barker AT (1999) The history and basic principles of magnetic nerve stimulation. Electroencephalogr Clin Neurophysiol Suppl 51:3–21

10. Tufail Y, Yoshihiro A, Pati S, Li MM, Tyler WJ (2011) Ultrasonic neuromodulation by brain stimulation with transcranial ultrasound. Nat Protoc 6:1453–1470

11. Zhao Y, Liao Q, Zhang G, Zhang Z, Liang Q, Liao X, Zhang Y (2015) High output piezoelectric nanocomposite generators composed of oriented BaTiO3 NPs@PVDF. Nano Energy 11:719–727

12. Wang X, Liu J, Song J, Wang ZL (2007) Integrated nanogenerators in biofluid. Nano Lett 7:2475–2479

13. Royo-Gascon N, Wininger M, Scheinbeim JI, Firestein BL, Craelius W (2013) Piezoelectric substrates promote neurite growth in rat spinal cord neurons. Ann Biomed Eng 41:112–122

14. Inaoka T, Shintaku H, Nakagawa T, Kawano S, Ogita H, Sakamoto T, Hamanishi S, Wada H, Ito J (2011) Piezoelectric materials mimic the function of the cochlear sensory epithelium. Proc Natl Acad Sci U S A 108:18390–18395

15. Ciofani G, Danti S, D'Alessandro D, Ricotti L, Moscato S, Bertoni G, Falqui A, Berrettini S, Petrini M, Mattoli V, Menciassi A (2010) Enhancement of neurite outgrowth in neuronal-like cells following boron nitride nanotube-mediated stimulation. ACS Nano 4:6267–6277

16. Marino A, Arai S, Hou Y, Sinibaldi E, Pellegrino M, Chang Y-T, Mazzolai B, Mattoli V, Suzuki M, Ciofani G (2015) Piezoelectric nanoparticle-assisted wireless neuronal stimulation. ACS Nano 9:7678–7689

17. Kovalevich J, Langford D (2013) Considerations for the use of SH-SY5Y neuroblastoma cells in neurobiology. Methods Mol Biol 1078:9–21

Chapter 9

Influence of External Electrical Stimulation on Cellular Uptake of Gold Nanoparticles

Samantha K. Franklin, Brandy Vincent, Sumeyra Tek, and Kelly L. Nash

Abstract

Metal nanoparticles, more specifically gold nanoparticles (AuNPs), have become increasingly popular in research due to their optical, thermal, and electronic properties. It is these properties that have made them excellent for usage in life science areas such as medical, biological imaging, and drug delivery. AuNPs have been applied to multimodal imaging techniques to assist in fluorescence, MRI, and CT and in diagnostics through surface-enhanced Raman spectroscopy (SERS). These techniques all rely on the interaction of nanoparticles with electromagnetic stimulation. Here, we describe an experimental approach to investigate the combinatorial effects of gold nanoparticles with ultra-short electrical pulses. Specifically, we investigate cell membrane interaction with AuNPs to obtain information at the cellular level of possible changes in the uptake mechanism of the AuNPs. Our goal is to study the possibility of designing AuNPs for use in biomedical studies such as neuron stimulation and brain mapping.

Key words Nanosecond electroporation, Gold nanoparticles, Cell membrane, Uptake, Biocompatible nanoparticles

1 Introduction: Gold Nanoparticles in Biomedical Applications

Gold nanoparticles are among the most commonly used particles for drug delivery, which can be used with proteins and molecules [1–3]. Their unique chemical and physical properties make their use in biological applications possible. Due to the strong electric fields at the surface of the noble metal, the absorption and scattering of the electromagnetic radiation is greatly enhanced, which produces heat to eliminate the surrounding tumor cells. The specificity of the photothermal therapy, i.e., to target only tumor tissues, is achieved by tagging the AuNPs with HER-2 antibodies, which now specifically bind only to the tumor cells, and a photosensitizer that will destroy the targeted area [4–8]. The main attractiveness to using gold nanoparticles is that they are inert and nontoxic in nature, increasing the possibility of use for drug delivery, targeting, and imaging. They have both advantageous and

Fidel Santamaria and Xomalin G. Peralta (eds.), *Use of Nanoparticles in Neuroscience*, Neuromethods, vol. 135, https://doi.org/10.1007/978-1-4939-7584-6_9, © Springer Science+Business Media, LLC 2018

Table 1
Advantages and disadvantages of gold nanoparticles in biological sciences

Advantages of AuNP	Disadvantages of AuNP
Biocompatibility: AuNPs show less or no toxicities in the in-vitro and in-vivo applications	Inefficient release: the primary disadvantage of using AuNPs for payload delivery is their dependence on micro-environmental physiological conditions of the cell as a stimulus, resulting in the drug release process being extremely inefficient
Ease of fabrication: chemical synthesis of AuNPs of different sizes and shapes is of ease	Low penetration depths of visible light: the tissue penetration depth is a huge limitation in treating deep-tissue carcinomas. (Note: The visible light tissue penetration depth is only 2–3 cm)
Ease of tailoring surface properties: ability to functionalize the nanoparticles to carry a wide range of therapeutic payloads is mandatory in treating a wide range of diseases	Specificity: use of AuNPs as contrasting agents also demands surface functionalization of the nanoparticles with cell-specific antibodies. Thus, only the carcinomas carrying the counter antibodies (or antigen) can be detected
Diagnostics: tumor detection based on the cell-specific biomarkers	

disadvantageous uses in biological applications which are outlined in Table 1.

Metal nanoparticles including gold (Au) have become increasingly popular in scientific research due to their optical, thermal, and electronic properties. It is these properties that have made them excellent for usage in life science areas such as medical, biological imaging, and drug delivery [9]. For example, gold nanoparticles have been combined with particles, such as Iron Oxide particles (Fe_3O_4) and other compounds for imaging [10–13]; more specifically, they have been applied to multimodal imaging techniques to assist in fluorescence, MRI, and CT [14–17]. Along with imaging, AuNPs are used in therapeutic applications. AuNPs can assist in photothermal therapy such as Nanoparticle Assisted Photothermal Therapy (NAPT) [15] and Plasmonic Assisted Photothermal Therapy (PPTT). The benefits of AuNPs with therapy applications, such as photodynamic therapy, are that they can be combined with imaging as well, so there is less preparation time during synthesis and application [18]. Metal nanoparticles (including AuNPs) can be highly potent thermal therapeutic agents, which come from their strong light absorption to heat conversion [19]. The photothermal effect (PTT) utilizes the formation of heat, due to thermal energy converted from vibrations of the photon energies, by employing light absorbing dyes for achieving the photothermal damage of tissue and cells and inducing cell death [18, 20–23]. Current PTT focuses on near-infrared (NIR), known as the "Optical Window" (650–900 nm), in which NIR can penetrate into human tissue with little absorption or scattering. Additional therapeutic approaches have found that AuNPs can serve as effective means to transfect cells with small interfering ribonucleic acid (siRNA) for gene therapy [24] and delivery systems for photodynamic therapy (PDT) compounds. PDT is a clinically approved form of cancer treatment that uses a photosensitive agent (photosensitizer) and a light source [6, 12, 25–28].

1.1 Methods of Gold Nanoparticle Synthesis

The most common method of producing gold nanoparticles is the Turkevich method, first reported in 1951, otherwise known as the sodium citrate (NaCit) synthesis. In this synthesis process, the reaction of small amounts of hot chlorauric acid with small amounts of sodium citrate produces relatively mono-disperse spherical gold particles around 10–20 nm in diameter [29]. Another method, the Brust Method, was developed in the 1990s and is capable of producing gold nanoparticles that are 2–6 nm in diameter in organic liquids that are not normally miscible in water. It involves the reaction of chlorauric acid with Tetracylammonium Bromide (TOAB) in toluene, as an anti-coagulant, and sodium borohydride as the reducing agent [30]. More recently, Eah et al. developed a facile method of creating naked gold nanoparticles in water by reducing $HAuCl_4$ in $NaBH_4$. The resulting particles were stably dispersed and were approximately 3.2–5.2 nm in diameter [31]. Among all these approaches, the resulting AuNPs can be toxic to cells and further work, beyond the synthesis, must be done to prepare the particles for biological application uptake, to ensure non-toxicity, and addition of functional surfaces. The tailoring of the NP surface is often accomplished by adding specific ligands to the nanoparticle, which result in exposed functional groups [32–37] that allow for use in specific applications due to changes of ionic charge state surrounding the nanoparticle [38–41]. The changes to their physical and chemical characteristics make them applicable in various biological systems for dispersibility, stability, and reactivity [37]. In a previous study, chemically synthesized block polymer carriers formed a gold nanocore where modification to the structure using three types of reactive groups enabled the attachment of a disease-specific drug for drug delivery [41]. The measured changes in agglomeration behavior of gold nanoparticles in various ionized states using amine or carboxyl surface modification have shown longer alkane chains created by specific bundling on the surface, resulting in asymmetric coatings [32]. Also, surface modified silica aerogels and core shell NPs have been used to remove organic molecules from water systems [36]. Prior to surface modification, the NPs were unstable in water (hydrophobic). With the use of microfluids and porogen templating, the surface change enabled the NPs to be dispersed in the water, while the core absorbed the surrounding organic molecules. Once the nanoparticle surface is modified, it allows for the ability to impart functionalities onto the nanoparticle by further allowing the introduction of chemical functional groups to the surface that are application specific.

Approaches to rendering nanoparticles functional vary for biological applications. These techniques range from both in situ methods to post synthesis treatment [42–46]. Often these methods involve the addition of small amino acids or protein-like molecules to the surface of the particle [47]. This imparts a unique and dynamic nano-bio-interface resulting in biocompatible particles, which make them excellent candidates for use in medical applications such as drug delivery and therapy techniques [48–54].

1.2 Stimulation of Biological Systems

Invasive stimulation in neurons has been studied to measure neuron response and aims to cure neurological disorders, such as Alzheimer's, Parkinson's disease, and neuropathic pains [55]. The currently practised neurosurgical method for brain stimulation is called deep brain stimulation (DBS) [56]. Deep brain stimulation uses implantable electrodes that are placed into the target area to mimic neural response and re-create the neural activity by electric stimulation; however, the electrodes are left in place for weeks to years for stimulation, depending on the type of disorder [57]. The main advantage of using DBS is that it is highly reversible, leaving no damage to the neural tissue [58]. Disadvantages extend from the cost of the surgery, to the invasiveness of the procedure, and potential side effects including memory loss, paralysis, and other permanent neurological complications [59, 60]. To overcome the disadvantages and complications with DBS, noninvasive methods are currently being sought to study these neurological disorders. Such noninvasive techniques use local magnetic pulses, which are generated through a coil that is placed on the scalp, which pass high electrical currents to the brain [61]. The electrodes, with saline sponges, are placed over the head, to stimulate brain tissue [62, 63].

1.3 Ultra-Short External Electrical Pulses and Cell Interaction

Electroporation is a technique in which an electric field is applied to cell membranes. The electric pulse results in increased permeability due to development of pores in the membranes [64]. Membrane electroporation describes the transient, reversible permeabilization of the membranes of cells, organelles, or lipid bilayers vesicles by electric field pulses [65]. The electroporation process is modulated considerably by the structure of membranes.

Significant difference between the values of the pulse amplitude is required to induce electroporation for different species [64]. Pulsed electric fields from μs to ms have found their use in many biological techniques, including gene transfection and gene transformation, introduction of exogenous molecules into cells, and hybridization [66].

Electric pulses in the nanosecond range have also been shown to create pores in the plasma membrane; however, these pores are much smaller (<2 nm) and are termed nanopores [67]. These nanopores are believed to be large enough for the passage of small inorganic ions, but too small to allow dyes such as propidium iodide (PI) into the cell. As shown by Pakhomov et al., the electrical breakdown of the plasma membrane that creates these nanopores happens within nanoseconds after the onset of the electrical pulse; however, the nanopores can persist for many minutes after a single nano-pulse exposure [67].

Traditionally, nanopore formation has been observed and characterized by the uptake of fluorescent dyes into the cell upon nsPEF exposure. In Fig. 1, the theoretical breakdown of the cell membrane upon exposure to ultra-short electrical pulses is depicted schematically. Substantial uptake of PI (diameter ~2 nm) is lacking, but other molecules with diameters around 1 nm, such as YO-PRO1 and Thallium, have been shown to enter the cell after nsPEF [68, 69]. While the lifetime of membrane permeabilization has been examined through patch-clamp or delayed addition of Thallium ion, it is

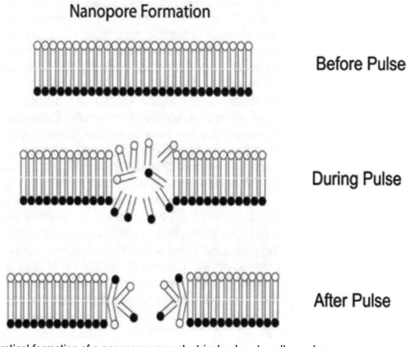

Fig. 1 Theoretical formation of a nanopore upon electrical pulses to cell membrane

difficult to determine how long the nanopores persist in most fluorescence-based approaches. In this work, we have chosen another approach to characterizing pore formation. 6-Propionyl-2-(N,N-Dimethylamine) Naphthalene, otherwise known as PRODAN, does not determine pore size, but it may determine pore formation. PRODAN was introduced by Weber and Farris in 1979. It has both electron donor and electron acceptor substituents resulting in a large excited-state dipole moment and extensive solvent polarity-dependent fluorescent shifts [70]. When PRODAN or its derivatives (Laurdan, acrylodan, and badan) are incorporated into membranes, their fluorescence spectra are sensitive to the physical state of the surrounding phospholipids [71]. For example, when temperature is increased across the membrane, the membrane shifts from more of a gel state to a more liquid state. This shift in membrane organization results in a spectral shift in PRODAN as the fluorescence shifts from blue to green fluorescence [68].

1.4 Combined Effects of Nanoparticles and Electromagnetic Stimulation

One focus that stems from these stimulation techniques is brain mapping. This allows for an imaging, whether direct (internal) or indirect (external), of various actions of the brain, such as functional and structural [72]. Neurons react by firing (signaling) to other neurons [73]. Understanding this neuronal activity plays an essential role in diagnostics and treatment of multiple neurological disorders noninvasively [74, 75]. Issues from current brain mapping techniques such as magnetoencephalography (MEG) and electroencephalography (EEG) [76, 77] are low spatial resolution and signal-to-noise ratio. While these applications are applied to 3D tissue system (in vivo), to further expand research in this area, it is advantageous to monitor interaction at a cellular level (in vitro). Few studies have utilized AuNP interaction with neurons, which can be studied for neural mapping and stimulation. One study by Benzanilla and colleagues showed that, when activated with green light, AuNPs would absorb and convert light energy into heat and activate un-modified neurons [78]. Studies have also used nanocomposites as therapeutic aids toward neural diseases such as Alzheimer's and Parkinson's disease, where the nanoparticles have been shown to inhibit Aβ fibrils [79, 80]. Gold nanoparticles have also been used with neural implant devices and showed threefold improvement in interfacial impedance with further improvement of electrical properties expected when using special shape gold nanoparticles [81]. It has been seen that with an applied external current, gold nanoparticles can alter intrinsic properties of neurons, by increasing their excitability [82]. However, there are still many types of toxic effects of nanoparticles on cells [83, 84]. Nanoparticles are generally synthesized as a multi-component system with additives such as polymers as capping agents [85] and other molecules for targeted drug delivery [86]. These gold nanoparticles typically contain a gold core shell and some surface functionalization. The left over chemicals from the synthesis can play a role in any observed toxicity [87]. This situation can be overcome by one-pot green synthesis techniques.

In general, cells will uptake the particles by endocytosis, where they consume some extracellular fluid, including material dissolved or suspended in the extracellular environment [88], a process that typically requires incubation of the particles for many hours before experimentation. nsPEFs are similar to electroporation, a reversible disturbance of the barrier function of membranes through the application of a μs–ms electrical pulse, to allow the passage of molecules such as DNA into the cell [89]. The loss of the barrier function of the membranes during electroporation can develop in less than microseconds for sufficiently high intensities of electrical treatments [90–92]. However, nsPEFs produce much smaller, nanometer-size holes in the plasma membrane [93–95].

In the present approach, we use nanosecond pulsed electric fields (nsPEF) to enhance the uptake of Chit-AuNPs vs. incubation methods. A similar technique was studied in 2005 by Gunderson et al. with the use of Quantum Dots (QDs) [96], where the uptake of the QDs due to pulsing was measured. This study revealed that larger QDs did not cross the membrane after pulsing and that uptake, due to pulsing, was dependent on particle size and membrane transport activity. By increasing the number of pulses, amplitude, or duration of the exposure, we can "controllably" increase the size of the nanopores. It is hypothesized that the uptake of particles with pulsing would be instant, due to prompt formation of the pore with pulsing. This would exceed incubation techniques, and decrease uptake time considerably. Successful induction of rapid nanoparticle uptake with nsPEF may prove very useful in studying the dynamics of membrane pores and delivery of intracellular nanosensors to measure cellular pathways and interactions.

2 Materials

2.1 Gold Nanoparticle Produced in the Presence of Chitosan

The use of chitosan has been purported as advantageous because of chitosan being a natural material with excellent physiochemical properties, environmentally friendly, and it possesses bioactivity that does not harm humans [97].Another beneficial property of chitosan is that it has the ability to form extremely strong, thin films on surfaces. This has been observed in applications such as wound dressing, creating a chitosan film in the form of a bandage [98]. In a study performed by Khan et al., they determined that the chitosan film would have strong tensile strength to allow normal movement from different parts of the body. The adhesive ability of chitosan comes from the property that it is positively charged, and readily binds to negatively charged surfaces [99]. In Fig. 2, chitosan is shown through a rendering in visual molecular dynamics (VMD) in

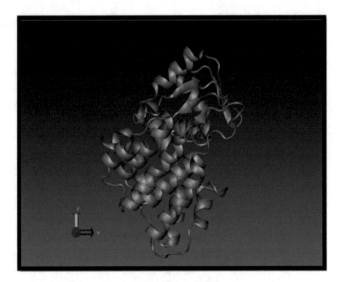

Fig. 2 Visual molecular dynamic rendering of chitosan

its neutral state. As the pH is adjusted, the α-helixes and β-sheets unfold, and the resulting exposure of the amino groups can be used for attachment, particularly to metal ions.

Medium molecular weight (MMW) chitosan, derived from Chitin, in a flake form was prepared in 500 mL deionized water, with the addition of 5 mL acetic acid, due to insolubility of chitosan in water [100]. Two weight percent (wt%) of MMW chitosan was prepared by the addition of different weights of flakes. 0.05% and 0.25% stock solution of chitosan were prepared under magnetic stirring, and adjusted to pH of 4 for use in synthesis. The solution was stored in a laboratory refrigerator when not in use, and has a degradation time of 4–6 months. For the first synthesis with the chitosan solution, $HAuCl_4$ was added for the formation of Chit + AuNPs. 10 μL/mL of $HAuCl_4$ was added to the stirring chitosan solution (0.05% and 0.25%) and allowed to mix for 30 min. The second formation was synthesized with the addition of tripolyphosphate (TPP). Chitosan-TPP-Au nanoparticles were formed with a chitosan-to-TPP weight ratio of 3:1. Sodium Tripolyphosphate, mixed with deionized water, was added dropwise to the solution. After 50 min of mixing, $HAuCl_4$ was added for the formation of the TPP-Au ions with chitosan and allowed to stir for an additional 30 min. The third formation of particles was with the addition of sodium chloride (NaCl). NaCl was added directly to the chitosan solution to produce a 150 mM solution, and allowed to stir for 50 min, before the addition of $HAuCl_4$. The solution was stirred for an additional 30 min. All the samples were then reduced by UV light in a Spectrolinker UV oven (Spectroline, Westbury, NY) for 20 min and then washed with a mixture of 1 part acetic acid and 3 parts deionized water. The samples were centrifuged at $21,130 \times g$ for 30 min. The supernatant liquid was removed and the pellet was resuspended in deionized water and stored at 4 °C prior to use. The reaction between chitosan and Au is sensitive to the concentration of chitosan present, so AuNPs prepared with high concentrations of chitosan have higher stability [101]. The formation of highly stable AuNPs can be attributed to the affinity of the gold ions to NH_2 groups within the chitosan structure prior to the photo-reduction. A colorimetric change can be observed after the photo-reduction process, as shown in Fig. 3. The pinkish color indicates formation of spherical gold nanoparticles, since the particles absorb in the green which yields a reddish-pink color [102]. As the UV reduction time increases, the resultant shade of the particle solution is dark, which implies a higher concentration of the particles. The 1-min solution contains less particles than the 20 min solution due to a shorter reaction time, which was supported by the polydispersity index (PdI) value given when dynamic light scattering (DLS) was measured. This effect can be seen for the Chit + Au, Chit + TPP, and Chit + NaCl. The purple shade seen in the reduced Chit + NaCl solutions indicates aggregation or rod formation. As particle size increases, the wavelength of the SPR related to the absorption of the sample shifts to a redder wavelength, leaving the red color absorbed. This leaves more blue light reflected (opposite the small gold spheres), resulting in a pale blue or purple color [102]. The shorter UV reduced samples have light purplish color, which would signify larger particles. As the reduction time is increased, there is additional interaction of the chitosan and gold, resulting in smaller particles.

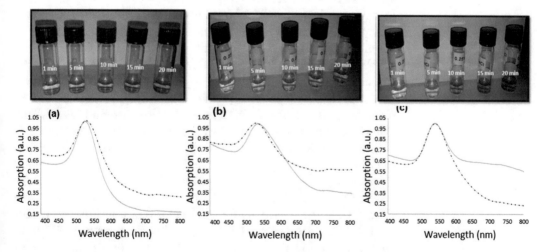

Fig. 3 UV-VIS absorption graphs of (a) 0.05% Chit+Au (solid, gray) and 0.25% Chit+Au 20 min UV (dashed, black) (b) 0.05% Chit+TPP (solid, gray) and 0.25% Chit+TPP 20 min UV (dashed, black) and (c) 0.05% Chit+NaCl (solid, gray) and 0.25% Chit+NaCl 20 min UV (dashed, black) with corresponding photo showing colorimetric change of samples after UV reduction above.

In Fig. 3a–c, we show the absorption data of the nanoparticles. For the Chit + Au absorption spectra (Fig. 3a), both the 0.05 and 0.25 wt% have an absorption peak at 525 nm. The absorption peak is the same for both the concentrations of chitosan due to the formation with only the $HAuCl_4$ addition before the 20 min UV reduction, which forms gold spheres. Absorption data for Chit + TPP (Fig. 3b) show a broad peak at ~540 nm. This is a 20 nm red shift, which indicates a change in the morphology of the resultant AuNPs. Chit + NaCl data reveal a narrower peak than the Chit + TPP solution, for both concentrations, suggesting more uniform AuNP spheres.

The 0.05 wt% concentration has a higher molecular weight than the 0.25 wt% after synthesis. The 0.05 wt% contains more chitosan in the solution. When the samples are synthesized, the same concentration of $HAuCl_4$ is added to both 0.05 wt% and 0.25 wt%, so there is more chitosan in the 0.05% for a higher concentration of AuNP formation. The broad peak of the 0.05% Chit + TPP (Fig. 3b) indicates a higher variation of shape and size formation in the solution, which can be confirmed by microscopy images (Fig. 3b). Due to the larger MW of the 0.05% chitosan solution and electrostatic repulsion that occurs with the addition of NaCl, the AuNPs form on the truncated chains, like a necklace, which give a resonance similar to rods. This is confirmed in Fig. 3c, where the spheres are in close proximity in the image, which will absorb like a gold nanorod. This explains the 0.05% absorption spectra in the NIR region in Fig. 3c.

The electron microscope images of AuNPs are taken after 20 min of UV reduction (Fig. 4a–f). The observation of AuNPs, synthesized with the three different conditions (AuNPs reduction in chitosan, in Chit + NaCl, and in Chit + TPP), is based on two different concentrations of chitosan solution (0.05 and 0.25 wt%). At a lower concentration of chitosan solution (0.05 wt%), the AuNPs reduced in chitosan are ≈20 nm in size (Fig. 4a), in the presence of NaCl in chitosan solution the AuNPs formation becomes smaller 8–15 nm (Fig. 4c) and in the presence of TPP in chitosan solution the AuNPs formed with different shapes like spherical, rod, triangle, and hexagonal and their average size is within 20–40 nm (Fig. 4b). In contrast at the higher concentration of chitosan solution (0.25 wt%) the formation of AuNPs is smaller compared to the lower concentration of chitosan solution AuNPs synthesis. The AuNPs reduced in chitosan are 8–10 nm in size (Fig. 4d), in the presence of NaCl in chitosan solution the AuNPs are much smaller 2–5 nm (Fig. 4e), and in the presence of TPP in chitosan solution the AuNPs are 5–10 nm (Fig. 4f) in size with spherical shapes.

0.05 wt% Chitosan

0.25 wt% Chitosan

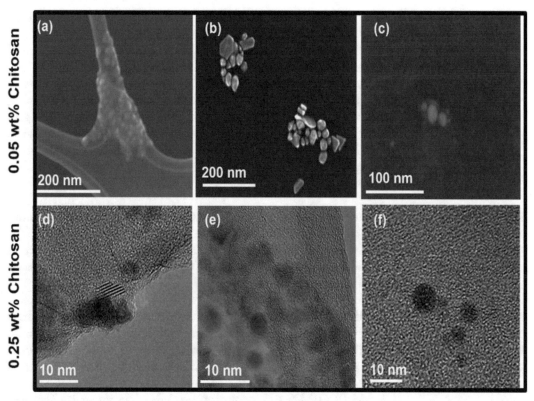

Fig. 4 SEM images of (**a**) 0.05% Chit + Au: ~20 nm spheres within a chitosan film (**b**) 0.05% Chit + TPP 20–40 nm gold structures (**c**) 0.05% Chit + NaCl 8–15 nm gold spheres and TEM images of (**d**) 0.25% Chit + Au 8–10 nm spheres (**e**) 0.25% Chit + TPP 5–10 nm spheres (**d**) 0.25% Chit + NaCl 2–5 nm spheres. All the samples are after 20 min of UV reduction

2.2 Viability Assay to Assess In Vitro Nanoparticle Cytotoxicity

NG108-15, Rat neuronal cells, and Chinese hamster ovary (CHO-K1) cells (ATCC# CCL-61, American Type Culture Collection, Manassas VA) were grown in a 75 mL flask with DMEM, 10% Fetal Bovine Serum (FBS), Penicillin Streptomycin (Pen/Strep), and hypoxanthine-aminopterin-thymidine (HAT), for NG108 cells, and were incubated at 37 °C and 5% CO_2. For viability measurements, CellTiter-Glo (CTG), from Promega Corporation, measures the number of viable cells based on the presence of Adenosine triphosphate (ATP), which is present in metabolically active cells [103]. This assay measures luminescence from a firefly luciferase that binds to ATP [104].

NG108 cells were plated with the Chit-AuNPs to measure viability. After 24 h, the plate was removed and the Promega CTG protocol was followed to measure the luminescence [103]. Luminescence was measured on a BioTek microplate reader and KC4 plate reader software was used to analyze well plate values.

2.3 Imaging Cellular Uptake of Nanoparticles

Lucifer Yellow (LY) VS, which forms covalent bonds with amino and suphydryl groups, is stable in water [105]. To image the particles loaded into the NG108 cells, first, 50–100 μL of cells were placed on a 15 mm coverslip in a 6-well plate.

The cells were washed with PBS and 100% ice-cold methanol was added. The cells were then incubated for 10 min at −20 °C, and washed, with PBS, again. Next, Chromeo Red, a red fluorescent cell stain, was added to the wells and they were incubated at room temperature, for 30 min, covered and protected from light. After the 30 min incubation time, the cells were washed, twice, with PBS. The same procedure was used to measure changes in nanoparticle fluorescence intensity during electroporation studies and analyzed with ImageJ software [106].

2.4 Cell Culture for Electroporation Studies

Jurkat cells were grown in RPMI media (with 1% Pen-Strep and 10% of Fetal Bovine Serum) in a 37 °C incubator with 5% CO_2. New media was added as needed and the cells were split as needed to maintain approximately 1×10^6 cells/mL (as determined optimum for cuvettes).

2.5 Fluorophore Loading for Electroporation Studies

PRODAN and its derivative, Laurdan, were ordered from Invitrogen, Eugene, OR. Laurdan was originally determined to be best for measuring membrane polarization, but after multiple scans with inaccurate results, PRODAN was used for the remainder of the experiments. The cells were grown, as previously stated, and a 2 mM PRODAN stock solution was added to the cells. This was 20 μL of 2 mM PRODAN per mL of cells. The cells were placed back into 37 °C for 45 min. Next, the cells were spun down (Jurkat cells are suspension cells) at 200 rpm for 4.5 min. The excess RMPI media was removed from the pellet of cells and 10 mL of DPBS was added and the cells were aspirated into the new solution. Only ~1 mL of cells were needed for spectrofluorometric, while the remaining 9 mL were placed into a well-plate and back into the incubator.

For high-speed imaging, the NG108 cells were used. The cells were allowed to adhere to a number 1 poly-L-lysine coated coverslip affixed to the bottom of a 35 mm culture dish for 24 h. Thirty minutes prior to the experimentation the growth medium was removed, the cells were washed with an outside buffer containing 2 mM $MgCl_2$, 5 mM KCL, 10 mM HEPES, 10 mM Glucose, 2 mM $CaCl_2$, and 135 mM NaCl at 135 mM, and adjusted to a pH of 7.4 resulting in an osmolality of 290–310 mOsm. The cells were then allowed to incubate for 30 min in 20 μL of 2 mM PRODAN stock solution added to outside buffer. The cells were then rinsed and the buffer replaced for imaging.

2.6 Fluorescence Acquisition for Electroporation Studies

Fluorescence was acquired using a PTI fluorescence spectrometer. Calibration was measured and adjusted using the Raman spectra of water. The dual emission monochromators were calibrated to each other using a time-based dark spot test. All lights to PMT (including overhead light) were shut off and a time-based scan was taken. A time-based scan outputs Intensity vs. Time. The intensity on either PMT1 or PMT2 was adjusted to overlay the other. A full scan was taken after to verify the calibration of the PMTs. The spectrometer allowed for UV excitation for PRODAN (359 nm) and a blue to green emission range. To measure the response of stained cells all light to the cells was blocked and the system shutter was closed. The only time light was applied to the cells was when the scan was taken. This was to ensure longevity of the cells because of PRODAN being light sensitive. A small stirrer was placed in the cuvette to ensure the cells from settling to the bottom of the cuvette. The excitation was set at 359 nm (as determined by Invitrogen for PRODAN membrane dye), emission range was 390-550 nm; recording every 2 nm steps, and bandpass was 1.5 nm.

A water bath was used to take scans at 20, 27, and 37 °C. When the cuvette was placed into the holder, it was allowed to reach the temperature of the holder. When temperature was increased during the run, time was taken to allow the cells to reach each temperature. When the cells were pulsed, an initial scan of the cells was taken at water bath temperature. The cells were placed into electroporation cuvettes and exposed. They were immediately removed from the electroporation cuvettes and placed back into the quartz cuvette and back into the holder and a scan was taken. This process took not more than 1 min and the cells were kept out of the light. A scan was taken every 3 min.

2.7 High-Speed Imaging System for Electroporation Studies

To examine the membrane response in high speed, we used a dual-fluorescence, EM-CCD-based system on an inverted microscope. An HBO arc lamp was filtered with a 360/12 nm bandpass filter to provide the excitation light. To limit photobleaching, the lamp was shuttered until just prior to exposure. A dual-view image splitter (Photometrics) was equipped with a 465 nm dichroic mirror to split the fluorescence images into two separate images, 405–465 nm and longer than 465 nm, onto an EM-CCD camera (Andor). An Opto-mask was inserted into the image pathway prior to the image splitter to crop the image to just the region-of-interest (single cell). Using the cropped frame transfer function of our EM-CCD camera, we were able to increase the acquisition rate from 56 frame per second (fps) full-frame to ~200 fps. A delay generator (Stanford Research Systems) was used to precisely trigger the beginning of the image acquisition and delivery of the electric pulse.

2.8 nsPEF Technique

A pulsing system, which can deliver multiple pulse widths (10, 30, 60, 200, 400, or 600 ns), was connected to a Zeiss LSM confocal microscope (Carl Zeiss Jena, Germany) and a 40x objective was used to image the cells and AuNPs. Two parallel electrodes, made of tungsten, 141 μm apart, used to apply the electric field, were positioned 50 μm above the cells, using a micromanipulator [94].

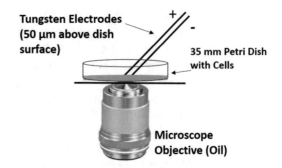

Fig. 5 Microscope stage set up for pulsing

The diagram in Fig. 5 displays the setup for pulsing. Viable cells were plated on 35 mm dish for application. Ten microliter of resuspended cells were added to 10 μL of Trypan Blue and placed in a Countess slide. Viability and count were measured with a Countess Automated Cell Counter (Invitrogen). One hundred microliter of cells were plated on 35 mm glass bottom dish (MaTek Corp. Ashland, MA) and allowed to incubate overnight. After 24 h incubation, the cells were washed with Dulbecco's phosphate-buffered saline (DPBS) and 2 mL of F12K media was added to the cells. Prior to pulsing, PI (2.5 mL) or LY labeled Chit + NaCl AuNPs (2 μL) were added, gradually, into the 35 mm dish. A 30 s delay was set in the Zen microscope software, and the images were taken every 30 s, after the 600 ns pulse width was initiated. Changes in fluorescence intensity were measured for cells positioned between the electrodes with ImageJ software [106].

3   Methods

3.1   Time-Dependent In Vitro Nanoparticle Cytotoxicity

Viability measurements of the NG108 cells incubated with the chitosan gold nanoparticles as well as sodium citrate (NaCit) particles for comparison. The particles were added to the NG108 cells and incubated for 24, 72, and 168 h. Their viability was measured using a luminescence assay CellTiter Glo (Fig. 6a). Data suggest there is no negative effect of the Chit-AuNPs and the NaCit NPs when incubated over the course of 7 days, when compared to the control. An increase in cell growth for 0.25% Chit + Au and 0.25% Chit + NaCl was observed on day 3. An increase in growth for the Chit + Au sample can be attributed to the effects of chitosan on the cell. In this case, this is solely a chitosan and gold particle, which leaves more chitosan present for cellular interaction, and due to its cross-linking chain and can stimulate cell response for growth [107]. The Chit + NaCl sample had the highest increase in cell growth on day 3. This growth can be due to the presence of NaCl. Neurons contain sodium and potassium gated ion channels, which allow

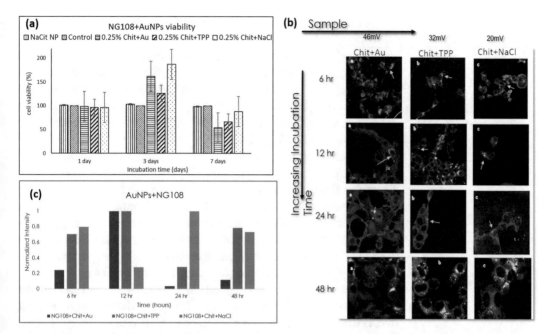

Fig. 6 (**a**) Viability of NG108 cells in the presence of Chit-AuNPs for 1, 3, and 7 days. (**b**) Incubation uptake of 0.25% Chit-AuNPs in NG108 cells. Cells were stained with Chromeo Red, and depicted as red, and the particles are depicted as yellow. (**c**) Quantized Data of NG108 uptake of Chit-AuNPs using ImageJ. Increased fluorescence intensity signifies an increase in AuNP uptake into cells

passage of ions through the cell membrane [108]. Excess sodium causes the sodium gates to open and the sodium diffuses into the cell and this changes the membrane potential of the cell from negative to positive, and then activates an action potential [109], which would create cell growth. While there is a decrease in viability on day 7 it is not significant cell death, as the cell viability is still above LD_{50}. This decrease can be attributed to the fact that NG108 cells being to split after approximately 15 h in the incubator and adherent cells, which undergo contact inhibition at higher densities, can result in a change of ATP per well at high densities and this results in a nonlinear relationship [103, 110]. CellTiter-Glo data reports that CHO-K1 cells continued to grow (metabolically active) and that the particles do not interfere with the adhesion process, which begins around 3 h after being tripsinized [111]. Figure 6a shows the growth of control cells and sample (0.25% Chit + NaCl AuNPs), which indicates that the addition of the gold nanoparticles does not affect the proliferation of the CHO K1 cells.

3.2 Time-Dependent Uptake of Nanoparticles

The chitosan synthesized gold nanoparticles were incubated in NG108 cells for 6, 12, 24, and 48 h and then imaged by confocal microscopy. Confocal images show uptake of the chitosan-based gold nanoparticles at across all incubation times, which can be seen with the arrows in Fig. 6b. The gold particles are depicted as a bright yellow in the cells, while the cells were stained with Chromeo Red.

ImageJ analysis of the collected images (Fig. 6c) shows increased uptake, based on sample type for 6 h incubation. This could be due to the stability of the particles in media during incubation. As the protein corona forms, they settle out of solution to the bottom of the sample where the cells are adhered, allowing for uptake in less time. After 12 h, there are not as many gold nanoparticles or cells present. This can be attributed to the cell phase. When the cells begin to split, they disperse while they grow [112]. There is a significant increase in gold nanoparticles in the cells after 48 h of incubation, for all three Chit-AuNP samples, as shown in Fig. 6b. These results are in agreement that gold nanoparticles accumulate inside dendritic cells when incubated for 24 and 48 h [113]. This accumulation of particles in the cells is expected due to the strong phagocytic capacity of dendritic cells [114].

These results led us to question if particles could potentially travel with cells during cell division. Studies in the literature suggest that as the cells split, they take loaded nanoparticles with them [115]. This was observed in all phases of cell division. Specifically, the results of these studies showed that uptake is not influenced or affected by specific phases of cell division and the nanoparticles were divided between newly formed daughter cells [116]. This supports the data in which we see an increased uptake of chitosan gold nanoparticles at hour 48 when compared to hour 6. NG108 cells continue to grow, and therefore continue to double, in the presence of Chit-AuNPs and the cells take the particles with them as they divide. This analysis provides evidence that not only do the particles remain internalized in the cells, but may also be transported with the mitosis process with little interfere with cell cycle function.

3.3 Measurement of Changes in Plasma Membrane Phospholipid Polarization Following Pulsed Electric Field Exposure

When an electric pulse is applied at a constant temperature, the same shift is expected to be seen because of the membrane becoming permeable from the electric field. Due to the sensitivity of PRODAN, as the membrane undergoes a phase shift, the fluorescence emission should shift. The emission of PRODAN and its derivatives is a dual peak. The intensity of the peak would shift depending on the phase of the membrane. This section uses this response to measure how an electrical pulse affects the phase of the membrane at a constant temperature. Additionally, since it is believed the formation and subsequent resealing of nanopores in the cell membrane is temperature dependent, we can monitor how the membranes respond and recover at different set temperatures.

To verify that the PRODAN and Laurdan dye was properly labeled to the cells, the images were captured using the DAPI filter on the Nikon Eclipse TE2000-E microscope. An example image is shown in Fig. 7a. Once fluorescent labeling was verified, the cells were transferred to a cuvette to be used in the fluorescence

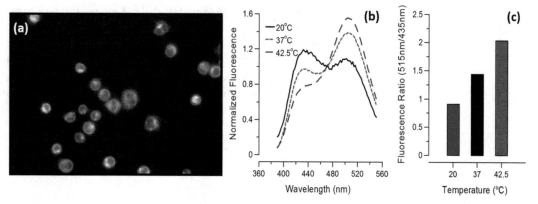

Fig. 7 (a) Image of PRODAN labeled Jurkat cells captured using the DAPI filter on the Nikon Eclipse TE2000-E microscope. Temperature effect of PRODAN labeled Jurkat cells. (b) Shift in the fluorescence spectra to longer wavelengths at higher temperatures. (c) The shift in fluorescence spectra is reflected in the ratio between the two peaks

spectrometer. While taking initial Laurdan scans, fluorescence peaks began to appear around the 440–480 nm range. These peaks were not fluorescent peaks due to shift in the membrane of the cells. To determine the product of error, a scan was taken with each ingredient of the cells, dye, and media. Scan by scan, a new chemical was added until the redundant peak was output; and then the error could be determined. The RPMI media alone (no cells or dye) had a peak at 440 nm and the Laurdan in ethanol and a peak at 480 nm. Different molar amounts of the stock solution did not deter this peak of 480 nm, so the membrane probe was changed from Laurdan to PRODAN. The RPMI media had a fluorescent peak due to the FBS in the solution. Due to the need of all ingredients in the growth media, the media was changed right before a scan was taken. DPBS was measured to have no fluorescent peak with cells and PRODAN, so it was used as cell media during scans.

As a first experiment, the fluorescence spectral shift of the PRODAN-labeled Jurkat cells was examined as a function of temperature. As shown in Fig. 7b, the fluorescence emission spectrum shifts from more blue fluorescence (maximum peak ~435 nm) to more green fluorescence (maximum peak ~515 nm) with increasing temperature. This response is because of the membrane becoming more fluid at higher temperatures. This shift in fluorescence can be quantified by taking the ratio of the two fluorescence peaks (515 nm/435 nm), as shown in Fig. 7c. With higher temperatures, the ratio increases, indicating a shift in emission from the blue to the green.

Next, the response of the cellular membranes to an applied electric field as a function of temperature was examined. The cells were held at a constant temperature of 27 or 37 °C and an electric field of 998 V for 100 µs was applied. For the first temperature-dependent scan, the temperature was kept constant at 27 °C while the pulse was applied. As shown in Fig. 8a, the ratio in fluorescence increases for the first spectra (1-min post exposure) and then remains fairly constant for several minutes. This increase in ratio, indicative of an increase in membrane fluidity, possibly indicates that the cellular membrane has been porated. Post-exposure, the ratio begins to

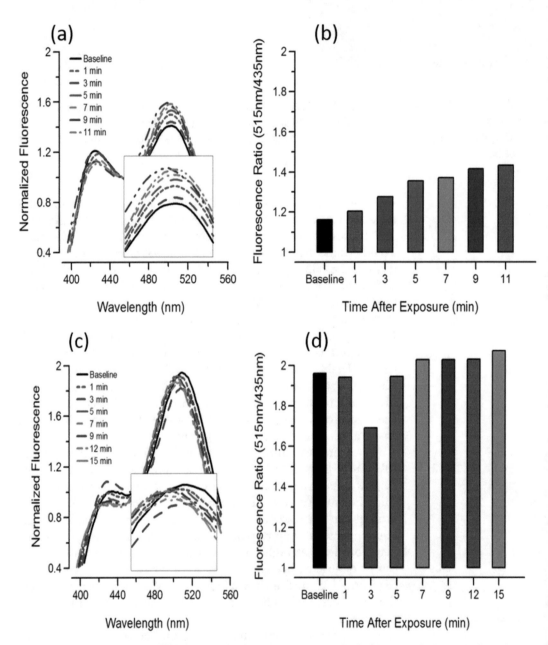

Fig. 8 (**a**) Spectral response of PRODAN-labeled Jurkat cells held at 27 °C with an applied electric field of 998 V for 100 μs. (**b**) Ratio spectral peaks at 27 °C. (**c**) Spectral response of PRODAN-labeled Jurkat cells held at 37 °C with an applied electric field of 998 V for 100 μs. (**d**) Ratio spectral peaks at 37 °C

increase even further. This delayed increase in membrane fluidity could indicate that the cells are swelling, as has been observed in microscope images of cells after electric pulse exposure [117]. This swelling appears to continue for minutes, as observed by the continued increase in membrane fluidity.

To examine the effect of temperature on poration of the membrane, the spectral response of the PRODAN-label cells to an applied electric field was then examined at higher temperature. The cells were held at a constant temperature of 37 °C and an electric field of 998 V for 100 μs was applied. As shown in Fig. 8c, as with the cells at 27 °C, a delayed increase in membrane fluidity is observed that could indicate that the cells are swelling. However, unlike the lower temperature results, the initial increase in fluorescence ratio is not observed. This result indicates any poration created at higher temperature can be difficult to observe because the cell membranes are already more fluid due to the swelling.

3.4 High-Speed Imaging of Membrane Response to Applied Electric Field

The previous experiments demonstrated two effects. First, upon exposure to a 100 μs electric pulse, an immediate increase in the membrane fluidity was observed at lower temperatures. Second, pulses shorter than 100 μs did not cause a change in the fluorescence ratio that was detectable by our fluorescence spectrometer setup. To further determine if a change in membrane organization was actually occurring with nsPEF, we used a high-speed imaging system to more dynamically record changes in membrane fluorescence. This system used a dual-view image splitter, equipped with a 465 nm dichroic mirror to split the fluorescence images into two separate images, 405–465 nm and longer than 465 nm, onto an EM-CCD camera. A representative of these two fluorescence images, along with the corresponding DIC image, is shown in Fig. 9a.

Using the fluorescence intensities from these two channels, we calculated the generalized polarization (GP) for the PRODAN

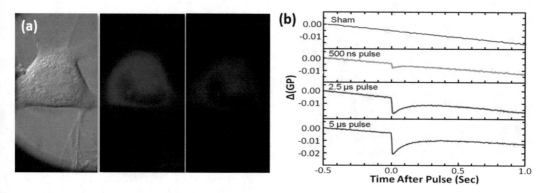

Fig. 9 (a) DIC and dual-channel fluorescence images of PRODAN-labeled NG-108 cells collected by high-speed imaging system. (b) Change in Generalized Polarization for PRODAN-labeled NG108 cells exposed to electrical pulses with various pulsewidths. The membrane becomes more fluid instantly upon exposure and recovers within 0.5 s

fluorescence [8]. The GP is a measure of membrane organization, with lower values indicating a more fluid membrane, and is defined as

$$GP = \frac{I_{(405-465)} - I_{(>465)}}{I_{(405-465)} + I_{(>465)}}. \tag{1}$$

As shown in Fig. 9b, the GP changes (i.e., membrane becomes more fluid) upon electric pulse exposure within the 5 ms acquisition limit of our current setup. This change also demonstrated a dose-dependent response with longer pulses more dramatically increasing the fluidity of the membrane. Despite the differences in the initial response, all exposures appeared to quickly recover within 500 ms. These results suggest that a disruption in the membrane, potentially due to the formation of nanopores, is occurring instantly upon electric pulse-exposure; however, this disruption in membrane organization is short-lived.

Using high-speed imaging we observe the effects of both 5 and 20 pulse exposures on the uptake of PI, as shown in Fig. 10. In Fig. 10a–c, we show that exposure to 20 pulses (high dose) likely opens large holes in the membrane resulting in the considerable uptake of PI, as measured by the fluorescence change in the graph on the right, as compared to the low dose (Fig. 10d–f). This increased uptake indicates pulse effects damaged the cell membrane to the point of irreversibility, which suggest the pore formation would allow AuNPs to flow into the cell, but due to the irreversible damage, it would render them insignificant for further studies. Fluorescence change at 5 pulses (low dose, Fig. 10d–f) shows a more gradual uptake of PI over 600 s, which indicates reversible pore formation, which again suggest the pore formation.

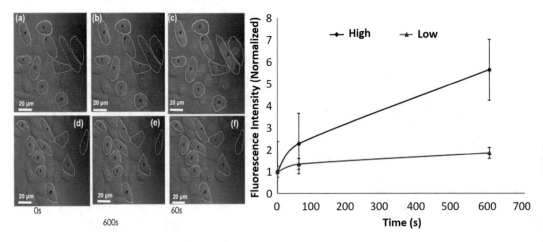

Fig. 10 (Left) Confocal images of CHO-K1 uptake of propidium iodide with high dose (600 ns; 20 pulses) (**a–c**) and low dose (600 ns; 5 pulses) (**d–f**) of nsPEF at 25.68 kV/cm. (Right) changes in fluorescence intensity (of ten cells in view) measured at 0, 60, and 600 s of propidium iodide exposed cells

However, the pores formed under these conditions would allow AuNPs to pass through the pores, and into the cell, before the membrane resealed.

3.5 AuNP Uptake by Cell Under Electrical Pulsing

While morphological changes to the membrane plasma are dependent on a number of pulses [118], in the previous section, the rapid increase in PI from the high dose indicated that although pores formed, the damage to the cells could not be repaired. The low dose (5 pulses) formed smaller pores, indicated by the low PI uptake, so we selected these parameters to measure the uptake of 0.25% Chit + NaCl AuNPs. Thus, two measurements were performed with 0.25% Chit + NaCl AuNPs in the presence of CHO K1 media, while applying nsPEF. For the first, the cells were preloaded with AuNPs 1 h prior to pulsing (Fig. 11a–c) to allow the particles to travel toward the surface of the cells. Second, AuNPs were added immediately before pulsing (Fig. 11d–f). Finally, uptake of AuNPs in CHO K1 cells was measured in the absence of pulsing and used as the control (Fig. 11g–i).

While the particles can be viewed in the confocal images, meaning that they are present around the cells during the pulsing, both pulsing studies show only a change in the presence of the nanoparticle over the 600 s observation period, as seen in the fluorescence change versus time graph in Fig. 11, while the fluorescence change

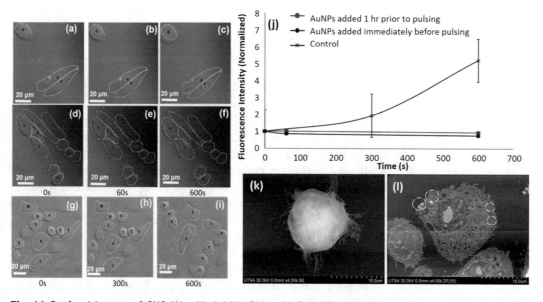

Fig. 11 Confocal images of CHO-K1 with 0.25% Chit + NaCl AuNPs loaded 1 h prior to pulsing (**a–c**); with AuNPs added immediately before pulsing (**d–f**); all exposed to 600 ns pulse width; 5 pulses; 20.6 kV/cm measured at 0, 60, and 600 s; and control (**g–i**) measured at 300 and 600 s. (**j**) Measure fluorescence intensity over time of LYVS labeled AuNPs. (**k**) Scanning electron microscopy (SEM) image of CHO-K1 cells after 1 h of incubation with AuNPs. (**l**) SEM image of cross-section of CHO-K1 cell after 1 h of incubation. Circles indicate clusters of AuNPs uptaken by cells are mostly found just inside the cell membrane

in the control sample increases. Fluorescence change shows no significant increase in particle uptake by CHO K1 cells in the presence of nsPEF with only slight differences in the uptaken amount of AuNPs in the cells when introduced 1 h before nsPEF versus just before the nsPEF treatment. The lack of enhanced uptake can be attributed to nsPEF interfering with the endocytosis process of the cells. Since AuNPs are naturally uptaken into the cells during endocytosis [118–121] uptake would be expected, especially if an opening, such as a nanopore, was formed. If nsPEF does, in fact, affect the endocytosis process, due to changing of the membrane, the AuNPs would not be endocytosed into the cells. This would explain why, in the control study, there is an opposite effect of the fluorescence intensity, signifying a constant increase in AuNP uptake. Additionally, further annalysis by electron microscopy was performed to determine where the uptaken nanoparticles are located within the cells. In Fig. 11k–l we show that after 1 h of incubation of AuNPs with the CHO-K1 cells, the nanoparticles have crossed or attached to the cell membrane. This supports the earlier observation by confocal imaging in Fig. 6b where the chitosan on the AuNPs may be enhancing the adhesion of the nanoparticles to the cell membrane. Furthermore, the almost constant fluorescence intensity of the labeled AuNP under the influence of nsPEF may be indicative that the adhesion of the AuNPs to the disordered membrane may be maintained sufficiently throughout the poration process.

4   Conclusions

This work demonstrates a method of examining poration and swelling in cellular membranes subjected to electrical pulses by using the polarity-sensitive dye, PRODAN. Upon exposure to a 100 µs electric pulse, an immediate increase in the membrane fluidity is observed at lower temperatures, likely indicating poration of the membrane. At longer times post-exposure, the fluorescence ratio continues to increase, likely indicating that the cell is swelling. This membrane disruption was shown to recover within 500 ms. Also, the utilization of nsPEF as a tool for AuNP uptake, using Chit-AuNPs, was studied. Luminescent viability, measured with CellTiter-Glo, confirmed that AuNPs did not inhibit cell growth; therefore, they could be used in this study. Parameters for CHO-K1 pulsing were measured by uptake of propidium iodide, and then applied to measure the enhanced AuNP uptake. While fluorescence data show uptake of PI in CHO-K1 cells, no significant enhancement of uptake 0.25% Chit + NaCl AuNPs, in CHO-K1 cells, occurs in the presence of an electric field (nsPEF) because of endocytosis being affected from the nsEP-induced membrane disruption [118]. Overall, in this work the use of the combined effects of nanoparticles with electrical stimulation provides unique opportunities for future studies related to basic biophysical studies of cells at the nanoscale to therapeutic applications for the treatment of disease states.

Acknowledgments

This research was supported in part by the Air Force Office of Scientific Research under grant numbers. FA9550-15-1-0109 and FA9550-15-1-0513. The authors would like to thank Dr. Jody Cantu for assistance with electroporation experiments and Dr. German Plascencia-Villa for assistance with electron microscopy of the cells.

References

1. Han G, Ghosh P, Rotello VM (2007) Functionalized gold nanoparticles for drug delivery. Nanomedicine (Lond) 2(1):113–123. https://doi.org/10.2217/17435889.2.1.113

2. Sato K, Hosokawa K, Maeda M (2003) Rapid aggregation of gold nanoparticles induced by non-cross-linking DNA hybridization. J Am Chem Soc 125(27):8102–8103. https://doi.org/10.1021/ja034876s

3. Thanh NT, Rosenzweig Z (2002) Development of an aggregation-based immunoassay for anti-protein A using gold nanoparticles. Anal Chem 74(7):1624–1628

4. Hainfeld JF, O'Connor MJ, Dilmanian FA, Slatkin DN, Adams DJ, Smilowitz HM (2011) Micro-CT enables microlocalisation and quantification of Her2-targeted gold nanoparticles within tumour regions. Br J Radiol 84(1002):526–533. https://doi.org/10.1259/bjr/42612922

5. Loo C, Lowery A, Halas N, West J, Drezek R (2005) Immunotargeted nanoshells for integrated cancer imaging and therapy. Nano Lett 5(4):709–711. https://doi.org/10.1021/nl050127s

6. Dolmans DE, Fukumura D, Jain RK (2003) Photodynamic therapy for cancer. Nat Rev Cancer 3(5):380–387. https://doi.org/10.1038/nrc1071

7. Li L, Nurunnabi M, Nafiujjaman M, Y-k L, Huh KM (2013) GSH-mediated photoactivity of pheophorbide a-conjugated heparin/gold nanoparticle for photodynamic therapy. J Control Release 171(2):241–250. https://doi.org/10.1016/j.jconrel.2013.07.002

8. Vankayala R, Lin C-C, Kalluru P, Chiang C-S, Hwang KC (2014) Gold nanoshells-mediated bimodal photodynamic and photothermal cancer treatment using ultra-low doses of near infra-red light. Biomaterials 35(21):5527–5538. https://doi.org/10.1016/j.biomaterials.2014.03.065

9. Tiwari P, Vig K, Dennis V, Singh S (2011) Functionalized gold nanoparticles and their biomedical applications. Nano 1(1):31–63

10. Menichetti L, Luigi P, Flori A, Kusmic C, De Marchi D, Sergio Casciaro FC, Lombardi M, Positano V, Arosio D (2013) Iron oxide-gold core-shell nanoparticles as multimodal imaging contrast agent. IEEE Sensors J 13(6):2341–2347

11. Topete A, Alatorre-Meda M, Iglesias P, Villar-Alvarez EM, Barbosa S, Costoya JA, Taboada P, Mosquera V (2014) Fluorescent drug-loaded, polymeric-based, branched gold nanoshells for localized multimodal therapy and imaging of tumoral cells. ACS Nano 8(3):2725–2738. https://doi.org/10.1021/nn406425h

12. Reju Thomas I-KP, Jeong YY (2013) Magnetic iron oxide nanoparticles for multimodal imaging and therapy of cancer. Int J Mol Sci 14(8):15910–15930

13. Jit Kang Lim SAMaRDT (2009) Stabilization of superparamagnetic iron oxide core–gold shell nanoparticles in high ionic strength media. Langmuir 25(23):13384–13393

14. Carril M, Fernández I, Rodríguez J, García I, Penadés S (2014) Gold-coated iron oxide glyconanoparticles for MRI, CT, and US multimodal imaging. Part Part Syst Charact

31(1):81–87. https://doi.org/10.1002/ppsc.201300239

15. Coughlin AJ (2013) Gold nanoconstructs for multimodal diagnostic imaging and photo-thermal cancer therapy. Dissertation, Rice University, ProQuest Dissertations and Theses (PQDT) Database

16. Kim D, Jeong YY, Jon S (2010) A drug-loaded aptamer-gold nanoparticle bioconjugate for combined CT imaging and therapy of prostate cancer. ACS Nano 4(7):3689–3696. https://doi.org/10.1021/nn901877h

17. Rengan AK, Jagtap M, De A, Banerjee R, Srivastava R (2014) Multifunctional gold coated thermo-sensitive liposomes for multimodal imaging and photo-thermal therapy of breast cancer cells. Nanoscale 6(2):916–923. https://doi.org/10.1039/c3nr04448c

18. Huang X, El-Sayed IH, Qian W, El-Sayed MA (2006) Cancer cell imaging and photo-thermal therapy in the near-infrared region by using gold nanorods. J Am Chem Soc 128(6):2115–2120. https://doi.org/10.1021/ja057254a

19. Nanophotonics Nanomedical Research Group (2007) Nanoplasmonics. http://www.nanot-rio.com/board/list.php?category=&board_num=11&rowid=19&go=&sw=&sn=&st=&sc=&page=1

20. Ahmed N, Fessi H, Elaissari A (2012) Theranostic applications of nanoparticles in cancer. Drug Discovery Today 17(17–18):928–934. https://doi.org/10.1016/j.drudis.2012.03.010

21. Boisselier E, Astruc D (2009) Gold nanoparticles in nanomedicine: preparations, imaging, diagnostics, therapies and toxicity. Chem Soc Rev 38(6):1759–1782. https://doi.org/10.1039/B806051G

22. Chen J, Keltner L, Christophersen J, Zheng F, Krouse M, Singhal A, Wang SS (2002) New technology for deep light distribution in tissue for phototherapy. Cancer J (Sudbury, MA) 8(2):154–163

23. Huang X, Jain P, El-Sayed I, El-Sayed M (2008) Plasmonic photothermal therapy (PPTT) using gold nanoparticles. Lasers Med Sci 23(3):217–228. https://doi.org/10.1007/s10103-007-0470-x

24. Heinemann D, Schomaker M, Kalies S, Schieck M, Carlson R, Escobar HM, Ripken T, Meyer H, Heisterkamp A (2013) Gold nanoparticle mediated laser transfection for efficient siRNA mediated gene knock down. PLoS One 8(3):e58604. https://doi.org/10.1371/journal.pone.0058604

25. Dev Kumar Chatterjeea LSF, Zhang Y (2008) Nanoparticles in photodynamic therapy: an emerging paradigm. Adv Drug Deliv Rev 50(15):1627–1637

26. Institute NC (2011) Photodynamic therapy for cancer. National Institutes of Health. http://www.cancer.gov/cancertopics/fact-sheet/Therapy/photodynamic. 2013

27. Liu TW, Huynh E, MacDonald TD, Zheng G (2014) Porphyrins for imaging, photodynamic therapy, and photothermal therapy. In: Chen X, Wong S (eds) Cancer theranostics. Academic, Oxford, pp 229–254. https://doi.org/10.1016/B978-0-12-407722-5.00014-1

28. Sakamoto JH, van de Ven AL, Godin B, Blanco E, Serda RE, Grattoni A, Ziemys A, Bouamrani A, Hu T, Ranganathan SI, De Rosa E, Martinez JO, Smid CA, Buchanan RM, Lee S-Y, Srinivasan S, Landry M, Meyn A, Tasciotti E, Liu X, Decuzzi P, Ferrari M (2010) Enabling individualized therapy through nanotechnology. Pharmacol Res 62(2):57–89. https://doi.org/10.1016/j.phrs.2009.12.011

29. Kimling J, Maier M, Okenve B, Kotaidis V, Ballot H, Plech A (2006) Turkevich method for gold nanoparticle synthesis revisited. J Phys Chem B 110(32):15700–15707. https://doi.org/10.1021/jp061667w

30. Daniel M-C, Astruc D (2003) Gold nanoparticles: assembly, supramolecular chemistry, quantum-size-related properties, and applications toward biology, catalysis, and nanotechnology. Chem Rev 104(1):293–346. https://doi.org/10.1021/cr030698+

31. Martin MN, Basham JI, Chando P, Eah SK (2010) Charged gold nanoparticles in non-polar solvents: 10-min synthesis and 2D self-assembly. Langmuir 26(10):7410–7417. https://doi.org/10.1021/la100591h

32. Bolintineanu DS, Lane JMD, Grest GS (2014) Effects of functional groups and ionization on the structure of alkanethiol-coated gold nanoparticles. Langmuir 30(37):11075–11085. https://doi.org/10.1021/la502795z

33. Lu Y-S, Kuo S-W (2014) Functional groups on POSS nanoparticles influence the self-assembled structures of diblock copolymer composites. RSC Adv 4(66):34849–34859. https://doi.org/10.1039/C4RA06193D

34. CytoViva I (2015) Identifying functional groups on nanoparticles used as drug delivery agents. CytoViva

35. Yong WYD, Zhang Z, Cristobal G, Chin WS (2014) One-pot synthesis of surface functionalized spherical silica particles. Colloids Surf A Physicochem Eng Asp 460:151–157. https://doi.org/10.1016/j.colsurfa.2014.03.039

36. Abbaspourrad A, Carroll NJ, Kim SH, Weitz DA (2013) Surface functionalized hydrophobic porous particles toward water treatment application. Adv Mater 25(23):3215–3221

37. Froimowicz P, Munoz-Espi R, Landfester L, Musyanovych A, Crespy D (2013) Surface-

functionalized particles: from their design and synthesis to materials science and bio-applications. Curr Org Chem 17:900–912

38. Storm G, Belliot SO, Daemen T, Lasic DD (1995) Surface modification of nanoparticles to oppose uptake by the mononuclear phagocyte system. Adv Drug Deliv Rev 17(1):31–48. https://doi.org/10.1016/0169-409X(95)00039-A

39. Evrim U (2013) Surface modification of nanoparticles used in biomedical applications. In: Aliofkhazraei M (ed) Modern surface engineering treatments. InTech, Rijeka

40. Hsu C-L, Wang K-H, Chang C-H, Hsu W-P, Lee Y-L (2011) Surface modification of gold nanoparticles and their monolayer behavior at the air/water interface. Appl Surf Sci 257(7):2756–2763. https://doi.org/10.1016/j.apsusc.2010.10.057

41. Li X, Guo J, Asong J, Wolfert MA, Boons G-J (2011) Multifunctional surface modification of gold-stabilized nanoparticles by bioorthogonal reactions. J Am Chem Soc 133(29):11147–11153. https://doi.org/10.1021/ja2012164

42. Castellanos LJ, Blanco-Tirado C, Hinestroza JP, Combariza MY (2012) In situ synthesis of gold nanoparticles using fique natural fibers as template. Cellulose 19(6):1933–1943. https://doi.org/10.1007/s10570-012-9763-8

43. Thanh NTK, LAW G (2010) Functionalisation of nanoparticles for biomedical applications. Nano Today 5:213–230

44. Sperling RA, Parak WJ (2010) Surface modification, functionalization and bioconjugation of colloidal inorganic nanoparticles. Phil Trans R Soc Lond A Math Phys Eng Sci 368(1915):1333–1383. https://doi.org/10.1098/rsta.2009.0273

45. Watthanaphanit A, Panomsuwan G, Saito N (2013) In situ preparation of gold nanoparticles in alginate gel matrix by solution plasma sputtering process. MRS Online Proc Lib Arch 1569:151–155

46. Wu W, He Q, Jiang C (2008) Magnetic iron oxide nanoparticles: synthesis and surface functionalization strategies. Nanoscale Res Lett 3(11):397–415. https://doi.org/10.1007/s11671-008-9174-9

47. Joanne Manson DK, Meenan BJ (2011) Polyethylene glycol functionalized gold nanoparticles: the influence of capping density on stability in various media. Gold Bull 44:99–105

48. Park W, Na K (2015) Advances in the synthesis and application of nanoparticles for drug delivery. Wiley Interdiscip Rev Nanomed Nanobiotechnol 7(4):494–508. https://doi.org/10.1002/wnan.1325

49. De Jong WH, PJA B (2008) Drug delivery and nanoparticles: applications and hazards. Int J Nanomedicine 3(2):133–149

50. Sun T, Zhang YS, Pang B, Hyun DC, Yang M, Xia Y (2014) Engineered nanoparticles for drug delivery in cancer therapy. Angew Chem Int Ed 53(46):12320–12364. https://doi.org/10.1002/anie.201403036

51. Svenson S, Prud'homme RK (eds) (2012) Multifunctional nanoparticles for drug delivery applications. Imaging, targeting, and delivery. Springer, New York, NY

52. Ahangari MSA (2014) Nanoparticle based drug delivery systems for treatment of infectious diseases. In: Sezer AD (ed) Nanotechnology and nanomaterials. Springer, New York, NY. CC BY 3.0 license

53. Mudshinge SR, Deore AB, Patil S, Bhalgat CM (2011) Nanoparticles: emerging carriers for drug delivery. Saudi Pharm J 19(3):129–141. https://doi.org/10.1016/j.jsps.2011.04.001

54. Dev A, Mohan JC, Sreeja V, Tamura H, Patzke GR, Hussain F, Weyeneth S, Nair SV, Jayakumar R (2010) Novel carboxymethyl chitin nanoparticles for cancer drug delivery applications. Carbohydr Polym 79(4):1073–1079. https://doi.org/10.1016/j.carbpol.2009.10.038

55. Pienaar IS, Gartside SE, Sharma P, De Paola V, Gretenkord S, Withers D, Elson JL, Dexter DT (2015) Pharmacogenetic stimulation of cholinergic pedunculopontine neurons reverses motor deficits in a rat model of Parkinson's disease. Mol Neurodegener 10(1):47. https://doi.org/10.1186/s13024-015-0044-5

56. Lee J-I (2015) The current status of deep brain stimulation for the treatment of Parkinson disease in the Republic of Korea. J Movement Dis 8(3):115–121. 10.14802/jmd.15043

57. Fontaine D, Lazorthes Y, Mertens P, Blond S, Géraud G, Fabre N, Navez M, Lucas C, Dubois F, Gonfrier S, Paquis P, Lantéri-Minet M (2009) Safety and efficacy of deep brain stimulation in refractory cluster headache: a randomized placebo-controlled double-blind trial followed by a 1-year open extension. J Headache Pain 11(1):23–31. https://doi.org/10.1007/s10194-009-0169-4

58. Groiss SJ, Wojtecki L, Südmeyer M, Schnitzler A (2009) Deep brain stimulation in Parkinson's disease. Ther Adv Neurol Disord 2(6):20–28. https://doi.org/10.1177/1756285609339382

59. Dams J, Siebert U, Bornschein B, Volkmann J, Deuschl G, Oertel WH, Dodel R, Reese JP (2013) Cost-effectiveness of deep brain stimulation in patients with Parkinson's disease. Movement Dis 28(6):763–771. https://doi.org/10.1002/mds.25407

60. Okun MS, Fernandez HH, Foote KD (2015) What are the risks of DBS? Center for Movement Disorders and Neurorestoration, Gainesville, FL

61. Davis NJ, van Koningsbruggen MG (2013) "Non-invasive" brain stimulation is not non-invasive. Front Syst Neurosci 7:76. https://doi.org/10.3389/fnsys.2013.00076

62. Gandiga PC, Hummel FC, Cohen LG (2006) Transcranial DC stimulation (tDCS): a tool for double-blind sham-controlled clinical studies in brain stimulation. Clin Neurophysiol 117(4):845–850. https://doi.org/10.1016/j.clinph.2005.12.003

63. Anwar Y (2010) Right or left? Brain stimulation can change the hand you favor. UC Berkeley Office of Communications, Berkeley, CA

64. Cheromordik LV (1992) Electropores in lipid bilayers and cell membranes. In: Chass DC (ed) Guide to electropation and electrofusion. Academic, San Diego, CA, pp 63–76

65. Neumann E (1992) Biophsyical considerations of membrane electroporation. In: Chass DC (ed) Guide to electropation and electrofusion. Academic, San Diego, CA, pp 77–90

66. Chang C, Hunt JR, Zheng Q, Gao PQ (1992) Electroporation and electrofusion using a pulsed radio-frequency electric field. In: Chass DC (ed) Guide to electropation and electrofusion. Academic, San Diego, CA, pp 303–326

67. Andrei G, Pakhomov JF (2007) Long-lasting plasma membrane permeabilization in mammilian cells by nanosecond pulsed electric field (nsPEF). Bioelectromagnetics 28:655–663

68. Bagatolli LAPT (1999) A model for the interaction of 6-lauroyl-2-(N,N-dimethlyamino) napthalene with lipid environments; implications for spectral properties. Photochem Photobiol 70:557–564

69. Bunker CB (1993) A photophysical study of solvatochromic probe 6-propionyl-2-(N,N-dimethylamo) naphthalene (PRODAN) in solution. Photochem Photobiol 58:499–499

70. Pakhomov AG (2009) Lipid nanopores can form a stable, ion channel-like conduction pathway in cell membrane. Biochem Biophys Res Commun 385(2):181–186

71. Bowman A (2010) Analysis of plasma membrane integrity by fluorescent detection of thallium uptake. J Membr Biol 236(1):15–26

72. Raichle ME (2003) Functional Brain Imaging and Human Brain Function. J Nuerosci 23(10):3959–3962

73. Potts PD, Hirooka Y, Dampney RA (1999) Activation of brain neurons by circulating angiotensin II: direct effects and baroreceptor-mediated secondary effects. Neuroscience 90(2):581–594

74. Lewis SJ, Dove A, Robbins TW, Barker RA, Owen AM (2003) Cognitive impairments in early Parkinson's disease are accompanied by reductions in activity in frontostriatal neural circuitry. J Neurosci 23(15):6351–6356

75. Phillis JW, Horrocks LA, Farooqui AA (2006) Cyclooxygenases, lipoxygenases, and epoxygenases in CNS: their role and involvement in neurological disorders. Brain Res Rev 52(2):201–243. https://doi.org/10.1016/j.brainresrev.2006.02.002

76. Chambers JA, Sanei S (2008) EEG signal processing. Wiley, Hoboken, NJ

77. Aston-Jones G (1981) Activity of norepinephrine-containing locus coeruleus neurons in behaving rats anticipates fluctuations in the sleep-waking cycle. J Neurosci 1:876

78. Jiang K (2015) New technique uses light to take genetics out of optics. UChicago News, 12 Mar, 2015

79. Liao Y-H, Chang Y-J, Yoshiike Y, Chang Y-C, Chen Y-R (2012) Negatively charged gold nanoparticles inhibit Alzheimer's amyloid-β fibrillization, induce fibril dissociation, and mitigate neurotoxicity. Small (Weinheim an der Bergstrasse, Germany) 8(23):3631–3639. https://doi.org/10.1002/smll.201201068

80. Liao YH, Chang Y-J, Yoshiike Y et al (2014) Negatively charged gold nanoparticles inhibit Alzheimer's amyloid-β fibrillization IFD, and mitigate neurotoxicity. In: Simon H (ed) Coated nanoparticles show Alzheimer's promise – chemistry world. Royal Society of Chemistry, London

81. Zhang H, Shih J, Zhu J, Kotov NA (2012) Layered nanocomposites from gold nanoparticles for neural prosthetic devices. Nano Lett 12(7):3391–3398. https://doi.org/10.1021/nl3015632

82. Jung S, Bang M, Kim BS, Lee S, Kotov NA, Kim B, Jeon D (2014) Intracellular gold nanoparticles increase neuronal excitability and aggravate seizure activity in the mouse brain. PLoS One 9(3):e91360. https://doi.org/10.1371/journal.pone.0091360

83. Crosera M, Bovenzi M, Maina G, Adami G, Zanette C, Florio C, Filon Larese F (2009) Nanoparticle dermal absorption and toxicity: a review of the literature. Int Arch Occup Environ Health 82(9):1043–1055. https://doi.org/10.1007/s00420-009-0458-x

84. Marquis BJ, Love SA, Braun KL, Haynes CL (2009) Analytical methods to assess nanoparticle toxicity. Analyst 134(3):425–439. https://doi.org/10.1039/b818082b

85. Orza A, Pruneanu S, Soritau O, Borodi G, Florea A, Bălici Ş, Matei H, Olenic L (2013) Single-step synthesis of gold nanowires using biomolecules as capping agent/template: applications for tissue engineering. Part Sci Technol 31(6):658–662. https://doi.org/1 0.1080/02726351.2013.831151

86. Dreaden EC, Austin LA, Mackey MA, El-Sayed MA (2012) Size matters: gold nanoparticles in targeted cancer drug delivery. Ther Deliv 3(4):457–478

87. Alkilany AM, Nagaria PK, Hexel CR, Shaw TJ, Murphy CJ, Wyatt MD (2009) Cellular uptake and cytotoxicity of gold nanorods: molecular origin of cytotoxicity and surface effects. Small (Weinheim an der Bergstrasse, Germany) 5(6):701–708. https://doi.org/10.1002/smll.200801546

88. Kimball JW (2014) Endocytosis. Creative commons attribution 3.0

89. Saunders JA, Smith CR, Kaper JM (1989) Effects of electroporation pulse wave on the incorporation of viral RNA inot tobacco protoplasts. BioTechniques 7:1124–1131

90. Benz R, Conti F (1981) Reversible electrical breakdown of squid giant axon membrane. Biochim Biophys Acta 645:115–123

91. Chernomordik LVSS, Pastushenko VF, Sokirko AV, Abidor IG, Chizmadzhev YA (1987) The electrical breakdown of cell and lipid membranes: the similarity of phenomenologies. Biochem Biophys Acta 902: 360–373

92. Zimmermann U (1982) Electric field mediated fusion and related electrical phenomena. Biochem Biophys Acta 694:227–277

93. Van Lehn RC, Ricci M, Silva PHJ, Andreozzi P, Reguera J, Voïtchovsky K, Stellacci F, Alexander-Katz A (2014) Lipid tail protrusions mediate the insertion of nanoparticles into model cell membranes. Nat Commun 5:4482. https://doi.org/10.1038/ncomms5482

94. Kneipp K, Haka AS, Kneipp H, Badizadegan K, Yoshizawa N, Boone C, Shafer-Peltier KE, Motz JT, Dasari RR, Feld MS (2002) Surface-enhanced raman spectroscopy in single living cells using gold nanoparticles. Appl Spectrosc 56(2):150–154

95. Bravo-Osuna I, Millotti G, Vauthier C, Ponchel G (2007) In vitro evaluation of calcium binding capacity of chitosan and thiolated chitosan poly(isobutyl cyanoacrylate) core–shell nanoparticles. Int J Pharm 338(1–2):284–290. https://doi.org/10.1016/j.ijpharm.2007.01.039

96. Sun Y, Vernier PT, Wang J, Kuthi A, Marcu L, Gundersen MA (2005) Materials research society symposium, 2005. pp 115–120

97. Kocak N, Sahin M, Akin I, Kus M, Yilmaz M (2011) Microwave assisted synthesis of chitosan nanoparticles. J Macromol Sci A 48(10):776–779. https://doi.org/10.1080/10601325.2011.603613

98. Azad AK, Sermsintham N, Chandrkrachang S, Stevens WF (2004) Chitosan membrane as a wound-healing dressing: characterization and clinical application. J Biomed Mater Res B Appl Biomater 69B(2):216–222. https://doi.org/10.1002/jbm.b.30000

99. NovaMatrix (2011) Chitosan. FMC biopolymer

100. Gan Q, Wang T, Cochrane C, McCarron P (2005) Modulation of surface charge, particle size and morphological properties of chitosan-TPP nanoparticles intended for gene delivery. Colloids Surf B Biointerfaces 44(2–3):65–73. https://doi.org/10.1016/j.colsurfb.2005.06.001

101. Wei D, Qian W (2006) Chitosan-mediated synthesis of gold nanoparticles by UV photoactivation and their characterization. J Nanosci Nanotechnol 6(8):2508–2514

102. Aldrich S (2015) Gold nanoparticles: properties and applications. Sigma-Aldrich, St. Louis, MO

103. Corporation P (2012) CellTiter-Glo luminescent cell viability assay tehnical bulletin, 2013

104. Riss TL, Moravec RA, Niles AL, Benink HA, Worzella TJ, Minor L (2004) Cell viability assays. In: Sittampalam GS, Nelson H et al (eds) Assay guidance manual. Eli Lilly & Company and the National Center for Advancing Translational Sciences, Bethesda, MD

105. Bailey MP, Rocks BF, Riley C (1983) Use of lucifer yellow VS as a label in fluorescent immunoassays illustrated by the determination of albumin in serum. Ann Clin Biochem 20(Pt 4):213–216

106. Schneider CA, Rasband WS, Eliceiri KW (2012) NIH Image to ImageJ: 25 years of image analysis. Nat Methods 9(7):671–675

107. Hu J, Huang L, Zhuang X, Zhang P, Lang L, Chen X, Wei Y, Jing X (2008) Electroactive aniline pentamer cross-linking chitosan for stimulation growth of electrically sensitive cells. Biomacromolecules 9(10):2637–2644. https://doi.org/10.1021/bm800705t

108. Marban E, Yamagishi T, Tomaselli GF (1998) Structure and function of voltage-gated sodium channels. J Physiol 508(Pt 3): 647–657. https://doi.org/10.1111/j.1469-7793.1998.647bp.x

109. Neurons and the nervous system. http://people.eku.edu/ritchisong/301notes2.htm

110. Gérard V, Rouzaire-Dubois B, Dilda P, Dubois JM (1998) Alterations of ionic membrane permeabilities in multidrug-resistant

neuroblastoma x glioma hybrid cells. J Exp Biol 201(Pt 1):21–31

111. Genlantis mammalian cells-CHO0K1. http://www.biocat.com/bc/pdf/C5021_2_00_Expresso%20Mammalian%20Cells%20CHO-K1_VKM080121.pdf

112. Tao Sun LL, Huang Q (2011) Research of neuron growth prediction and influence of its geometric configuration. Appl Math 2: 904–907

113. Villiers C, Freitas H, Couderc R, Villiers M-B, Marche P (2010) Analysis of the toxicity of gold nano particles on the immune system: effect on dendritic cell functions. J Nanopart Res 12(1):55–60. https://doi.org/10.1007/s11051-009-9692-0

114. Savina A, Amigorena S (2007) Phagocytosis and antigen presentation in dendritic cells. Immunol Rev 219:143–156. https://doi.org/10.1111/j.1600-065X.2007.00552.x

115. Capco DG, Yongsheng C (eds) (2014) Nanomaterial. Impacts on cell biology and medicine, 1st edn. Springer, Dordrecht

116. Krpeti Z (2014) Nanomaterials: impact on cells and cell organelles. In: Capco DG, Yongsheng C (eds) Nanomaterial. Impacts on cell biology and medicine, 1st edn. Springer, Dordrecht, pp 135–156

117. Nesin OM (1808) Manipulation of cell volume and membrane pore comparison following single cell permeabilization with 60- and 600-ns electric pulses. Biochim Biophys Acta Biomembr 3:792–801

118. Thompson GL, Roth CC, Dalzell DR, Kuipers M, Ibey BL (2014) Calcium influx affects intracellular transport and membrane repair following nanosecond pulsed electric field exposure. J Biomed Opt 19(5):055005. https://doi.org/10.1117/1.jbo.19.5.055005

119. Freese C, Uboldi C, Gibson M, Unger R, Weksler B, Romero I, Couraud P-O, Kirkpatrick C (2012) Uptake and cytotoxicity of citrate-coated gold nanospheres: comparative studies on human endothelial and epithelial cells. Part Fibre Toxicol 9(1):23

120. Malugin A, Ghandehari H (2010) Cellular uptake and toxicity of gold nanoparticles in prostate cancer cells: a comparative study of rods and spheres. J Appl Toxicol 30(3):212–217. https://doi.org/10.1002/jat.1486

121. Yang A, Jie Z, Tao K, Xiaoping W, Roa W, El-Bialy T, Xing J, Jie C The effect of surface properties of gold nanoparticles on cellular uptake. In: Life science systems and applications workshop, 2007. LISA 2007. IEEE/NIH, 8–9 Nov 2007, pp 92–95. doi:https://doi.org/10.1109/LSSA.2007.4400892

Chapter 10

Estimating the Effects of Nanoparticles on Neuronal Field Potentials Based on Their Effects on Single Neurons In Vitro

Michael Busse, Narsis Salafzoon, Annette Kraegeloh, David R. Stevens, and Daniel J. Strauss

Abstract

The application of nanoparticles in medicine requires a rigorous examination of their safety in order to determine and predict their benefits and potential side effects.

The aim of this study was to examine in vitro effects of coated silver-nanoparticles (cAg-NPs) on the excitability of single neuronal cells and to integrate those findings into an in silico model to predict their effects on neuronal circuits and finally on field potentials generated by those circuits.

As a first step, patch-clamp experiments were performed on single cells to investigate the effects of nano-sized silver particles surrounded by an organic coating. The parameters that were altered by exposure to those nanoparticles were then determined through using the Hodgkin & Huxley model of the sodium current. As a next step, to predict possible changes in network signaling due to the applied cAg-NPs, those findings were integrated into a well-defined neuronal circuit of thalamocortical interactions in silico. The model was then extended to observe neural fields originating from activity of neurons exhibiting Hodgkin & Huxley type action potentials. As a last step, the loop between field potentials and its generators was closed to investigate how the neural field potentials influence the spike generation in neurons that are physically located within these fields, if this feedback causes relevant changes in the underlying neuronal signaling within the circuit, and most importantly if the cAg-NPs effects on single neurons of the network are strong enough to cause observable changes in the generated field potentials themselves.

Key words Coated silver nanoparticles, Patch clamp recordings, Neuromodulatory effect, Nonviral vectors, Modeling, Neuronal circuit model, Llinás model

1 Introduction

The development of nanomaterials in biology and medicine has led to high expectations and hopes for treatment of hitherto untreatable diseases.

Nanoparticles (NPs) with a wide range of sizes, morphologies, and surface features have potential for implementation as pathologic or therapeutic agents.

Fidel Santamaria and Xomalin G. Peralta (eds.), *Use of Nanoparticles in Neuroscience*, Neuromethods, vol. 135, https://doi.org/10.1007/978-1-4939-7584-6_10, © Springer Science+Business Media, LLC 2018

The application of nanoparticles has been developed through neuroscience and pharmacological research in the last decades.

A major goal of nanoparticle research in medicine is to overcome existing drug delivery barriers such as biomembranes and the blood-brain barrier by employing nonviral vectors functioning at the nano-to-micro scale. Those systems could be utilized for the systemic delivery of drugs or genes to target cells for therapy of cancer, inflammation or for the intended modulation of neural activity in the brain tissue, e.g., see [1, 2].

Studies in biological systems show that the absorption of nano–sized particles and their potential to induce responses in the cell is affected by physico-chemical parameters [3]. Exploring the nanoparticle-membrane interaction processes on the molecular scale, i.e., receptor-binding, endocytosis, and signaling activation is crucial to understand how NPs of different material, size, and geometry interact with cells.

Several nanotoxicology studies have used the patch-clamp method to examine the influence of nano-sized CuO, ZnO, and Ag on single neurons in vitro. Effects of silver nanoparticles (Ag-NPs) on the amplitude and the time course of the sodium current (I_{Na}) were observed. However, the underlying mechanism of the changes in I_{Na} was not defined [4–6].

The main goal of this work was to identify the effects of NPs on neuronal network activity. For this purpose, patch-clamp recordings in neuroendocrine cells were carried out and the effects of coated silver-nanoparticles (cAg-NPs) on excitability were examined.

In order to model the observed changes in I_{Na} an established model of dynamic changes in membrane conductance was employed. Based on the fitting results, the observed changes in I_{Na} parameters in individual neurons were computationally reproduced.

Furthermore, we computationally investigated the effects of cAg-NPs exposed neurons at the:

(a) Cellular level, using the Hodgkin & Huxley model (HH-model) [7].

(b) Network level, using well explored model of thalamocortical activity by Llinás [8–11].

(c) Field potential effects that could feedback to the cellular level by using the neural field model of Amari [12].

The in vitro measurements have shown that the application of cAg-NPs to chromaffin cells reduces the amplitude of sodium currents without an appreciable shift in either activation voltage or null potential. Those changes were rapid, partially reversible and dose-dependent ((Fig. 1)). It was not possible to figure out if cAg-NPs may produce mechanical effects on the ion channels

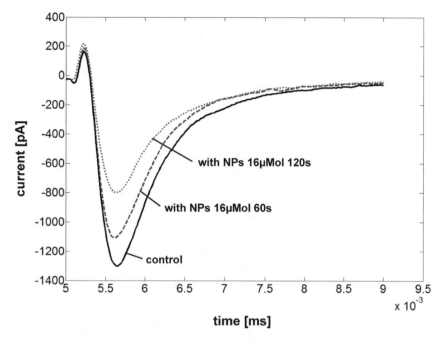

Fig. 1 Local application of cAg-NPs to chromaffin cells. Records of I_{Na} before and after local application of cAg-NPs after 60 s and 120 s respectively (adopted from [25])

leading to lower conductivity nor if there are fewer channels that reach the open state. In any case, the net effects of cAg-NPs on neuronal feedback circuits will be reduced excitability of affected cells, and these effects could have been simulated. Transferring those findings to neuronal circuits via modeling of the thalamo-cortical network resulted in the possibility of making initial predictions as to what effects cAg-NPs' suppression of sodium currents will have on thalamocortical circuits. This model extension predicted that an alteration of the properties of thalamic neurons, as they were found by the HH-model fits of I_{Na} after treatment with cAg-NPs, ends in large alterations of network signaling behavior. It can be expected that NPs brought into contact with few cells of a neuronal feedback circuit will extensively alter network rhythms of large neuronal populations in vivo. As the in silico thalamocortical circuit model was expanded to observe related field potentials spreading over a spatial cortical patch by incorporating the idea of back propagating the field potentials effecting single neuron activity to this new model, the emerging field potentials of a basic two-dimensional two-layer approach were also found to be widely diversified after assuming cAg-NPs presence in network thalamic neurons. The field potentials seem to have strong effects on the action potential generation of neurons that are exposed to those fields as well. The results presume that as a consequence of cAg-NPs affected thalamocortical network cells, the emerging in vivo neural field potentials will be found to be extensively diversified. This model may also subserve as a basic approach to estimating the spatiotemporal dynamics of cortical field potentials on a very small cortical patch that may be electrophysiologically measurable.

2   Materials

2.1   *Coated Silver Nanoparticles*

Thermal decomposition of silver oleate and stabilization by oleyamine was used in a first step to prepare hydrophobic cAg-NPs with a mean diameter of 5 ± 2 nm (Fig. 2).

The particles were capped with an amphiphilic polymer (poly(maleic-acid-anhydride-alt-1-octadecen)) coupled to polyethylenglycol (750 Da) for the transfer into the aqueous phase and steric stabilization [14]. After dispersion in ultrapure water, particles were filtered through a sterile 0.22 μm membrane. A total silver concentration of 1.3 mM, including a fraction of 0.4 μM free Ag^+, was determined by ICP-OESO (Horiba Jobin Yvon GmbH, Munich, Germany) measurements and titration using a silver ion selective electrode [15].

The measured zeta-potential (Zetasizer Nano, Malvern Instruments, Malvern, UK) of the nanoparticles was ~69 mV in pure water. Because the solution ingredients bind to the negatively charged particle surfaces, the absolute value of the zeta-potential

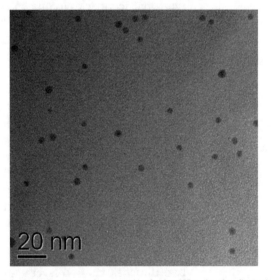

20 nm

Fig. 2 Transmission electron micrograph of cAg-NPs. As the first objective, the electrophysiological effects of cAg-NPs on single excitable cells were determined. For this purpose, particle agglomeration in salt containing medium had to be prevented by coating of Ag-NPs with a polymer shell, which makes it suitable for drug delivery

was significantly reduced when dispersed in 10× PBS buffer (−6 mV) and RPMI medium (−16 mV). A Dyna Pro Titan instrument (Wyatt Technology Europe GmbH, Dernbach, Germany) was further used to determine the hydrodynamic diameter of the particles via dynamic light scattering in various media. In pure water, a mean diameter of 13 ± 2 nm was determined. This value slightly increased to 16 ± 4 nm in the presence of Roswell Park Memorial Institute (RPMI) cell culture medium (Life Technologies, Carlsbad, CA, USA) and decreased to 9 nm ± 1 nm in 10× phosphate-buffered saline (PBS, Life Technologies), indicating binding of molecules to the particle surface as well as particle stability due to steric stabilization by the polymer shell.

2.2 Chromaffin Cells

Neuroendocrine chromaffin cells are well characterized electrophysiologically and are well suited for voltage-clamp analysis of membrane currents due to their small size and spherical shape [16–18]. We have focused on voltage-gated sodium currents (Na_v 1.7) which initiate action potentials and propagation in excitable cells [19] (*see* **Note 1**).

The similar findings on hippocampal neurons (Na_v 1.2) [4, 6] are consistent with the observation that Na_v isoforms in chromaffin cells (Na_v1.7) and in hippocampal cells (Na_v1.1 and 1.2) have an amino acid sequence similarity of about 95% and show nearly identical electrophysiological behavior [19–24].

For cell preparation, the collected adrenal glands from 1- to 3-day-old mice were digested with 20 units of papain (Worthington Biochemical Corp., Lakewood, NY, USA) at 37 °C for 25–30 min. After trituration, the cells were plated on 25 mm cover glasses and then incubated at 37 °C and 8% CO_2. The chromaffin cells were kept in the culture medium (ITS-X, DMEM with GlutaMax and 100 U Penicillin/Streptomycin; Invitrogen, Life Technologies GmbH, Darmstadt, Germany) prior to recordings 12–48 h later.

3   Methods

3.1 Patch-Clamp Measurements

Patch-clamp recordings on chromaffin cells were carried out to evaluate the effects of cAg-NPs on the voltage gated sodium channels (isoform Na_v 1.7). Therefore, borosilicate glass pipettes (1.1 mm inner diameter, 1.5 mm outer diameter, 10 cm long, GB 150F 8P, Science Products GmbH) were pulled on a P-97 Flaming/Brown type micropipette puller. The pipette tip was coated with an insulator to decrease capacitance and improve noise characteristics. We utilized dental wax. A polishing rig (microforge) with a platinum heating filament controlled by a foot pedal was utilized to polish the pipette tip (1 μm tip openings in diameter). A brief heat pulse (1–2 s) is sufficient to remove wax from the tip of the pipette and smooth the glass.

The measurements in the whole-cell configuration were carried out with these 3–6 MΩ pipettes using an EPC-9 patch-clamp amplifier controlled by PULSE software (HEKA Instruments Inc., Lambrecht, Germany). Sodium current was measured on both cAg-NPs exposed and naive cells at a holding potential of −70 mV at room temperature.

The measurement procedure was the following:

First, four control cells were recorded in a new dish and each recording comprised around ten depolarizations to −10 mV with the step duration of 30 ms. While the last control cell patch was still active the cAg-NPs dispersion was pipetted into the chamber and then effects on the patched cell could be measured.

Afterward, the sodium currents of 3–5 more cells from the treated dish were collected. This procedure was repeated for concentrations of 13 µMol, 16 µMol, 43 µMol, 130 µMol, and 1.3 mMol respectively. As a diluent, the extracellular solution (Hepes buffered saline solution) was utilized. Altogether, the current traces of 70 control and 45 cAg-NPs exposed chromaffin-cells were recorded. The extracellular solution contained 145 mM NaCl, 2.4 mM KCl, 10 mM HEPES, 1.2 mM $MgCl_2$, 2.5 mM $CaCl_2$, 10 mM glucose (pH 7.5). The pipette solution (intracellular solution) contained 135 mM Cs-aspartate, 10 mM Cs-HEPES, 5 mM Cs-EGTA, 3 mM $CaCl_2$, 1 mM $MgCl_2$, 2 mM Mg-ATP, 0.3 mM Na_2-GTP, (pH 7.2).

The patch clamp measurements show that the application of cAg-NPs to chromaffin cells reduces the amplitude of sodium currents. The changes were rapid, partially reversible, and dose-dependent.

3.2 Hodgkin-Huxley Fitting Employing Differential Evolution

As a next step, the HH-model of dynamic changes in membrane conductance [7] was computationally fitted to the measured patch-clamp data.

Since all the parameters in the original model's equations were at that time fixed based on notorious squid's giant axon, it was necessary to adjust all the model parameters to be capable of adapting the electrophysiological behavior of chromaffin cells.

A way to adjust these parameters is to fit the HH-model output, e.g., a particular ion flux, to a measured ion flux by mutating the model's particular parameters. First, this fitting process is completed for the recorded ionic sodium currents and the HH-model became capable of exactly simulating those, then another fitting procedure for the I_{Na} after cAg-NPs addition is performed from the recorded patch clamp data to determine which parameters might be modified under the influence of cAg-NPs.

Those factors directly link to physiological mechanisms involved in action potential generation. The basic Hodgkin & Huxley equations used [26] are given by

$$G_{Na} = G_{Namax} m^3 h \qquad (1)$$

$$\frac{dm}{dt} = \alpha_m (1 - m) - \beta_m m \qquad (2)$$

$$\frac{dh}{dt} = \alpha_h (1 - h) - \beta_h h \qquad (3)$$

for the conductance of sodium ions under the experimental conditions of Hodgkin & Huxley [7] with a maximum sodium conductance of

$$G_{Namax} = 120 \, \text{mS} / \text{cm}^2 \qquad (4)$$

The corresponding empirical theorems for the transfer rate coefficients are

$$\alpha_m = \frac{2.5 - 0.1 V'}{e^{2.5 - 0.1 V'} - 1} \frac{1}{\text{ms}} \qquad (5)$$

$$\beta_m = \frac{4}{e^{V'/18}} \frac{1}{\text{ms}} \qquad (6)$$

$$\alpha_h = \frac{0.07}{e^{V'/20}} \frac{1}{\text{ms}} \qquad (7)$$

$$\beta_h = \frac{1}{e^{30 - V'/10} + 1} \frac{1}{\text{ms}} \qquad (8)$$

with $V' = V_m - V_r$ [mV], where V_m is the actual membrane voltage and V_r is the resting voltage. Using voltage clamp, for a voltage step, the transfer rate coefficients α_m, β_m, α_h, and β_h change instantly to new values, (see [25] for steady state). As in steady state, the transfer rate coefficients in Eqs. 5–8 are constant, so the primary differential equation can be readily solved for m and h, giving

$$m(t) = m_\infty - (m_\infty - m_0) e^{\frac{-t}{\tau m}}, \qquad (9)$$

where

$$m_\infty = \frac{\alpha_m}{\alpha_m + \beta_m} \qquad (10)$$

represents the value of m and

$$\tau_m = \frac{1}{\alpha_m + \beta_m} \qquad (11)$$

defines the time constant in [s] in steady state. The mathematical term of h is similar to the m in Eq. 9. By applying voltage clamp, a voltage step initiates an exponential change in m (and h) from its initial value of m_0 or h_0 (at $t = 0$) toward the steady state value of m_∞ or h_∞ (at $t = \infty$). Finally, the sodium current I_{Na} that has to be fitted is then given by

$$I_{Na} = G_{Na} \cdot (V_m - V_{Na}) \frac{nA}{cm^2}, \tag{12}$$

where V_{Na} [mV] expresses the resulting Nernst- or also called reversal-potential following the equation of Nernst, simplified for this issue to

$$V_{Na} = \varphi_i - \varphi_0 \tag{13}$$

with ϕ_i = intracellular and ϕ_0 = extracellular potential (ion concentration dependent).

Now, the Differential-Evolution (DE) algorithm, a technique that originates from the genetic annealing algorithm, which is introduced by [27], could optimize the real parameters and real-valued functions such as the 13 parameters in the HH-model in order to fit them (Eqs. 1–12) to the measured patch-clamp data (see **Note 2**). The DE-algorithm was implemented to minimize the distance between the measured sodium current and the one iteratively calculated by the HH-model. As a consequence, the parameters for the best fit were evaluated. The 13 parameters, which give the freedom for the curve fitting process, represent all the empirical coefficients in Eqs. 1–12 that were estimated by Hodgkin & Huxley in their experiments in 1952 [7]. The following equations indicate again the applied Hodgkin & Huxley formulations, but now with the 13 free parameters expressed by $\xi.1 - \xi.13$:

$$\alpha_h = \frac{\xi.1}{e^{V'/\xi.2}} \frac{1}{ms} \tag{14}$$

$$\beta_h = \frac{1}{e^{\xi.3-V'/\xi.4} + \xi.5} \frac{1}{ms} \tag{15}$$

$$\alpha_m = \frac{\xi.6 - \xi.7.V'}{e^{\xi.6-\xi.7.V'} + \xi.8} \frac{1}{ms} \tag{16}$$

$$\beta_m = \frac{\xi.9}{e^{V'/\xi.10}} \frac{1}{ms} \tag{17}$$

$$G_{Na} = \xi.11.m^3 h \tag{18}$$

$$I_{Na} = G_{Na} \cdot (V_m - \xi.12) \tag{19}$$

and lastly the point of time for the start of the depolarization t_{stim} as $\xi.13$.

3.3 Computational Upscaling

Computational models in neuroscience have become useful tools to explore the functions of complex nervous systems and to understand why they operate in ways observed experimentally.

The modeling of biological neuron started in early twentieth century and it has been always aimed to discover different scales and levels of the brain's dynamic processes that are not known yet from single-cell dynamics in the microscopic level, up to large-scale neuronal population activity. New developments of cognitive sciences and computational models in brain have a crucial role for making links between biological mechanisms and the behavioral and cognitive phenomena that they produce.

3.3.1 Thalamocortical Interactions

The major route for afferents to the neocortex and extrasensory regions of the brain is through thalamus. Traditionally, thalamus was considered the sensory gateway to the cortex which was an oversimplification, because the cortex receives input from both the sense-specific nuclei and the nonspecific thalamic nuclei, which carry multimodal connections to the cortex, probably controlling overall arousal [28].

These thalamocortical connections create bidirectional neuronal loops between the thalamus and the cortex. Based on the Llinás model, generation of cognition is depending on both of these systems and damage to the specific systems induces loss of particular modality while damage to the nonspecific thalamus leads to deep disturbances of the consciousness.

Their statement implies that these two systems can only generate a cognitive experience synced, based on the summation of nonspecific and specific activity along the dendritic tree of the cortical element, by coincidence detection at the pyramidal neuron [10, 11].

Thalamic nuclei receive reciprocal connections from the cortical areas that they project to, though the number of corticothalamic fibers is significantly greater than the number of thalamocortical axons [28, 29].

The intrinsic electrical properties of neurons and the dynamic events resulting from their connectivity cause global resonant states, which will be changed by just small alterations in the neural signaling characteristics.

Consequently, distributed neural representations of simultaneous perceptual events or features could be related to each other within the thalamocortical system. Binding input from different sensory modalities into a single cognitive event is assumed to be a consequence. The underlying mechanism has been proposed as temporal binding, a process based on the synchronization of neural signals [30].

Based on different studies some specific types of cognitive functions are closely related to synchronized neuronal oscillations at both low (4–7/8–13 Hz) and high (18–35/30–70 Hz) frequencies [12, 31–35]. The activity patterns of these oscillations are formed within one or more bounded areas (corresponding to cortical columns) [10, 11, 36]. A group of the cortical neurons can produce repetitive, high-frequency burst discharges known as chattering cells, which can generate burst with intraburst frequencies of 300–750 Hz and interburst frequencies of 8–80 Hz [37–39]. The firing patterns of those chattering cells in burst mode contain more information to evaluate the effect of NPs on the circuit's activity. That is why we focused on the thalamocortical activity of those chattering cells.

3.3.2 The Llinás-Model as Basis for the Developed Corticothalamic Network Algorithm

The Llinás et al. model exposing the thalamocortical circuit is depicted in Fig. 3.

For the purpose of this work, the focus was on two types of thalamic cells: first specific thalamic cells, also known as thalamic core cells due to their focused projection to an individual cortical area and second intralaminar nonspecific thalamic cells, also known as thalamic matrix cells as they project across larger neocortical areas in a more dispersed way [28]. The model development includes both thalamocortical resonant loops and the system operates on the basis of thalamocortical resonant columns that can support global cognitive experiences. In this context, the specific system provides the content that relates to the external world while the nonspecific system would give rise to the temporal conjunction (binding circuit).

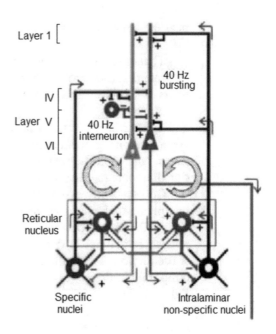

Fig. 3 Thalamocortical circuit. The Llinás et al. model (Adopted from [10])

This well-explored and accepted model served as the theoretical fundament for the investigations on NPs' induced signaling modifications in neuronal circuits (*see* **Note 3**).

3.3.3 Applied Thalamocortical Network

The simplified in silico model that we used to predict in vivo behavior is based on the principle of thalamocortical interaction and binding as elucidated before.

The applied circuit is based on kinetic models of pyramidal neurons (here: chattering cells, PY) [40–43], inhibitory cortical interneurons (IN) [43, 44], thalamic cells (TC) including specific thalamic cells (STC) and nonspecific thalamic cells (NSTC) [45–47], and reticular thalamic neurons (RTN) [40, 43, 48, 49] (*see* **Note 4**).

Each single-neuron model is considered one compartment (except PY, modeled as two-compartments) and is represented by coupled differential equations according to an extended HH type Scheme [7].

All currents in this model have the unit $\mu A/cm^2$ and all conductances have the unit mS/cm^2 [49]. After using Kirchoff's Law, the current balance equations for the ionic currents for the somatic and dendritic membrane potentials and voltage change across the membrane of each compartment are obtained [50].

The synaptic connectivity scheme shown in Fig. 4 is used in the developed simplified thalamocortical network model. In this model, specific thalamic inputs are represented by a thalamic neuron (STC) that projects to both PY neuron and inhibitory IN located in cortical layer IV after sending an axon collateral to the RTN neuron. Another thalamic neuron (NSTC) represents intralaminar, nonspecific thalamic inputs and projects to neocortical layer I after sending axon collaterals to the RTN neuron. STC and NSTC neurons produce excitatory postsynaptic potentials that are mediated by fast excitatory receptors in the model. RTN neurons project with inhibitory characteristic to specific and intralaminar nucleus neurons. Both the inhibitory fast and slow receptors in the thalamic neurons mediate the inhibitory postsynaptic potentials of those cells. The RTN neurons also have reciprocal inhibitory synaptic connectivity and fast inhibitory receptors mediate the corresponding activity. The cerebral cortex was considered a simple network model of inhibitory IN and excitatory PY neurons [8–11]. Although this is a highly simplified representation of the neocortex's multilayered structure, no additional complexity was required for the theoretical modeling. Consequently, the pyramidal neurons in the layers IV and V are described by single neurons that receive inputs and project to both STC and NSTC cells and have axon collaterals to the RTN neurons. Four PY neurons were included, two of which receive input from the STC to provide specific sensory input to the cortex. This structure indicates the specific resonant loop on the left. The nonspecific resonant loop

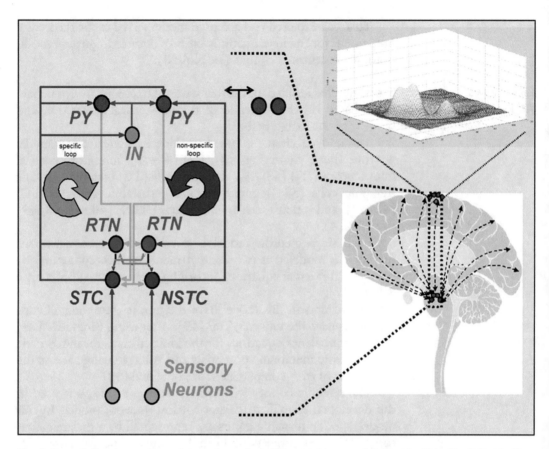

Fig. 4 Computational upscaling scheme, from single HH-model to neural field potentials. The left side illustrates the developed model based on the concept of Llinás [9, 10]; The blue, green, brown, and red arrows correspond to the AMPA, AMPA + NMDA, GABA$_A$, and GABA$_A$ + GABA$_B$ receptors respectively. The figures on the bottom right and top right illustrate the idea of a cortico-thalamic column within a very small cortical patch and its 3D Amari field potential respectively (adopted from [13])

is represented on the right side, where three PY neurons receive their inputs in a more diffuse way from NSTC to provide the multimodal connectivity to the cortex. Two PY neurons that receive inputs and project back to the NSTC are shown as two black spots. In this model, both the slow and fast excitatory receptors mediate the corticofugal excitation of the PY. In the circuit, all PY neurons receive axons from the cortical inhibitory IN, in which the inhibitory postsynaptic potential elicited by this cell is also mediated by both the fast and slow excitatory receptors. Lastly, the essential sensory inputs to activate STC and NSTC are provided by including two sensory neurons in the present model (see Fig. 4). Thus, the synaptic projections from those sensory neurons activate the fast receptors of the thalamic nuclei. Every neuron of the simplified thalamocortical model receives various synaptic inputs that are modeled as the sum over all synaptic currents that each cell receives [51, 52]. Accordingly, every neuron is described by the generic membrane equation

$$C_m \frac{dV_i}{dt} = -\overline{g}_L \left(V_i - V_L \right) - \sum_j I_{ji}^{int} - \sum_k I_{ki}^{syn} \qquad (20)$$

where C_m is the specific membrane capacity and V_i is the postsynaptic membrane potential. I_{ji}^{int} and I_{ki}^{syn} signify the intrinsic (ionic) and synaptic currents. The generic form of the intrinsic currents (generalization of Eq. 12) is represented by

$$I_{ji}^{int} = \overline{g}_j \, m_j^M h_j^N \left(V_i - E_j \right) \qquad (21)$$

Here, i connotes the postsynaptic neuron and j represents for the specific ionic type. Further, $\overline{g}_j$ is the maximal conductance, m the time and voltage-dependent activation variable, h is the corresponding (time and voltage-dependent) inactivation variable, and finally $(V_i - E_j)$ is the difference between membrane potential and reversal potential of each ion. The following generic equation represents the synaptic currents in the system:

$$I_{ki}^{syn} = \overline{g}_{ki} s_{ki} \left(V_i - E_{ki} \right) \qquad (22)$$

where k_i designates the synaptic junction from the presynaptic neuron k to the postsynaptic neuron i, g_{ki} is the maximal conductance of the postsynaptic receptors, and E_{ki} is the reversal potential. The fraction of open receptors is specified by s_{ki} according to the simple two-state scheme.

$$\begin{aligned} (\text{closed}) + T\left(V_k\right) &\overset{\alpha}{\rightarrow} (\text{open}) \\ (\text{closed}) + T\left(V_k\right) &\overset{\beta}{\leftarrow} (\text{open}) \end{aligned} \qquad (23)$$

For computational efficiency, a reduced transmitter release model is used, assuming that all the intervening reactions in the release process are relatively fast and thus, can be considered in steady state (instantaneous) [53]. Consequently, the stationary relationship between the transmitter concentration $[T]$ and presynaptic voltage is described by a simple sigmoidal function [53]

$$[T]\left(V_{pre}\right) = \frac{T_{max}}{1 + \exp\left(-\left(V_{pre} - V_p\right) / K_p\right)} \qquad (24)$$

where T_{max} is the maximal concentration of the transmitter in the synaptic cleft, V_{pre} is the presynaptic voltage, K_p gives the steepness, and V_p sets the value at which the function is half activated. This form, in conjunction with simple kinetic models of postsynaptic channels, provides a model of synaptic interaction based on autonomous differential equations with only one or two variables [54, 55].

In the model, all initial conditions and set parameters representing a single cell or synaptic connection originate from electrophysiological measurements on the specific neurons taken from cited references. Please see [40, 42, 53, 54, 56] for a detailed description of the particular equations and the corresponding cell- and transmitter specific parameters. According to [51], different values of I_h and I_{KL} would cause heterogeneity in the intrinsic properties of the cells. Therefore, the intrinsic conductance of g_h and g_{KL} for the STC and the NSTC and g_{KL} for the RTN neuron models are slightly different in the present model.

To simulate the effects of cAg-NPs in contact with thalamic cells, i.e., STC, NSTC, and RTN, respectively, the changes of the intrinsic currents that were identified by fitting the patch-clamp data to the HH-model were applied to those cells.

The kinetics and voltage dependence of the modeled currents are very similar to the currents measured in patch clamp experiments (see Sect. 2.2). MATLAB was used as a simulation environment in which the differential equations were solved by employing a fourth-order Runge-Kutta method.

3.3.4 Linking Hodgkin-Huxley Neuronal Circuit Activity to Neural Field Potentials

After creating an in silico model considering the cell membrane dynamics of the involved cortical and thalamic neurons, the study of neural networks as a population of a few cortical neurons is noticeable. According to the Amari model and through mathematical analyses, which are known as neural field equations, the mechanism of formation and interaction of firing patterns and their response to input stimuli in homogenous fields is considered [12].

Amari's general idea to express and estimate the oscillation of neural field, which consists of m types of excitatory and inhibitory neurons and m layers, under natural assumptions on the connectivity and firing rate function (commonly referred to as Mexican hat connectivity), is mathematically noted as

$$\tau_i \frac{\partial u_i(x,t)}{\partial t} = -u_i + \sum_{j=1}^{m} \int w_{ij}(x,x';t-t') Z_j(x',t') dx' dt' + h_i + s_i(x,t). \quad (25)$$

To develop the current study the neural field model that consists of the two-layer architecture is introduced. The excitatory (top) layer represents the mean firing rate of four PY neurons, and the inhibitory (sub-) layer is substituting IN neurons, located over a wider spatial area possessing unspecific all-to-all connections (lateral inhibition). A neural field F is calculated by the extended 2D Amari model through Eq. 26:

$$\frac{du(x,t)}{dt} = \frac{1}{\tau}\left(-u(x,t) + h_u\right.$$

$$+ \int_F w_u(x-x')f\left[u(x',t)\right]dx' - \int_F w_v(x-x')f\left[v(x',t)\right]dx' + S_{\text{PY1som}}(x,t) \tag{26}$$

$$+ S_{\text{PY2som}}(x,t) + S_{\text{PY3som}}(x,t) + S_{\text{PY4som}}(x,t)$$

$$+ S_{\text{PY1den}}(x,t) + S_{\text{PY2den}}(x,t) + S_{\text{PY3den}}(x,t) + S_{\text{PY4den}}(x,t)$$

Here, $\dfrac{du(x,t)}{dt}$ is the rate of each neuron's change of fieldactivation level across the spatial dimension x as a function of time t. τ is the time scale of the dynamics and $u(x, t)$ is defined as the activation in the field at each position $x = (x1, x2)$ at time t and is the first factor that advances the rate of change of activation and due to its negative term, the activation changes toward the activation level h_u, that is relevant for the threshold function f which is an output function in the form of a nonlinear sigmoid function and is monotonically nondecreasing, saturating to a constant for large u.

$$f(u) = \frac{1}{1 + \exp\left[-\beta(u - u_0)\right]} \tag{27}$$

This determines that in both the cases only the field parts that are sufficiently activated are contributing to intrafield interactions. Here, β is determined as the sigmoidal function's slope and shows the degree to which neurons close to threshold contribute to the activation dynamics. u_0 is the inflection point, namely threshold point.

Based on the Amari model introduced in 1977, the field in this model also consists of excitatory and inhibitory layers, in which the inhibitory neurons only target the excitatory neurons. Moreover, the excitatory neurons have very narrow fan-out connections to the inhibitory neurons so that the excitatory neurons at place x excite the inhibitory neurons at place x only [12].

This local excitation and lateral inhibition is respectively defined by $\int w_u(x - x')f[u(x', t)]dx'$ and $\int w_v(x - x')f[v(x', t)]dx'$ in Eq. 26.

The intrafield interaction between the neurons of one layer takes the shape of a convolution over the threshold field $f(u)$ with a homogeneous convolution kernel w_u. The interfield interaction between the neurons placed at two layers follows the same scheme with the threshold field $f(v)$ and convolution kernel w_v.

Furthermore, the connectivity w that is deduced from Gaussian kernel and is often referred to as lateral inhibition explains excitatory behavior over small distances, inhibitory over medium distances and either inhibitory or zero over larger distances (global inhibition). This interaction mode is widely applied in modeling, known as Mexican hat function:

$$w(x - x') = \exp\left[-0.5\frac{(x - x')^2}{\sigma^2}\right], \tag{28}$$

where σ denotes the width of the excitatory part of the kernel.

A detailed description of these functions and their parameters can be found in [24].

According to the Amari field model, a local excitation pattern is elicited by external time-invariant input stimuli, that is applied to the cortical neurons at position x. After excitation of the cortical neurons, the oscillating activity of both, the dendritic tree's characteristic low-pass activity, and also the somatic spiking activity are taken as mean firing rate inputs to the neural field model. To do that, it is necessary to map the 1D neuron model firing output (soma + dendrite) to a 2D Gaussian spatiotemporal signal distribution since the average value that stands at $x = (x_0, y_0)$ has two coordinates:

$$f(x,y) = a.\exp\left(-\frac{(x - x_0)^2}{2\sigma_x^2} + \frac{(y - y_0)^2}{2\sigma_y^2}\right) \tag{29}$$

where a is the height of the curve's peak that is substituted by the amplitude of somatic and dendritic activity calculated. x_0 is the location parameter or a mean value that is replaced by a coordinate of the firing neurons for chosen coordinates, and σ^2 is the squared scale parameter that corresponds to the variance of the distribution.

Therefore, each stimulus in this model can be depicted for the somatic part as

$$S_{\text{PYsom}} = \frac{1}{C_m}(-\bar{g}_L(V_s - E_L) - I_{\text{Na}}(V_s,h) - I_K(V_s,n) - I_{\text{Ca}}(V_s) - I_{\text{AHP}}(V_s,[\text{Ca}^{2+}]_i)) \tag{30}$$

$$-\frac{g_c}{p}(V_s - V_D) + C_m V_m \exp(-\frac{(x - x_0)^2}{2\sigma_x^2} + \frac{(y - y_0)^2}{2\sigma_y^2})$$

and for the dendritic part as

$$S_{\text{PYden}} = \frac{1}{C_m}(-\bar{g}_L(V_D - E_L) - I_{\text{Ca}}(V_D) - I_{\text{AHP}}(V_s,n) - I_{\text{AHP}}(V_s,[\text{Ca}^{2+}]_i)) \tag{31}$$

$$-\frac{g_c}{1 - p}(V_D - V_s) + I_{\text{syn}} \exp(-\frac{(x - x_0)^2}{2\sigma_x^2} + \frac{(y - y_0)^2}{2\sigma_y^2})$$

As stated, the field in the current study is generated by a set of four pyramidal neurons. So eight inputs must be considered here, four to the dendritic and four to the somatic compartments.

3.3.5 Backpropagation of the Neural Field Potentials on the Cortical Neurons

After calculating the neural field at position x and time t by the Amari neural field model on the membrane potential of neurons by means of the single neuron's Hodgkin-Huxley formalism embedded in the neural circuit model of Llinás and his colleagues, the effect of neural field oscillations is investigated through the Nernst model.

$$C_m . V_m = I_m \tag{32}$$

In this way, it is possible to convert the neural field potentials V_m to electric current I_m at a given spatial point of a modeled neuron and to use that as an additional feedback input for the thalamocortical circuit in order to calculate the effects of the generated field potentials back on each pyramidal neuron. Because changes in the neural signaling properties will change the resonant states in the network, it is expected that this feedback is mandatory to gain a valid multiscale model.

Since we separated the PY neurons into a dendritic and a somatic compartment, the potential equations for somatic part are given by

$$\frac{dV_s}{dt} = \frac{1}{C_m} \left(\begin{array}{c} -\bar{g}_L (V_s - E_L) - I_{Na}(V_s,h) - I_K(V_s,n) - I_{Ca}(V_s) \\ -I_{AHP}(V_s,[Ca^{2+}]_i) - \frac{g_c}{p}(V_s - V_D) + I_{app} \end{array} \right), \tag{33}$$

and for dendritic part are given by

$$\frac{dV_D}{dt} = \frac{1}{C_m} \left(-\bar{g}_L (V_D - E_L) - I_{Ca}(V_D) - I_{AHP}(V_D,[Ca^{2+}]_i) - \frac{g_c}{1-p}(V_D - V_s) + I_{syn} \right), \tag{34}$$

where C_m is the membrane capacitance, V_S and V_D are the membrane potentials of somatic and dendritic compartment, respectively, and $\bar{g}_L (V_s - E_L)$ and $\bar{g}_L (V_D - E_L)$ depict the corresponding leakage conductance term for each compartment. The current flow between the soma and the dendrite is proportional to $(V_S - V_D)$ in A/cm², with coupling conductance $g_c = 2$ mS/cm² and the parameter $p =$ somatic area/total area $= 0.5$. The cell can be either excited by an injected current I_{app} (in µA/cm²) to the soma or by synaptic inputs to the dendrite [41].

According to recent research in rat cortical pyramidal neuron stimulation, it has been found that extracellular fields cause changes in the exposed cell's somatic membrane potentials [57]. Therefore, the feedback current caused by the neural fields is only modeled in the somatic part of the two-compartment model as I_{app}.

After successfully simulating the cortical field potentials the field activity was simulated again through the presence of Ag-NPs in the referred thalamic neurons (RTN1, RTN2, STC, NSTC).

4   Conclusion

The in vitro measurements that served as basis for this study have shown a rapid suppression of sodium currents of the chromaffin cells after exposure to cAg-NPs without an appreciable shift in either activation voltage or null potential. Those changes were found to be rapid, partially reversible, and dose-dependent.

Transferring those findings to neuronal circuits via modeling of the thalamocortical network predicted that an alteration of the properties of RTN, STC, or NSTC neurons, as they were found by the HH-model fits of I_{Na} after treatment with cAg-NPs, end in large alterations of network signaling behavior. It can be expected that NPs brought into contact with few cells of a neuronal feedback circuit will extensively alter network rhythms of large neuronal populations in vivo.

The in silico thalamocortical circuit model was additionally expanded to observe related field potentials spreading over a small spatial cortical patch. Incorporating the idea of back propagating the field potentials effecting single neuron activity to this new model showed that the field potentials seem to have strong effects on the action potential generation of neurons that are exposed to those fields as well.

Also, NPs brought into contact with thalamic cells of the circuit actually lead to diversified neuronal field activity of much higher spatial domains. The results presume that as a consequence of cAg-NPs affected thalamocortical network cells, the emerging in vivo neural field potentials will be found to be extensively diversified.

In the future, two-dimensional multi-electrode array as well as voltage sensitive dye measurements with high spatiotemporal resolution will be carried out on rat auditory cortex. The experimental data will be applied to validate and also to extend the model with the focus on underlying neuronal mechanisms. Besides, cAg-NPs effects on other aspects of membrane excitability will be investigated and these findings will be included into the model. The experimental data will be applied to compare and also to extend the model with the focus on recovering estimates of the underlying mechanisms by means of excitable cells exposed to NPs. Furthermore, different NPs types, sizes, as well as their coatings' material and surface structure will be investigated in respect of their stability and impact on neuronal cells and tissues.

5   Notes

1. The in vivo effects of such NPs on any network will likely be complicated by additional effects. The changes in voltage-dependent sodium currents are introduced in this model. However, we have not examined whether cAg-NPs may produce mechanical effects on the ion channels leading to lower conductivity or if there are fewer channels that reach the open state.

2. Various problems in applied mathematics have target functions that are noncontinuous, nonlinear, non-differentiable, noisy, multi-dimensional or have many local minima, constraints or stochasticity. Fitting the steady-state HH model (Eqs. 1–7) to the measured patch clamp data requires the solution of a high dimensional inverse problem: since the essential HH-Model-equations that need to be taken into account for the fitting consist of 13 independent parameters (constants evaluated by Hodgkin & Huxley for the squid axon) that have to be estimated, this problem cannot be solved analytically.

3. The developed in silico model enables the investigation of potential effects of NPs which enter the CNS.

   The simulations examine the systemic influences of cAg-NPs in neuronal systems. Though our computational model of the thalamocortical network is highly simplified and does not consider inputs or projections to other involved brain regions, it indicates that reduced excitability of a few neurons in such a circuit has distinct effects on network activity. The model is able to predict possible consequences of cAg-NPs introduction to neuronal feedback circuits.

4. Those findings on Na_v 1.7 channels were employed for the modeling approach, in which the measured effects of cAg-NPs on chromaffin cell voltage-gated sodium currents were devolved to voltage-gated sodium channels of thalamic neurons (STC, NSTC, RTN).

References

1. Zhang Y, Satterlee A, Huang L (2012) In vivo gene delivery by nonviral vectors: overcoming hurdles? Mol Ther 20(7):1298–1304

2. Lamarre B, Ryadnov MG (2011) Self-assembling viral mimetics: one long journey with short steps. Macromol Biosci 11(4):503–513

3. Chithrani B, Ghazani A, Chan W (2006) Determining the size and shape dependence of gold nanoparticle uptake into mammalian cells. Nano Lett 6:662–668

4. Xu L, Zhao J, Zhang T, Ren G, Yang ZL (2009) In vitro study on influence of nanoparticles of CuO on Ca1 pyramidal neurons of rat hippocampus potassium currents. Environ Toxicol 24:211–217

5. Zhao J, Xu L, Zhang T, Ren G, Yang ZL (2009b) Influences of nanoparticles zinc oxide on acutely isolated rat hippocampal Ca3 pyramidal neurons. Neurotoxicology 30:220–230

6. Liu Z, Ren G, Zhang T, Yang ZL (2009) Action potential changes associated with the inhibitory effects on voltage-gated sodium current of hippocampal ca1 neurons by silver nanoparticles. Toxicology 264:179–174

7. Hodgkin AL, Huxley AF (1952) A quantitative description of membrane current and its application to conduction and excitation in nerve. J Physiol 117:500–544

8. Llinás R, Ribary U (1993) Coherent 40-Hz oscillation characterizes dream state in humans. Proc Natl Acad Sci U S A 90:2078–2081

9. Llinás R, Ribary U, Joliot M, Wang XJ (1994) Content and context in temporal thalamocortical binding. In: Buzski G et al (eds) Temporal coding in the brain. Springer, Berlin

10. Llinás R, Ribary U, Contreras D, Pedroarena C (1998) The neuronal basis for consciousness. Philos Trans R Soc Lond B 353:1841–1849

11. Llinás R, Leznik E, Urbano FJ (2002) Temporal binding via cortical coincidence detection of specific and nonspecific thalamocortical inputs: a voltage-dependent dye-imaging study in mouse brain slices. Proc Natl Acad Sci 99(1):449–454

12. Amari S (1977) Dynamics of pattern formation in lateral inhibition type neural fields. Biol Cybern 27:77–87

13. Busse M, Kraegeloh A, Arzt E, Strauss D (2011) Modeling the influences of nanoparticles on neural field oscillations in thalamocortical networks. Conf Proc IEEE Eng Med Biol Soc 2011:136–139

14. Pellegrino T, Manna L, Kudera S, Liedl T, Koktysh D, Rogach AL, Keller S, Rdler J, Natile G, Parak WJ (2004) Hydrophobic nanocrystals coated with an amphiphilic polymer shell: a general route to water soluble nanocrystals. Nano Lett 127:703–707

15. Koch M, Kiefer S, Cavelius C, Kraegeloh A (2012) Use of a silver ion selective electrode to assess mechanisms responsible for biological effects of silver nanoparticles. J Nanopart Res 14(2):646. https://doi.org/10.1007/s11051-011-0646-y

16. Fenwick EM, Marty A, Neher E (1982) A patch-clamp study of bovine chromaffin cells and of their sensitivity to acetylcholine. J Physiol 331:577–597

17. Kobayashi H, Shiraishi S, Yanagita T, Yokoo H, Yamamoto R, Minami S, Saitoh T, Wada A (2002) Regulation of voltage-dependent sodium channel expression in adrenal chromaffin cells: involvement of multiple calcium signaling pathways. Ann N Y Acad Sci 971:127–134

18. Tischler AS (2002) Chromaffin cells as models of endocrine cells and neurons. Ann N Y Acad Sci 971:366–370

19. Goldin AL, Barchi RL, Caldwell J, Hofmann F, Howe JR, Hunter JC et al (2000) Nomenclature of voltage-gated sodium channels. Neuron 28:365–368

20. Catterall WA, Goldin AL, Waxman SG (2005) International union of pharmacology. xlvii. nomenclature and structure-function relationships of voltage-gated sodium channels. Pharmacol Rev 57:397–409

21. Goldin AL (2001) Resurgence of sodium channel research. Annu Rev Physiol 63:871–894

22. Lorincz A, Nusser ZA (2010) Molecular identity of dendritic voltage-gated sodium channels. Science 328(5980):906–909

23. Royeck M, Horstmann M, Remy S, Reitze M, Yaari Y, Beck HA (2008) Role of axonal nav1.6 sodium channels in action potential initiation of ca1 pyramidal neurons. J Neurophysiol 100(4):2361–2380

24. Toledo-Aral JJ, Moss BL, He Z, Koszowski AG, Whisenand T et al (1997) Identification of pn1, a predominant voltage-dependent sodium channel expressed principally in peripheral neurons. Proc Natl Acad Sci U S A 94:1527–1532

25. Busse M, Stevens D, Kraegeloh A, Cavelius C, Vukelic M, Arzt E, Strauss D (2013) Estimating the modulatory effects of nanoparticles on neuronal circuits using computational upscaling. Int J Nanomed 8:3559–3572

26. Malmivuo J, Plonsey R (1995) Bioelectromagnetism. Oxford University Press, New York, NY

27. Price KV, Storn RM, Lampinen JA (2005) Differential evolution: a prictical approach to global optimization. Springer, Berlin

28. Shepherd GM (2001) The synaptic organization of the brain, 5th edn. Oxford University Press, Inc., Oxford

29. Jones EG (2002) Thalamic circuitry and thalamocortical synchrony. Philos Trans R Soc Lond B Biol Sci 357(1428):1659–1673

30. Singer W (1999) Neuronal synchrony: a versatile code for the definition of relations? Neuron 21(1):49–65

31. Hughes SW, Errington A, Lorincz ML, Kkesi KA, Juhsz G, Orbn G, Cope DW, Crunelli V (2008) Novel modes of rhythmic burst firing at cognitively-relevant frequencies in thalamocortical neurons. Brain Res 1235:12–20

32. Gray CM, Singer W (1989) Stimulus-specific neuronal oscillations in orientation columns of cat visual cortex. Proc Natl Acad Sci U S A 86:1698–1702

33. Ribary U, Ioannides AA, Singh KD, Hasson R, Bolton JPR, Lado F, Mogilner A, Llinás R (1991b) Magnetic field tomography of coherent thalamocortical 40-Hz oscillations in humans. Proc Natl Acad Sci U S A 8:11037–11041

34. Singer W (1993) Synchronization of cortical activity and its putative role in information

processing and learning. Annu Rev Physiol 55:349–374

35. Gregoriou GG, Gotts SJ, Zhou H, Desimone R (2009) High-frequency, long-range coupling between prefrontal and visual cortex during attention. Science 324:1207–1210

36. Ribary U, Cappel J, Yamamoto T, Suk J, Llinás R (1991a) Anatomical localizaiton revealed by meg recordings of the human somatosensory system. Electroencephalogr Clin Neurophysiol 78:185–196

37. Gray CM, McCormick DA (1996) Chattering cells: superficial pyramidal neurons contributing to the generation of synchronous oscillations in the visual cortex. Science 274:109–113

38. Steriade M, Timofeev I, Drmller N, Grenier F (1998) Dynamic properties of corticothalamic neurons and local cortical interneurons generating fast rhythmic (30–40 Hz) spike bursts. J Neurophysiol 79:483–490

39. Brumberg JC, Nowak LG, McCormick DA (2000) Ionic mechanisms underlying repetitive high-frequency burst firing in supragranular cortical neurons. J Neurosci 20:4829–4843

40. Wang XJ, Golomb D, Rinzel J (1995) Emergent spindle oscillations and intermittent burst firing in a thalamic model: specific neuronal mechanisms. Proc Natl Acad Sci U S A 92:5577–5581

41. Wang XJ (1998) Calcium coding and adaptive temporal computation in cortical pyramidal neurons. J Neurophysiol 79:1549–1566

42. Golomb D, Shedmi A, Curt R, Ermentrout GB (2006) Persistent synchronized bursting activity in cortical tissues with low magnesium concentration: a modeling study. J Neurophysiol 95:1049–1067

43. Golomb D, Wang XJ, Rinzel J (1996) Propagation of spindle waves in a thalamic slice model. J Neurophysiol 75(2):750–769

44. Wang XJ, Buz'saki G (1996) Gamma oscillation by synaptic inhibition in a hippocampal interneuronal network model. J Neurosci 16(20):6402–6413

45. Destexhe A, Bal T, McCormick DA, Sejnowski TJ (1996a) Ionic mechanisms underlying synchronized oscillations and propagating waves in a model of ferret thalamic slices. J Neurophysiol 76(3):2049–2070

46. Bazhenov M, Timofeev I, Steriade M, Sejnowski TJ (1998) Cellular and network models for intrathalamic augmenting responses during 10-Hz stimulation. J Neurophysiol 79:2730–2748

47. Golomb D, Amitai Y (1997) Propagating neuronal discharges in neocortical slices: computational and experimental study. J Neurophysiol 78:1199–1211

48. Destexhe A, Contreras D, Sejnowski TJ, Steriade M (1994a) A model of spindle rhythmicity in the isolated thalamic reticular nucleus. J Neurophysiol 72(2):803–818

49. Destexhe A, Contreras D, Steriade M, Sejnowski TJ, Huguenard JR (1996b) In vivo, in vitro, and computational analysis of dendritic calcium currents in thalamic reticular neurons. J Neurosci 16:999–1016

50. Ferguson KA, Campbell SA (2009) A two compartment model of a ca1 pyramidal neuron. Can Appl Math Q 17(2):293–307

51. Destexhe A, Contreras D, Steriade M (1998a) Mechanisms underlying the synchronizing action of corticothalamic feedback through inhibition of thalamic relay cells. J Neurophysiol 79:999–1016

52. Bazhenov M, Timofeev I, Steriade M, Sejnowski TJ (2002) Model of thalamocortical slow-wave sleep oscillations and transitions to activated states. J Neurosci 22(19):8691–8704

53. Destexhe A, Mainen ZF, Sejnowski TJ (1998b) Kinetic models of synaptic transmission. In: Koch C, Segev I (eds) Methods in neuronal modeling, 2nd edn. The MIT Press, Cambridge

54. Wang XJ, Rinzel J (1992) Alternating and synchronous rhythms in reciprocally inhibitory model neurons. Neural Comput 4:84–97

55. Golomb D, Wang XJ, Rinzel J (1994) Synchronization properties of spindle oscillations in a thalamic reticular nucleus model. J Neurophysiol 72(3):1109–1126

56. Destexhe A, Mainen ZF, Sejnowski TJ (1994b) An efficient method for computing synaptic conductances based on a kinetic model of receptor binding. Neural Comput 6:14–18

57. Anastassiou CA, Perin R, Markram H, Koch CK (2011) Ephaptic coupling of cortical neurons. Nat Neurosci 14(2):217–223

Chapter 11

The Application of In Vivo Extracellular Recording Technique to Study the Biological Effects of Nanoparticles in Brain

Yanyan Miao, Han Zhao, Jutao Chen, Ming Wang, and Longping Wen

Abstract

With the ability to penetrate the brain blood barrier (BBB), many types of nanoparticles have the chance to interact with the central nervous system, eliciting various and sometimes unexpected biological effects. Thus, understanding the effects of nanoparticles in central nervous system and the underlying mechanisms is critically important for the biomedical applications of engineered nanomaterials in the brain. Various techniques have been developed to study the electrophysiology of neurons and neuronal communication, providing an insight into the molecular mechanisms of learning and memory. Here, we describe the in vivo extracellular recording techniques that are used to record the long-term potential (LTP), short-term potential (paired-pulse facilitation, PPF), and the basal synaptic transmission (input–output curve, I/O curve) in dentate gyrus of hippocampus in the brain. Dentate gyrus plays a critical role in learning and memory. Thus, intrahippocampal infusion of nanoparticles in this area and the subsequent in vivo extracellular recording may help explore the function of nanoparticles in brain.

Key words Extracellular recording technique, Nanoparticles, Learning and memory, Long-term potential, Paired-pulse facilitation, Input–output curve

1 Introduction

Nanomaterials exhibit many size-dependent physical and chemical properties, which make them unique and indispensable in a wide array of areas ranging from catalysis to biology [1]. With the extensive application in biology, especially in the central nervous system (CNS), the diverse and oftentimes unexpected biological effects elicited by engineered nanomaterials have started to attract wide attention [2–6]. The central nervous system is a very complex structure, and the neurons are specialized to communicate with each other through integration and propagation of electrical events, which occurs at the synapse, a structure that allows information flow from one neuron (pre-synaptic neuron) to another

Fidel Santamaria and Xomalin G. Peralta (eds.), *Use of Nanoparticles in Neuroscience*, Neuromethods, vol. 135, https://doi.org/10.1007/978-1-4939-7584-6_11, © Springer Science+Business Media, LLC 2018

neuron (postsynaptic neuron) [7]. The electrical activities can be measured by various electrophysiological techniques, providing a quantifiable readout of the underlying synaptic activity in real time [8]. Consequently, electrophysiological studies are fundamental for understanding synaptic plasticity, the ability of synapses to change their strength in response to their own activity and for mechanisms of learning and memory [9, 10].

Extracellular recording techniques are the approach of choice for studying the electrical activities of neurons in anesthetized or awake animals. With the placement of extracellular electrodes in the vicinity of cells, this in vivo electrophysiological approach can record the extracellular field potentials, which are the electrical potentials produced by a single neuron or population of neurons [11]. The application of the extracellular recording technique in the studies of neuronal activity, learning, and memory has provided many important clues of the way in which the brain processes information.

The hippocampus, which contains the dentate gyrus (DG area), CA fields, and subicular complex, is a brain region in the medial temporal lobe that is critical for learning and memory [12–14]. Long-term potentiation (LTP, a form of synaptic plasticity), a long-lived increase in synaptic strength following high-frequency stimulation (HFS) of synapses, was first demonstrated as a phenomenon in the dentate area of hippocampus by using the in vitro extracellular recording technique in 1973 [15]. Recently, LTP has been widely considered the molecular mechanisms and the main experimental model of learning and memory [16, 17]. Paired-pulse facilitation (PPF) is known as a phenomenon in which the amplitude of the field excitatory postsynaptic potentials (fEPSPs) or population spikes (PS) evoked by one stimulus is larger when the stimulus rapidly follows a prior stimulus. PPF represents the form of short-term synaptic plasticity, crucial in the information processing in synapses and meaning the ability of the brain to remember information for a short time [18]. In addition, neurons that are in the basal state also can communicate with each other at low rates. And this basal synaptic transmission, estimated by the input–output curve (I/O curve, EPSP slope vs. stimulus intensity) of extracellularly recorded field EPSPs, is important for the information processing in brain [19, 20]. Consequently, targeted manipulation of the biological process may help facilitate synaptic plasticity events for improving learning and memory.

Fullerene C60, a classic engineered nanomaterial with unique physicochemical properties [21, 22], has attracted intense attention from biomedical areas due to their outstanding biological activities, such as anti-viral [23], anti-bacteria [24], autophagy induction [25], inhibition of allergic responses [26], and ion channel blockage [27]. Fullerene C60 is reportedly capable of translocating through the blood–brain barrier (BBB) [28, 29], suggesting the possibilities of interaction with the neurons and modulation their functions. Currently, fullerene C60 and its derivatives have been shown to have potential neuroprotective abilities. For instance, C60 derivatives may exert neuroprotective functions by blocking glutamate receptors [30]. A single introcerebroventricular injection of C60 was shown to improve the performance of cognitive task, and prevent the cognitive impairment by amyloid β_{25-35} [31]. Hexasulfobutylated C60 exerts the ability to significantly ameliorate total volume of infarction in the focal cerebral ischemia [32]. Carboxy-modified C60 can significantly prevent iron-induced oxidative stress in rat brain, and serve as free radical scavengers and prevent excitotoxic neuronal death [33, 34]. In our previous studies, we have found that fullerene nanoparticles can specifically bind to and elicit persistent activation of hippocampal Ca^{2+}/calmodulin-dependent protein kinase II (CaMKII), a kinase critical for synaptic plasticity, but the functional consequence of that modulation is unknown [35]. Thus an in vivo electrophysiological approach will be an ideal method to study the process involving the modulation of learning and memory by fullerene nanoparticles, which are delivered into hippocampus through intrahippocampal (i.h.) infusion.

2   Materials

Prepare all of the solutions by using ultrapure water (pH 6.7; Milli-Q, Bedford, MA).

2.1   Components and Equipment Used in Nanoparticle Preparation

1. Fullerene C60 (99.9% pure, BuckyUSA).

2. The solvent used to make Nano C60 nanocrystals: HPLC grade tetrahydrofuran (THF) (*see* **Note 1**).

3. Rotary evaporator (BC-R202, Shanghai Biochemical Equipment Co., Ltd., China, http://www.shbioc.cn): Rotatory bottle, vertical condenser, condensate-collecting flask and constant temperature bath pan (*see* **Note 2**).

4. 2XZ(S)-2 type rotary vane vacuum pump: Motor power 370 W, ultimate pressure $\leq 6 \times 10^{-2}$ Pa, pumping speed 2 L/s, rotational speed $16{,}000 \times g$.

2.2   Components Used in Stimulation and Recording

1. Artificial cerebrospinal fluid (ACSF): 124 mM NaCl, 2.0 mM KCl, 2.0 mM $CaCl_2$, 2.0 mM $MgSO_4$, 1.25 mM Na_2HPO_4, 26 mM $NaHCO_3$ and 10 mM glucose, add ultrapure water to a volume of 90 mL. Mix and adjust pH to 7.3–7.4. Make up to 100 mL with water (Osmotic pressure: 300–310 mOsm), followed by bubbling with 95% O_2/5% CO_2. Then filter through

a 0.22 μm Millex-GP filter (Millipore). Store at 4 °C in a sterile bottle (*see* **Note 3**).

2. Anesthetic: Dissolve 10 g urethane in 100 mL ultrapure water (10% urethane solution). Store at 4 °C (*see* **Note 4**).

3. 70% ethanol (analytical grade reagents) in ultrapure water.

2.3 Equipment for Intrahippocampal Stimulation, Recording, and Infusion

1. Stereotaxic apparatus (Cat. 51503D Digital Dual New Standard Stereotaxic, Rat and Mouse, STOELTING.) including two ear bars and tooth holder (Fig. 1a).

2. Homeothermic Monitor with 8 in. heating pad (Fig. 1b) (Harvard Apparatus).

3. A concentric bipolar stimulating electrode (Cat. CBBRE75, FHC) used for stimulation.

4. A conjoined electrode/cannula assembly for intrahippocampal recording and infusion in DG area: Two individually insulated wire electrodes (coated diameter of 230 μm, recording electrodes) and stainless steel tubing of the guide cannula (26 G, Cat. C315G/SPC, PlasticsOne). An inner cannula (31 G, Cat. C316I/SPC, PlasticsOne) inserted along the guide cannula.

5. Hamilton syringe (10 μl, Cat. 20,779, Sigma). PE50 polyethylene tubing (Cat. C232CT, PlasticsOne). Ultra Micro Pump (WPI, Sarasota, FL, USA).

6. Master-8-cp stimulator (AMPI, Israel). AC Amplifier Model 1800 (A-M system, USA). Axon™ digidata 1550 low-noise data acquisition system (Molecular Devices, USA) (Fig. 2). The recording software is Clampex 10.4 and analysis software of waveform is clampfit 10.4 (Axon, USA).

2.4 Animal Preparation

Adult Sprague-Dawley (SD) rats weighing 180–250 g have free access to food and water, and should be maintained in a pathogen-free environment with a 12 h light-dark cycle at a constant temperature (23 ± 0.5 °C). All efforts are made to minimize animals suffering and the number of animals used, while all animal treatments are strictly in accordance with the National Institutes of Health Guides for the Care and Use of Laboratory Animals (NIH publication NO. 80-23, revised 1996).

3 Methods

3.1 Aqueous Suspension of Nanoparticles Preparation

Nanoparticles dispersed and stabilized in aqueous suspensions are applied in the intrahippocampal infusion. The aqueous suspension of nano-C60 is prepared following a modified method of Fortner et al. [36].

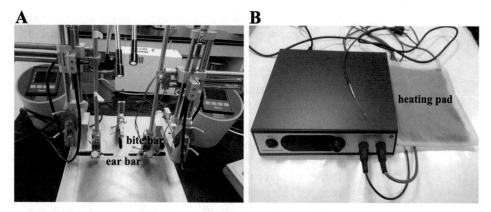

Fig. 1 Stereotaxic apparatus (**a**) and homeothermic monitor (**b**)

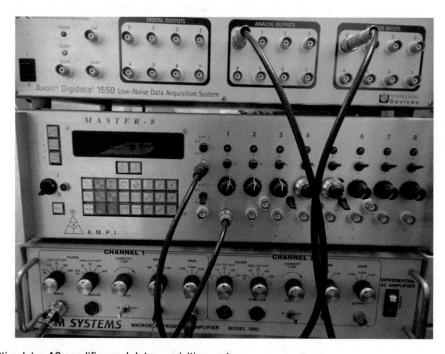

Fig. 2 Stimulator, AC amplifier, and data acquisition system

1. Add 10 mg C60 powder and a magnetic stirring bar to 500 mL previously unopened THF in a brown bottle (*see* **Note 5**). Cover the cap and then wrap the foil around the bottle. Stir the solution by using a magnetic stirrer at an ambient temperature (20–25 °C) in the dark for 24 h (*see* **Note 6**).

2. Remove the insoluble C60 by filtration through a 0.45 μm nylon membrane under vacuum, collect the C60 dissolved in THF in a receiving bottle made of borosilicate glass (*see* **Note 7**).

3. Put a clean and dried magnetic stirring bar in the receiving bottle and vigorously stir the C60/THF solution by using a magnetic stirrer.

4. Add an equal amount of Milli-Q water (pH 6.7; 500 mL) in the vigorously stirred solution at a constant rate of 1 L/min and continue to stir for 1–10 min until well mixed (*see* **Note 8**).

5. Transfer the C60/THF/water solution to a rotatory bottle in the constant temperature bath pan. Remove the THF through evaporation at 55 °C and the majority of water at 80 °C by using a rotary evaporator (*see* **Note 9**). The amount of the resulted Nano C60 aqueous is about 50 mL. Transfer the resulted Nano C60 in a 50 mL conical centrifuge tube, wrapped by foil and stored at 4 °C.

6. Determine the accurate concentration of obtained nano-C60 aquenous by HPLC [36]. Characterize the particles by transmission electron microscopy (TEM), performed on a JEOL-2010 high-resolution transmission electron microscopy (HRTEM, Tokyo, Japan) at 200 kV. The particle size is about 100 nm.

3.2 Stimulating and Recording Electrode Insertion

1. Anesthetize the SD rats with 1.8 g urethane per kg of body weight through intraperitoneal injection (10% urethane, 1.8 g/kg, i.p.) and provide supplemental injections as needed to maintain surgical level of anesthesia (*see* **Note 10**).

2. 20–30 min after anesthesia, place the rat in a stereotaxic apparatus. Insert each ear bar into the animal's ear cannals, the animals teeth are positioned on bite bar (tooth holder) and the nose is held firmly in the nose brace (*see* **Note 11**). Keep the head fixed and restrict movement any side (*see* **Note 12**) (Fig. 3a). Insert an automatic heating pad under the body and monitor the rectal temperature by a rectal thermometer. Maintain the rectal temperature at 37 ± 0.5 °C (*see* **Note 13**).

3. Clip the hair carefully from surgical site at dorsum of the rat skull with a scissor and avoid skin damage. Remove the loose hair by using a cotton swab soaked with 70% ethanol.

4. Incise the skin along the midline of the scalp (2–3 cm in length, start from rostral to caudal) by using ophthalmic scissor and forcep. The skin is gently moved aside to expose the skull surface and clean the skull with cotton swab (Fig. 3b).

5. Keep the head in a flat skull position. To achieve the top surface of the skull is flat from bregma to lambda, the micromanipulator in the stereotax is used to measure the three-dimensional coordinates in the *x*, *y*, and *z* axes. Use the lambda as zero point, move the micromanipulator to the bregma and observe the z axis, if the skull is not flat, just raise

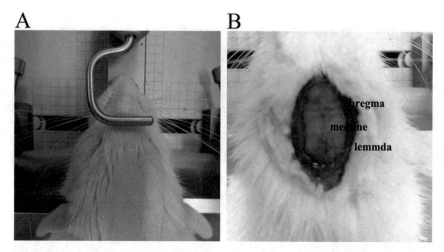

Fig. 3 Head fixation (**a**) and skull exposure (**b**)

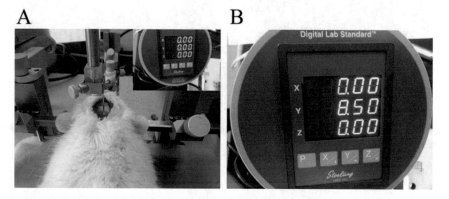

Fig. 4 Set the lambda as zero point (**a**), raise or lower the bite bar until the flat skull position is achieved (**b**)

or lower the bite bar until the flat skull position is achieved (Fig. 4a, b).

6. Use a dental drill to make two small holes (2–3 mm diameter, position: 8.0 mm posterior to bregma, 4.0 mm lateral to the midline, and 4.0 mm posterior to bregma, 2.0 mm lateral to the midline) (Fig. 5a–c) in the skull above DG area of the left dorsal hippocampus (*see* **Note 14**). Clean up the blood with a sterile cotton swab or thin tissue bud.

7. Remove the dura, the thin meningeal of tissue surrounding the brain.

8. Move the stimulating electrode to the positions on the skull surface based on the coordinates. Set midpoint of Bregma as zero point. Lower the electrode slowly until it just touches the brain and record the dorsal-ventral coordinate, set *z* axis as zero (Fig. 6a). Continue to lower the electrode slowly to the calculated position. Place the stimulating electrode in the

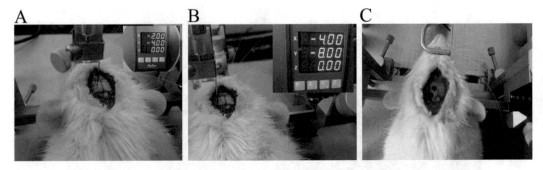

Fig. 5 The position of drilling holes of stimulation (**a**) and recording (**b**), and finished two holes (**c**)

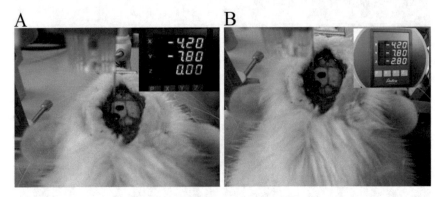

Fig. 6 When stimulating electrode touches the brain, set *z* axis zero (**a**). Place the stimulating electrode in the medial perforant path (**b**)

medial perforant path (MPP, coordinates with the skull surface flat at Bregma: 7.8 mm posterior, 4.2 mm lateral, and 2.8 mm ventral) (Fig. 6b).

9. Set midpoint of bregma as zero point (Fig. 7a). Swing the conjoined electrode/cannula assembly used for recording field excitatory postsynaptic potentials and intrahippocampal infusion in the DG area to the positions on the skull surface based on the coordinates. When recoding electrode touches the brain, the recorded waveform becomes a flat line, set *z* axis as zero (Fig. 7b). Continue to lower the electrode/cannula slowly to the calculated position (coordinates with the skull surface flat: 3.8 mm posterior to bregma, 2.2 mm lateral to the midline, and 3.0–3.5 mm ventral) (Fig. 7c) (see **Note 15**).

10. All instruments should be grounded with wires to avoid 50 Hz disturbance affecting the waveform.

11. Adjust the depth of the conjoined electrode by the *z* axis in the micromanipulator to obtain a maximum response.

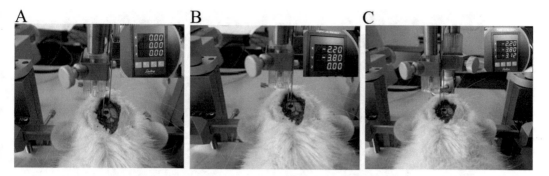

Fig. 7 Set midpoint of bregma as zero point (**a**). When recoding electrode touches the brain, recording waveform becomes a flat line, set *z* axis as zero (**b**). Lower the electrode/cannula slowly to the calculated position (**c**)

3.3 Acute Intrahippocampal Infusion in DG Area

1. Connect the inner cannula (31 G) with Hamilton syringe (10 μL) through PE50 polyethylene tubing. Inject 4 ng/μL Nano C60, ultrapure water, and ACSF respectively to the inner cannula by Hamilton syringe and make two air bubbles in the tubing (*see* **Note 16**). The Ultra Micro 4 pumps at 5 μL/min until a bead of liquid comes out of the inner cannula (*see* **Note 17**).

2. Insert the inner cannula along the guide cannula and the position is about 300 μm above the DG recording site.

3. Set the rate of the ultra-micro pump and infuse the liquids (5 μL) into the DG area at a rate of 250 nL/min for 20 min (*see* **Note 18**) (Fig. 8).

3.4 Input/Output (I/O) Curves Recording

1. Perform the experiments in a room with stable temperature (26–28 °C).

2. Use the same and clean stimulation electrode across the experiments (*see* **Note 19**).

3. Wait for 10 min after drug infusion (*see* **Note 20**).

4. Apply systematic variation of the stimulus current (0.1–1.0 mA) in steps of 0.1 mA and stimulus pulses at 0.05 Hz to evaluate synaptic potency. Each current level records three times. The duration of the pulse is 0.2 ms, and the interval is 20 s (*see* **Note 21**).

5. Measure the slope of the excitatory postsynaptic potential (EPSP) and amplitude of the population spike (PS) by Clampfit 10.4 analysis software to evaluate the strength of a field potential. The EPSP slope (mV/ms) is measured on the rising phase of EPSP by measuring the amplitude at a fixed latency (0.5 ms) from EPSP onset (Fig. 9). The PS amplitude (mV) is the averaging distance from the negative peak to the preceding peak and following positive peak (Fig. 9). Three responses at each current are averaged.

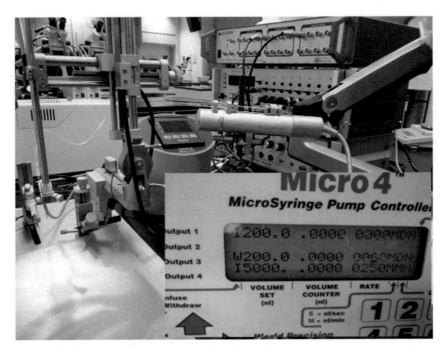

Fig. 8 Set the rate of the ultra-micro pump and infuse the liquids (5 µL) into the DG area at a rate of 250 nL/min for 20 min

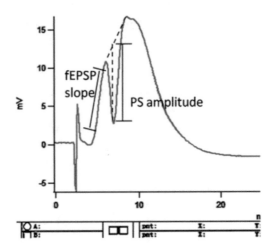

Fig. 9 Schematic diagram of EPSP slope and PS amplitude analysis

3.5 Paired Pulse Facilitation Recording

1. After the I/O curve determination, record the paired-pulse facilitation (PPF) to measure the short-term potentiation.

2. Choose the stimulus intensity to elicit 50% of the maximal PS.

3. Set the stimulus pulses at 0.05 Hz. The duration of each pulse is 0.2 ms. Then deliver pairs of identical stimuli with inter-pulse intervals (IPI) ranging from 10 to 700 ms. The pair of

pulses is given repeatedly every 20 s (0.05 Hz). Repeat three times for each IPI.

4. Measure amplitude of the PS (the distance from the negative peak to the line drawn on the top of the two positive peaks). Three responses are averaged at each IPI. Express the results as a ratio of the amplitude of the second population spike (PS2) to the amplitude of the first population spike (PS1) vs interval (Fig. 10).

3.6   Long-Term Potential Recording

1. After I/O curve and PPF determinations, measure the long-term potentiation (LTP).

2. Stable stimulation response is necessary when the following stimulation and recording are started.

3. Apply the stimulus intensity eliciting 50% of the maximal PS to obtain stable recording of LTP for 20 min with stimulus pulses at 0.05 Hz (baseline) (*see* **Note 22**), followed by high-frequency stimulation (HFS: 11 trains of 11 pulses at 250 Hz separated by 1 s). And then perform the post-tetanic recordings for 1 h with single pulses at 0.05 Hz to assess the induction of LTP (*see* **Note 23**).

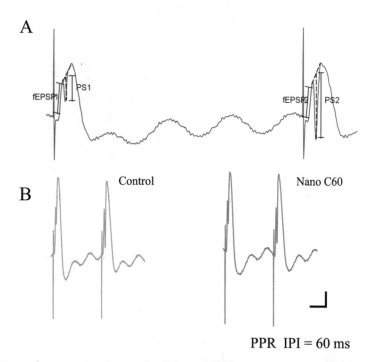

Fig. 10 Expressing PPF as a ratio of the amplitude of the second population spike (PS2) to the amplitude of the first population spike (PS1) vs interval (**a**). The representative sample recordings of PPR (IPI = 60 ms) in the DG area in rats after intrahippocampal infusion of ACSF (Control) and Nano C60 were shown (**b**). Scale bars = 3 mV and 5 ms

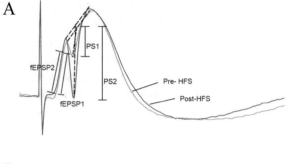

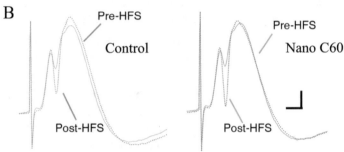

Fig. 11 Waveform before and after HFS (**a**). The representative sample recordings of field potentials in the DG area before and after LTP induction in rats after intra-hippocampal infusion of ACSF (Control) and Nano C60 were shown (**b**). Scale bars = 3 mV and 5 ms

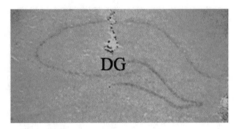

Fig. 12 Recording site of hippocampal DG area

4. Measure the slope of the EPSP (the initial positive slope) and amplitude of the PS before and after HFS (Fig. 11). Average the evoked responses in every 5 min and express the LTP as the percent changes of EPSP and PS from baseline (fEPSP and PS before HFS).

3.7 Recording Site Assay

After eletrophysiological experiments, hippocampal tissues are removed and postfixed in 4% paraformaldehyde overnight at 4 °C, and embedded in paraffin,10 μm thick sections are cut using a vibratome (leica CM 1900, Germany) and processed for HE staining. The recording site in DG area of hippocampus is shown in Fig. 12.

<table>
<tr><td>

3.8 Statistical Analysis

</td><td>

Represent all of the data as mean ± SEM. Use the two-way analysis of variance (ANOVA) with the Turkey test to analyze the date from electrophysiological measurement studies, Probabilities of less than 0.05 were considered significant.

</td></tr>
</table>

4 Notes

1. Make sure the THF is new and unopened before Nano C60 preparation.

2. Thoroughly clean the rotatory bottle. First, use a soft brush to clean any residue or contamination in the bottle in a cleaning solution of detergent and water. Rinse with tap water to remove the detergent. Second, soak the bottle in a 1% HCL or HNO_3 solution for 6–12 h and then rinse with tap water. Third, add the anhydrous alcohol or THF in the bottle followed by shaking and sonication. Lastly, rinse the bottle with deionized water and allow it to dry.

3. Store the ACSF at 4 °C within 3 days.

4. 10% urethane solution is stored at 4 °C within a month.

5. The saturation of C60 in THF is 9 mg/L.

6. The aim of the 24 h stirring time is to get the solution saturated with soluble C60.

7. Upon saturation, a transparent lavender solution is collected through filtration. The solution is stored in the dark for later use. Make sure the receiving bottle is completely dry before using.

8. As the water is added to the solution, the mixed solution is changed to a transparent yellow solution.

9. When there is no liquid condensed in the condenser at 55 °C, gently heat the solution (65, 75, 80 °C) to ensure the THF removal from the solution.

10. Twenty to thirty minutes after anesthesia, the rat is still mobile or responsive to pain, give subsequent doses of urethane at 0.3 g/kg (intraperitoneal).

11. Just rest the ear bar on a solid place, not too far into the head.

12. The pinnae should lie flat on the ear bars, and no head wobble is happened when pressing on the neck. Make sure the rat is still anesthetized.

13. Keep the animals in a stable temperature, so all recording data comes from health and normal condition.

14. Use micromanipulator of stereotaxic apparatus to make sure the position of the hole. Drill the hole carefully and avoid disrupting the brain tissue.

15. Bregma is regarded as an origin of coordinates, frontal margin of coronal suture is regarded as x axis, and sagittal suture is regarded as y axis. Lower the electrode/cannula slowly to get the right position of the z axis.

16. Before injecting the Nano C60 into the hippocampus, we should first infuse the solution through the tube, so we can get a precise amount of solution when pumping the Nano C60 solution.

17. Pump at a speed so the solution can just come through the tube.

18. Pump carefully and slowly, so the solution could be injected completely into the hippocampus.

19. Use 75% ethanol to clean the stimulation electrode and recording electrode.

20. Place the stimulating electrode in the right site, and wait to make sure that the rat is in a stable stage.

21. When given a pulse stimulation, as the intensity increased, the response becomes obvious. The depth of the electrode should be adjusted into the site when we get an obvious and evident response.

22. The intensity of the stimulation is decided in the previous recording, so the baseline is recorded before the high-frequency stimulation.

23. Monitor the animal closely. Make sure the animal has no seizure, otherwise the experiment should be terminated and the data should be excluded.

5   Summary

Local infusion of nanoparticles directly to different areas of hippocampus in intact rats allows the nanoparticles to interact with hippocampal neurons and researchers the chance to investigate the in vivo behavior of nanoparticles in the brain under controlled conditions. The intrahippocampal infused nanoparticles should ideally be dispersed in water or ACSF (no more than 5 μL) to avoid the solvent interference in the experimental outcomes. Furthermore, slow constant-rate infusion may minimize any damage to the hippocampus and be sufficient for the nanoparticles distribution. After nanoparticles infusion, the results of extracellular recording techniques, such as LTP, PPF, and I/O curve, reflect the physiological changes associated with learning and memory, and are easy to measure. Thus, the extracellular recording techniques can be considered powerful tools for quickly and effectively uncovering the effects of nanoparticles on the neural activity and synaptic transmission in hippocampus in the brain, which may further expand our knowledge of the way how nanoparticles function on thoughts, feelings, and behaviors.

Acknowledgment

This work was supported by grants from the National Natural Science Foundation of China (31500813, 31170965), MOST (2012CB932502), Natural Science Foundation of Guangdong Province (2017A030313134), and the fundamental research funds for the central universities (No. 17lgpy106).

References

1. Salata O (2004) Applications of nanoparticles in biology and medicine. J Nanobiotechnol 2(1):3

2. Mout R, Moyano DF, Rana S, Rotello VM (2012) Surface functionalization of nanoparticles for nanomedicine. Chem Soc Rev 41(7):2539–2544

3. Yang K, Feng L, Shi X, Liu Z (2013) Nanographene in biomedicine: theranostic applications. Chem Soc Rev 42(2):530–547

4. Lee HJ, Park J, Yoon OJ et al (2011) Amine-modified single-walled carbon nanotubes protect neurons from injury in a rat stroke model. Nat Nanotechnol 6(2):121–125

5. Park SY, Park J, Sim SH et al (2011) Enhanced differentiation of human neural stem cells into neurons on graphene. Adv Mater 23(36):H263–H267

6. Steketee MB, Moysidis SN, Jin XL et al (2011) Nanoparticle-mediated signaling endosome localization regulates growth cone motility and neurite growth. Proc Natl Acad Sci U S A 108(47):19042–19047

7. Rutecki PA (1992) Neuronal excitability: voltage-dependent currents and synaptic transmission. J Clin Neurophysiol 9(2):195–211

8. Jeggo R, Zhao FY, Spanswick D (2014) Electrophysiological techniques for studying synaptic activity in vivo. Curr Protoc Pharmacol 64:11.11.1–11.11.17

9. Takeuchi T, Duszkiewicz AJ, Morris RG (2014) The synaptic plasticity and memory hypothesis: encoding, storage and persistence. Philos Trans R Soc Lond Ser B Biol Sci 369(1633):20130288

10. Neves G, Cooke SF, Bliss TV (2008) Synaptic plasticity, memory and the hippocampus: a neural network approach to causality. Nat Rev Neurosci 9(1):65–75

11. Buzsaki G, Anastassiou CA, Koch C (2012) The origin of extracellular fields and currents—EEG, ECoG, LFP and spikes. Nat Rev Neurosci 13(6):407–420

12. Jonas P, Lisman J (2014) Structure, function, and plasticity of hippocampal dentate gyrus microcircuits. Front Neural Circuits 8:107

13. Lysetskiy M, Foldy C, Soltesz I (2005) Long- and short-term plasticity at mossy fiber synapses on mossy cells in the rat dentate gyrus. Hippocampus 15(6):691–696

14. Jarrard LE (1993) On the role of the hippocampus in learning and memory in the rat. Behav Neural Biol 60(1):9–26

15. Bliss TV, Gardner-Medwin AR (1973) Long-lasting potentiation of synaptic transmission in the dentate area of the unanaestetized rabbit following stimulation of the perforant path. J Physiol 232(2):357–374

16. Shors TJ, Matzel LD (1997) Long-term potentiation: what's learning got to do with it? Behav Brain Sci 20(4):597–614

17. Lynch MA (2004) Long-term potentiation and memory. Physiol Rev 84(1):87–136

18. Tamura R, Nishida H, Eifuku S et al (2011) Short-term synaptic plasticity in the dentate gyrus of monkeys. PLoS One 6(5):e20006

19. Manahan-Vaughan D, Schwegler H (2011) Strain-dependent variations in spatial learning and in hippocampal synaptic plasticity in the dentate gyrus of freely behaving rats. Front Behav Neurosci 5:7

20. Navarrete M, Araque A (2011) Basal synaptic transmission: astrocytes rule! Cell 146(5):675–677

21. Kroto HW, Heath JR, Obrien SC et al (1985) C-60 – buckminsterfullerene. Nature 318:162–163

22. Zhang QM, Yi JY, Bernholc J (1991) Structure and dynamics of solid C60. Phys Rev Lett 66:2633–2636

23. Marchesan S, Da Ros T, Spalluto G et al (2005) Anti-HIV properties of cationic fullerene derivatives. Bioorg Med Chem Lett 15(15):3615–3618

24. Lyon DY, Adams LK, Falkner JC et al (2006) Antibacterial activity of fullerene water suspensions: effects of preparation method and particle size. Environ Sci Technol 40(14): 4360–4366

25. Zhang Q, Yang WJ, Man N et al (2009) Autophagy-mediated chemosensitization in cancer cells by fullerene C60 nanocrystal. Autophagy 5(8):1107–1117

26. Ryan JJ, Bateman HR, Stover A et al (2007) Fullerene nanomaterials inhibit the allergic response. J Immunol 179(1):665–672

27. Park KH, Chhowalla M, Iqbal Z et al (2003) Single-walled carbon nanotubes are a new class of ion channel blockers. J Biol Chem 278:50212–50216

28. Kubota R, Tahara M, Shimizu K et al (2011) Time-dependent variation in the biodistribution of C(6)(0) in rats determined by liquid chromatography-tandem mass spectrometry. Toxicol Lett 206(2):172–177

29. Moussa F, Pressac M, Genin E et al (1997) Quantitative analysis of C60 fullerene in blood and tissues by high-performance liquid chromatography with photodiode-array and mass spectrometric detection. J Chromatogr B Biomed Sci Appl 696(1):153–159

30. Jin H, Chen WQ, Tang XW et al (2000) Polyhydroxylated C(60), fullerenols, as glutamate receptor antagonists and neuroprotective agents. J Neurosci Res 62:600–607

31. Podolski IY, Podlubnaya ZA, Kosenko EA et al (2007) Effects of hydrated forms of C60 fullerene on amyloid 1-peptide fibrillization in vitro and performance of the cognitive task. J Nanosci Nanotechnol 7(4–5):1479–1485

32. Huang SS, Tsai SK, Chih CL et al (2001) Neuroprotective effect of hexasulfobutylated C60 on rats subjected to focal cerebral ischemia. Free Radic Biol Med 30(6):643–649

33. Lin AM, Chyi BY, Wang SD et al (1999) Carboxyfullerene prevents iron-induced oxidative stress in rat brain. J Neurochem 72(4): 1634–1640

34. Dugan LL, Lovett EG, Quick KL et al (2001) Fullerene-based antioxidants and neurodegenerative disorders. Parkinsonism Relat Disord 7(3):243–246

35. Miao Y, Xu J, Shen Y et al (2014) Nanoparticle as signaling protein mimic: robust structural and functional modulation of CaMKII upon specific binding to fullerene C60 nanocrystals. ACS Nano 8(6):6131–6144

36. Fortner JD, Lyon DY, Sayes CM et al (2005) C60 in water: nanocrystal formation and microbial response. Environ Sci Technol 39(11): 4307–4316

Chapter 12

Using the Whole Cell Patch Clamp Technique to Study the Effect of Nanoparticles in Hippocampal Neurons

Xiaochen Zhang and Zhuo Yang

Abstract

Several types of nanoparticles are considered toxic to the central nervous system. Patch-clamp is one of the most indispensable techniques in the study of neuroscience, especially in the field of neurophysiology. Here, we describe the experimental details using the whole-cell patch clamp mode in the study of nanoparticles in hippocampal slices of the rat, including the generation of giga-seals and cell clamped, recording of neuronal spontaneous discharge and neuronal evoked action potentials, recording of sodium current and potassium current, and recording of glutamatergic synaptic transmission as an example. Our goal is to provide readers with guidelines on how to take the advantage of patch-clamp in the study of nanoparticles in neuroscience.

Key words Nanoparticles, Patch-clamp, Hippocampal slices, Giga-seals

1 Introduction

Nanoparticles are used more and more widely in people's daily life due to the development of nanotechnology [1–3]. Nanoparticles have been used in the electronic industry for many years due to their unique physical and chemical functional properties brought by their nanoscale level sizes [4]. Nowadays, many kinds of nanoparticles have shown extraordinary applications in the biomedical fields such as biosensors, tissue engineering, and drug-delivery systems [5–7]. However, many studies have reported that some widely used nanomaterials may harm human health [8–11]. Therefore, it is necessary to determine the biological toxicity and safety of these nanomaterials.

Neurons are the basic cellular units of the central nervous system (CNS) and are responsible for brain function. They are vulnerable to external stimuli such as hypoxia and ischemia. Nanoparticles can also harm neurons due to their special physical and chemical properties [12–15]. Multiple types of nanoparticles have the ability to pass through biological membranes including

Fidel Santamaria and Xomalin G. Peralta (eds.), *Use of Nanoparticles in Neuroscience*, Neuromethods, vol. 135, https://doi.org/10.1007/978-1-4939-7584-6_12, © Springer Science+Business Media, LLC 2018

the blood-brain barrier (BBB) [16–19]. Thus, nanoparticle toxicity on the CNS should be assessed.

The whole cell patch-clamp technique was developed by Dr. Erwin Neher and Dr. Bert Sakmann in 1975 [20]. This technique can be used in the study of single or multiple ion channels in cells. The giga-seals formed between the microelectrode tips and the cell membrane provide a method to record minute currents. In the voltage clamp configuration it is then possible to record the current through ion channels. In the current clamp mode it is possible to record the voltage activity, including spiking of the cell in response to electrical stimulation and pharmacological manipulations. Many reports have suggested that different types of nanoparticles have various effects on CNS detected by patch-clamp [15, 21]. Here, we introduce the whole-cell mode application in the study of nanoparticles in hippocampal slices of rats as an example to represent some experiment details in patch-clamp techniques for the purpose of providing some assist to neurophysiological workers.

2   Materials

2.1   Major Solution Preparation

1. Artificial cerebrospinal fluid (ACSF) contained in mM: NaCl 125, NaH_2PO_4 1.25, $NaHCO_3$ 25, D-glucose 10, KCl 1.25, $MgCl_2$ 2.0, $CaCl_2$ 2.0, pH 7.4 and maintain temperature at 25 °C (see Note 1).

2. Dissection buffer also called cutting buffer contained in mM: sucrose 220, KCl 2.5, $MgCl_2$ 6, $CaCl_2$ 1, NaH_2PO_4 1.23, $NaHCO_3$ 26, D-glucose 10, pH 7.4, store at 4 °C (see Note 2).

3. The standard pipette solution for current-clamp experiments contained in mM: K-gluconate 135, $MgCl_2$ 2, HEPES 10, EGTA 10, Mg-ATP 2, pH 7.2 (see Note 3).

4. The standard pipette solution for sodium currents contained in mM: CsCl 140, $MgCl_2$ 2, HEPES 10, EGTA 10, Mg-ATP 2, pH 7.2 (see Note 4).

5. The standard pipette solution for potassium currents contained in mM: KCl 140, $MgCl_2$ 2, HEPES 10, Mg-ATP 2, EGTA 10, pH 7.2 (see Note 3).

2.2   Nanoparticle Solution Preparation

1. Stock solution of nanoparticles is prepared in ultrapure water (prepared by purifying deionized water to attain a sensitivity of 18 MΩ cm at 25 °C) and dispersed by ultrasonic vibration for 20 min (see Note 5). Almost all of nanoparticles need this process in experiments, we use MWCNTs as an example here. Dispersion and characterization of the MWCNTs were characterized DLS, which can be seen in our previous study. The TEM image was shown in Fig. 1 [22].

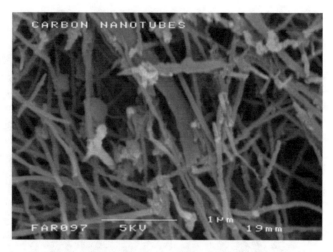

Fig. 1 SEM micrographs of MWCNTs. Reproduced from [22] with permission from [Elsevier]

2. To investigate the effect of different concentrations of nanoparticles, stock solutions are diluted into ACSF to working concentrations.

2.3 Animal Preparation

Healthy male Wistar rats or Sprague Dawley (SD) rats postnatal days 14–18 are used for the experiment.

2.4 Patch Electrodes

Patch electrodes are made of borosilicate glass with tip electrical resistance value of 4–8 MΩ by using multistage micropipette puller (P-97, Sutter Instruments, USA). The specification of borosilicate glass is 1.5 mm × 0.9 mm × 100 mm (see in Sect. 3.2).

2.5 Major Experiment Apparatus

Major experiment apparatus are shown in Table 1.

3 Methods

3.1 Slice Preparation

1. The rats are anesthetized with evaporated ether and decapitated quickly. The scalp of animal is cut with ophthalmic scissors in order to expose the skull. Remove the skull to expose the brain. Cut off the metencephalon and forebrain along the coronal plane. Quickly remove and immerse the remaining part in an ice cold, oxygenated (95% O_2 and 5% CO_2) dissection buffer for 30 s (see Note 6).

2. Use a flat bottom lab spoon taking the freezed brain out of the dissection buffer and place it with the forebrain up and hindbrain down. Absorb the liquid around the brain using a filter paper and fix the brain on the specimen holder. Then, pour the dissection buffer in and make the circumference

Table 1
Major experiment apparatus

Experiment apparatus	Mode	Manufacturer
Vibratome	VT1000S	Leica, Germany
CCD camera	710 M–TI-FW	DVC, USA
Patch-clamp amplifier	EPC-10	HEKA, Germany
Micropipette puller	P-97	Sutter, USA
Micromanipulator	MP-285	Sutter, USA
Stimulator	Master-8	A.M.P.I, Israel
Isolator	ISO-Flex	A.M.P.I, Israel
Microscopy	BX51WI	Olympus, Japan

Table 2
Pulling parameters of P-97

Step	Heat	Pull	Velocity	Delay	Pressure
1	505	0	31	1	500
2	505	0	31	1	500
3	505	0	28	1	500
4	525	20	31	1	500

around the brain moist and cold. After that, hippocampal slices (400 μm) are cut using the vibratome (VT1000S, Leica, Germany, see Notes 7 and 8).

3. Slowly wash the brain slices with ACSF two times in a culture dish. Subsequently, transfer the slices into a holding chamber with ACSF at 37 °C. The ACSF is saturated with 95% O_2 and 5% CO_2. One hour later, the slices can be used for electrophysiological recordings (see Note 9).

3.2 Patch Electrode Preparation

1. Borosilicate glass is used to prepare the patch electrodes. First, immerse borosilicate glass with ethyl alcohol and then ultrasonic vibrate for 15 min to clean up the grease and dust. After that, the cleaned borosilicate glass is put into an oven for desiccation.

2. Patch electrodes are pulled with a multistage micropipette puller (P-97, Sutter Instruments, USA) using the cleaned borosilicate glass. The pulling process has four steps in total. The tip electrical resistance value is 4–8 MΩ (see in Note 10). The detailed parameters for pulling can be seen in Table 2.

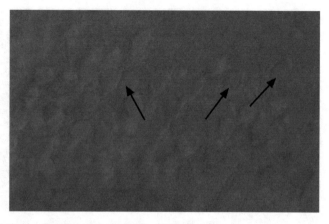

Fig. 2 Hippocampal CA1 pyramidal cell layer of rat. The black arrows denote pyramidal cells

3. Well-prepared patch electrodes need to fill pipette solution as a conducting medium. The solution liquid level is about 1 cm in the patch electrodes. Ensure the liquid level can contact with the silver probe (see Note 11).

3.3 Electro-physiological Recordings

1. The generation of giga-seals and cell clamped.

    (a) One the slice is transferred to a recording chamber (filled with 1 ml ACSF) and placed on the stage of an upright infrared differential interference contrast microscopy (BX51WI, Olympus). A U-shape cover net is used (rc-26 g, Warner, USA) to fix the slice at the bottom of the recording chamber to avoid the slice floating (see Note 12).

    (b) Use a perfusion drug delivery system (DAD-8/16VCP, DL Naturegene Life Sciences, Inc) to delivery nanoparticle solutions (see Sect. 2.2) into a recording chamber (see Note 13).

    (c) Switch on microscopy and make the hippocampus at central of the field. Then, switch on the patch-clamp amplifier, CDD, micromanipulator, and computer in turn.

    (d) Use the low power lens to find CA1 region of hippocampus and change to a high power lens. Find a clear image of cells on a monitor by adjusting the focal length (see in Fig. 2, see in Notes 14 and 15).

    (e) Put the objective cell at the middle of the monitor. Make a prepared patch electrode (filling with pipette solution) mounting on an electrode holder. Use the micromanipulator to move the electrode to central field of vision at low power lens. Then, change to a high power lens and make the tip of electrode on the objective cell.

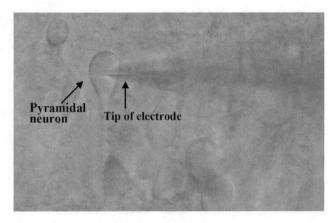

Fig. 3 The forming of giga-seals. Tip of patch electrode is contacted with the pyramidal cell under the whole-cell mode

(f) Before the electrode gets into the ACSF, the pipette solution should be given a positive pressure in order to prevent cell debris or contaminant ACSF blocking the electrode. Use the micromanipulator to move the electrode located in the center of the cell body or a little right shift. Keep pressing the cell membrane gently until the cell membrane has refractive changes and slightly concave. Release the positive pressure and give a negative pressure when the resistance value increases about 1–2 MΩ. The resistance value should increase to GΩ level rapidly to generate the giga-seals (see in Fig. 3). Turn the mode to "whole-cell mode" when the baseline is stable. Set the V-membrane to −70 mV and compensate the capacitor fast (C-fast, see in Notes 16 and 17).

(g) After the giga-seal is generated, give the objective cell a short but strong negative pressure to break the cell membrane. Meantime, compensate the capacitor low (C-low). Keep the resistance value at hundreds of MΩ or GΩ level and record currents.

2. Recording of neuronal spontaneous discharge and neuronal evoked action potentials.

(a) When recording the neuronal spontaneous discharge, change the mode to current clamp mode (C-C mode). A 0 pA current is given. Observe the membrane potential and spontaneous discharge of the objective cell. Recording 5–10 min in the case of normal ACSF perfusion is set as control group. Then, switch the perfusion to ACSF containing nanoparticles for 5 min (see in Note 18). After that, record another 5–10 min as a nanoparticle group. The pipette solution used in this experiment is the standard pipette solution for current-clamp experiments

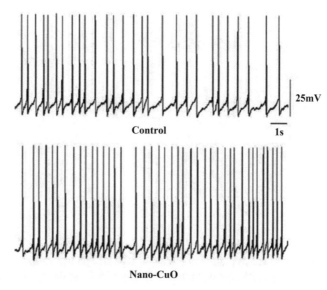

Fig. 4 The representative pattern of the neuronal spontaneous discharge with nano-CuO or without nano-CuO

(see in Sect. 2.1, item 3). The representative pattern can be seen in Fig. 4.

(b) When recording the neuronal evoked action potentials, C-C mode is used. Evoked action potentials include two types: single-action potential (sAP) and repeated firing (RF). To get sAP, a depolarizing current pulse is given, the duration is 5 ms and stimulation intensity is 100 pA. A depolarizing current pulse with a duration of 500 ms and stimulation intensity 50 pA is given to get RF. The recording process is as same as previous (see in Sect. 3.3.2, **step 1**). The representative pattern can be seen in Fig. 5 [23].

3. Recording of sodium current.

(a) In order to record the sodium current (I_{Na}), TEA-Cl (25 mM), 4-AP (3 mM), $CdCl_2$ (0.2 mM) are put in ACSF to block potassium channels and calcium channels. Change the mode to voltage clamp mode (V-C mode) and use the standard pipette solution for sodium currents (see in Sect. 2.1, item 4) as a pipette solution. The sodium currents are measured during depolarizing voltage steps from a holding potential of −70 mV. Give a depolarizing potential from −90 to +50 mV, the increment is 10 mV and pulse width is 20 ms to activate inward currents, which are completely and reversibly blocked by bath application of 1 μM tetrodotoxin (TTX). Therefore, these inward currents are thus considered as voltage-gated sodium currents (I_{Na}). The recording process is as same as previous (see in Sect. 3.3.2, **step 1**).

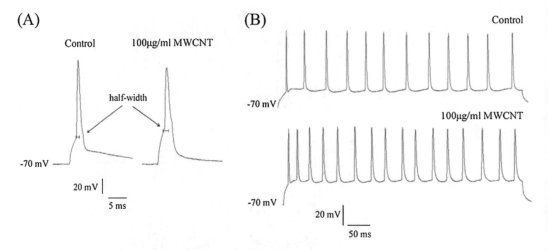

Fig. 5 The representative pattern of effects of MWCNTs on the evoked action potential properties. (**a**) Single-action potentials and (**b**) repeated firing. Reproduced from [23] with permission from [Elsevier]

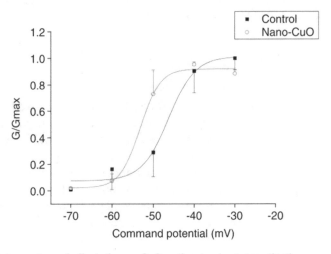

Fig. 6 The representative pattern of effect of nano-CuO on the steady-state activation curves of I_{Na}. Reproduced from [21] with permission from [Wiley]

(b) To research the activation kinetics of I_{Na}, formulas are introduced to calculate. First, currents are converted to conductance (G) using the formula $G = I/(V_m - V_r)$. V_m is membrane potential and V_r is reversal potential. The peak conductance value for each test potential is normalized to G_{max} and plot against the test potential to produce voltage-conductance relationship curves, which are fitted using Boltzmann eq. $G/G_{max} = 1/\{1 + \exp.[(V_m - V_{1/2})/k]\}$, where $V_{1/2}$ is the voltage at which conductance being half-maximal, and k is a slope factor. The activation of I_{Na} is a S-shape. The representative pattern can be seen in Fig. 6 [24].

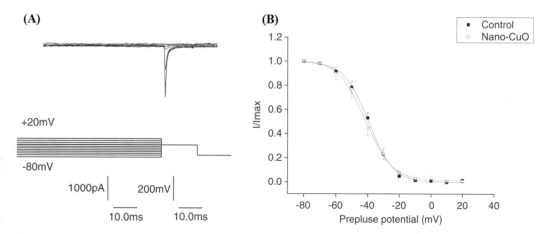

Fig. 7 (**a**) The representative pattern of effect of nano-CuO on the inactivation kinetics of I_{Na}. Peak amplitudes for I_{Na} currents were normalized and plotted vs command potentials. (**b**) The data were fitted with Boltzmann function, Each point represents mean ± SEM ($n = 8$). Reproduced from [21] with permission from [Wiley]

(c) To research the inactivation kinetics of I_{Na}, cells are held at −70 mV and currents are elicited with a 20 ms test pulse to −10 mV, preceded by 30 ms prepulse to potentials between −100 and −30 mV in 5 mV increments. The recording process is as same as previous (see in Sect. 3.3.2, **step 1**). The representative pattern can be seen in Fig. 7 [24].

(d) To study the time course of recovery of sodium channels from inactivation, a double-pulse protocol is applied as follows. Cells are held at −70 mV. A 50 ms conditioning depolarizing pulse of −40 mV is applied to inactivate the sodium channels fully, and then a 50 ms test pulse of −40 mV is applied after a series of −90 mV intervals varying from 2 to 36 ms in 2 ms increments. The peak value of I_{Na} evoked by the conditioning pulse is designated as I_1, while the peak value of I_{Na} evoked by the test pulse is designated as I_2. The ratio of I_2 to I_1 represents the recovery of I_{Na} from inactivation. The plot of I_2/I_1 vs. the duration of the +90 mV intervals is well fitted with a monoexponential function: $I/I_{max} = A + B \exp.(-t/\tau)$, where t is the recovery interval of the conditioning prepulse and τ is the time constant for the recovery from inactivation of I_{Na}. The recording process is same as previous (see in Sect. 3.3.2, **step 1**). The representative pattern can be seen in Fig. 8 [24].

4. Recording of potassium currents.

(a) In order to record potassium currents, TTX (0.001 mM) and CdCl$_2$ (0.2 mM) are put in extracellular solution (ACSF) to block sodium channels and calcium channels. When recording transient outward potassium current (I_A),

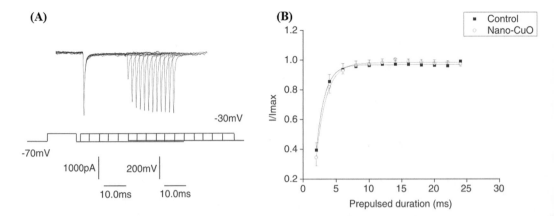

Fig. 8 (a) The representative pattern of effect of nano-CuO on recovery from inactivation of I_{Na}. The peak value of I_{Na} evoked by the conditioning pulse was designated as I_{max}, while the peak value of I_{Na} evoked by the test pulse was designated as I. The ratio of I to I_{max} represents the recovery of I_{Na} from inactivation. **(b)** Each point represents mean ± SEM ($n = 8$). Reproduced from [21] with permission from [Wiley]

the objective cell is clamped at −70 mV, a depolarizing potential from −90 to +50 mV, in which 10 mV increments and 80 ms pulse width are given to activate I_A, at the same time, TEA-Cl (25 mM) is added to block delayed rectifier potassium current (I_K). On the other hand, 4-AP (3 mM) is added to block I_A when recording I_K. The I_K is measured during depolarizing voltage steps from a holding potential of −50 mV. Give a depolarizing potential from −90 to +50 mV, in which increment is 10 mV and pulse width is 300 ms to activate I_K. The pipette solution used here is as same as the standard pipette solution for sodium currents (see in Sect. 2.1, item 5). The recording process is as same as previous (see in Sect. 3.3.2, **step 1**).

(b) To study the effects of nanoparticles on the activation kinetics of I_A and I_K, formulas are introduced to calculate (see in Sect. 3.3.3, **step 2**). The representative pattern can be seen in Fig. 9 [23].

(c) To research the inactivation kinetics of I_A, neurons are held at −70 mV and currents are elicited with an 80 ms test pulse to +50 mV proceeded by 80 ms prepulses to potentials between −110 and +10 mV at 10 mV increments. Peak amplitudes for I_A currents are normalized and plotted vs. prepulse potentials. The curves are well fit with Boltzmann equation described in Sect. 3.3.3, **step 2**. The representative pattern can be seen in Fig. 10 [23].

(d) To study the effects of nanoparticles on the recovery from inactivation of I_A, the cells are held at −70 mV, and an 80 ms conditioning depolarizing pulse of +50 mV is applied

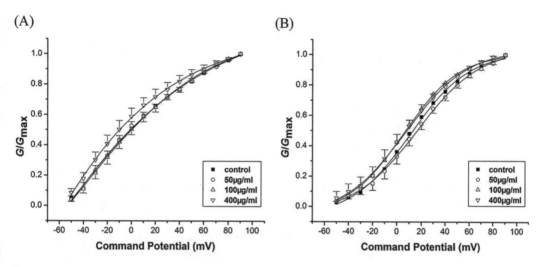

Fig. 9 The representative pattern of effects of MWCNTs on the steady-state activation curves of (**a**) I_A and (**b**) I_K. Each point represents mean ± SEM ($n = 7$). Reproduced from [23] with permission from [Elsevier]

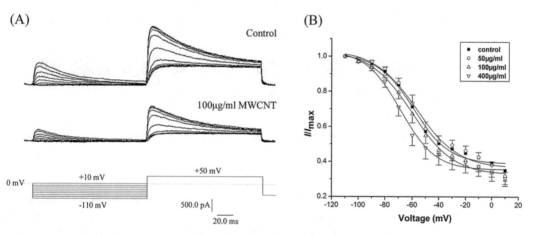

Fig. 10 (**a**) The representative pattern of effects of MWCNTs on the steady-state inactivation of I_A. (**b**) Data were presented as mean ± SEM ($n = 7$). Reproduced from [23] with permission from [Elsevier]

to inactivate the transient outward potassium channels fully, then an 80 ms test pulse of +50 mV is applied after a series of −80 mV intervals varying from 5 to 265 ms. The following steps are the same as Sect. 3.3.3, **step 4**. The representative pattern can be seen in Fig. 11 [23].

5. Recording of glutamatergic synaptic transmission.

   (a) To study spontaneous excitatory postsynaptic current (sEPSC), bicuculline (10 µM) is added into bath to block spontaneous inhibitory currents. The sEPSCs are totally blocked by bath application of CNQX (20 µM) and D-APV (50 µM), confirming that sEPSCs recorded here

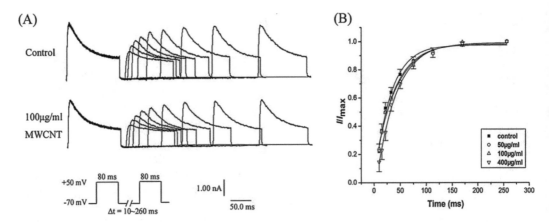

Fig. 11 (a) The representative pattern of effects of MWCNTs on the recovery from inactivation of I_A. **(b)** Data were presented as mean ± SEM ($n = 7$). Reproduced from [23] with permission from [Elsevier]

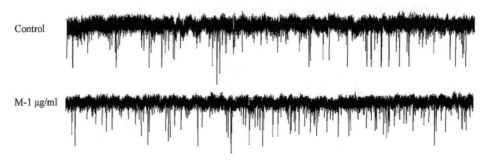

Fig. 12 The representative pattern of effects of MWCNTs on the amplitude and frequency of sEPSCs. Reproduced from [22] with permission from [Elsevier]

are definite glutamate currents. The sEPSCs are recorded under a holding potential of −70 mV in voltage-clamp mode. The pipette solution used here is as same as the standard pipette solution for sodium currents (see in Sect. 2.1, item 4). The representative pattern can be seen in Fig. 12 [22].

(b) To study action potential independent miniature EPSC (mEPSCs), TTX (1 μM) and bicuculline (10 μM) are applied to the bath to eliminate the influence of action potential and inhibitory transmission. The mEPSCs are recorded under a holding potential of −70 mV in voltage-clamp mode. The pipette solution used here is as same as the standard pipette solution for sodium currents (see in Sect. 2.1, item 4). The representative pattern can be seen in Fig. 13 [22].

(c) In order to collect the evoked excitatory postsynaptic currents (eEPSCs), a bipolar stainless-steel stimulating electrode is extra placed on the Schaffer collateral pathway

Fig. 13 The representative pattern of effects of MWCNTs on the amplitude and frequency of mEPSCs. Reproduced from [22] with permission from [Elsevier]

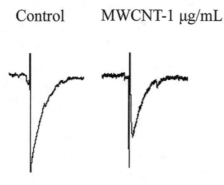

Fig. 14 The representative pattern of effects of MWCNTs on the eEPSCs. Reproduced from [22] with permission from [Elsevier]

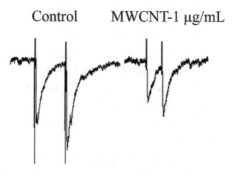

Fig. 15 The representative pattern of effects of MWCNTs on the PPF. Reproduced from [22] with permission from [Elsevier]

100–200 µm away from the objective pyramidal cell. The eEPSCs are elicited by using about eighteen 40 mV constant depolarizing pulse with 20 s interval. In the meantime, bicuculline (10 µM) is added into bath to block spontaneous inhibitory currents. The representative pattern can be seen in Fig. 14 [22].

In order to obtain paired-pulse ratio (PPR), eEPSCs are recorded in response to pairs of a 40 mV stimuli separated by an interval of 100 ms. The PPR is defined as the ratio of amplitude of the second eEPSC to the first one. The representative pattern can be seen in Fig. 15 [22].

4   Notes

1. ACSF is strongly suggested to be prepared freshly each time, if you want to store it, the condition is 0–4 °C. Using NaOH adjusts pH = 7.4.

2. Dissection buffer should be frozen at −20 °C 12 h before use. Break the dissection buffer into pieces and put it into vibratome for the hippocamal slice preparation.

3. Using KOH adjusts pH = 7.2. Filter the standard pipette solution using a 0.2 μm filter (Pall Corporation).

4. Using CsOH adjusts pH = 7.2. Filter the standard pipette solution using a 0.2 μm filter (Pall Corporation).

5. The suspension is stirred on vortex agitator before every use in order to prevent the aggregation of nanoparticles.

6. The whole process should be skilled, quick, and gentle. Do your best to avoid damaging and compressing the brain. To ensure the vitality of brain slices, it is advisable to complete the whole process within 1 min.

7. The transfer process of the brain should be as quickly as possible without damaging brain. Try to make the location of the brain straight in order to obtain uniform and regular slices. After fixing the brain onto a specimen holder, it is better to make the brain immersed into the dissection buffer as quickly as possible to prevent the exsiccation of the brain.

8. Increase the blade lateral vibration frequency and try to slow down the speed of the blade moving forward when using the vibratome in order to decrease the damage to the brain. The speed of slicing is 2 mm/s. For the purpose to decrease the hypoxia time, the whole process should be controlled within 15 min.

9. Control the oxygenated speed and make the generation of bubbles uniform. Avoid the accumulation of bubbles under the slices and turnover of slices.

10. In the actual operation, special attention should be paid to interfere brought by airflow and vibration. Pulling parameters can be appropriately adjusted according to changes of room temperature.

11. Over filling is not desirable, it can increase the ground capacitance of patch electrodes and disturb the recording results. Make sure that there are no bubbles in patch electrodes; otherwise, there will be open circuit.

12. The method of fixing brain slice is cover net method. The U-sharp or circular cover net is made by platinum wire.

13. Perfuse normal ACSF when recording the cells in control group, and perfuse ACSF containing nanoparticles when recording the cells in nanoparticle treatment group. The whole process is done at room temperature.

14. Cells in different layers can be seen by adjusting the focal length. In order to make the image clearly, use the infrared light to replace ordinary light. The differential interference contract images are more stereoscopic on the monitor.

15. Don't choose the surface cells because the cell activity is low. Neither choose the cells in deep layers even though the cells are in a good condition. The reason is that the nanoparticle effects will be delayed on such cells. Choose the cells in the two or three layers as the objective cells. These pyramidal cells should have smooth surface, full form, distinct outline, and good refraction.

16. The generation of giga-seals is the key of whole-cell patch clamp. If the seal resistance cannot reach the $G\Omega$ level, the measurement current should have a big error because of the leakage current. Order to successfully formed giga-seals, the patch clamp table and micromanipulation system must be stable and shockproof, the cells should have high vitality.

17. The patch electrodes are also very important. The tip of patch electrodes should be clean and smooth. Well-prepared patch electrodes should be sealed saved. The pipette solution plays a critical role also. High content of divalent cations solution and slightly hypotonic solution can improve the success rate of forming giga-seals. If the pipette solution contains calcium ions, the phosphate buffer system should be avoided.

18. The exposed time of nanoparticles is according to different nanoparticles used in respective experiment. Make appropriate adjustment in your experiment.

19. Signals are low-pass filtered at 3 k Hz and digitized at 10 kHz. The series resistance is compensated at least 60% in all recording process. The experiments are all conducted at room temperature (22–25 °C).

20. All the recording parameters used in electrophysiological recordings can be appropriately adjusted according to your experiment conditions.

References

1. Becheri A, Dürr M, Nostro PL, Baglioni P (2008) Synthesis and characterization of zinc oxide nanoparticles: application to textiles as UV-absorbers. J Nanopart Res 10(4):679–689

2. Kim G-S, Seo H-K, Godble V, Kim Y-S, Yang O-B, Shin H-S (2006) Electrophoretic deposi-

tion of titanate nanotubes from commercial titania nanoparticles: application to dye-sensitized solar cells. Electrochem Commun 8(6):961–966

3. Das M, Saxena N, Dwivedi PD (2009) Emerging trends of nanoparticles application

in food technology: safety paradigms. Nanotoxicology 3(1):10–18

4. Ahlbom A, Bridges J, De SR, Hillert L, Juutilainen J, Mattsson MO, Neubauer G, Schüz J, Simko M, Bromen K (2008) Possible effects of electromagnetic fields (EMF) on human health--opinion of the scientific committee on emerging and newly identified health risks (SCENIHR). Toxicology 246(2–3): 248–250

5. Shuangyun L, Wenjuan G, Ying GH (2008) Construction, application and biosafety of silver nanocrystalline chitosan wound dressing. Burns 34(5):623–628

6. Kim KJ, Sung WS, Bo KS, Moon SK, Choi JS, Kim JG, Dong GL (2009) Antifungal activity and mode of action of silver nano-particles on Candida albicans. Biometals 22(2):235–242

7. Richardson RT, Thompson B, Moulton S, Newbold C, Lum MG, Cameron A, Wallace G, Kapsa R, Clark G, O'Leary S (2007) The effect of polypyrrole with incorporated neurotrophin-3 on the promotion of neurite outgrowth from auditory neurons. Biomaterials 28(3):513–523

8. Han YG, Xu J, Li ZG, Ren GG, Yang Z (2012) In vitro toxicity of multi-walled carbon nanotubes in C6 rat glioma cells. Neurotoxicology 33(5):1128–1134

9. Mohd Imran K, Akbar M, Govil P, Naqvi SAH, Chauhan LKS, Iqbal A (2011) Induction of ROS, mitochondrial damage and autophagy in lung epithelial cancer cells by iron oxide nanoparticles. Biomaterials 33(5):1477–1488

10. Jing X, Xu P, Li Z, Jie H, Zhuo Y (2012) Oxidative stress and apoptosis induced by hydroxyapatite nanoparticles in C6 cells. J Biomed Mater Res A 100(3):738–745

11. Xu P, Jing X, Liu S, Zhuo Y (2012) Nano copper induced apoptosis in podocytes via increasing oxidative stress. J Hazard Mater 241–242(4):279–286

12. Bardi G, Malvindi MA, Gherardini L, Costa M, Pompa PP, Cingolani R, Pizzorusso T (2010) The biocompatibility of amino functionalized CdSe/ZnS quantum-dot-Doped SiO 2 nanoparticles with primary neural cells and their gene carrying performance. Biomaterials 31(25):6555–6566

13. Wu J, Sun J, Xue Y (2010) Involvement of JNK and P53 activation in G2/M cell cycle arrest and apoptosis induced by titanium diox-ide nanoparticles in neuron cells. Toxicol Lett 199(3):269–276

14. Liu S, Xu L, Zhang T, Ren G, Yang Z (2010) Oxidative stress and apoptosis induced by nanosized titanium dioxide in PC12 cells. Toxicology 267(1):172–177

15. Xu LJ, Zhao JX, Zhang T, Ren GG, Yang Z (2009) In vitro study on influence of nano particles of CuO on CA1 pyramidal neurons of rat hippocampus potassium currents. Environ Toxicol 24(3):211–217

16. Hoet PH, Brüske-Hohlfeld I, Salata OV (2004) Nanoparticles – known and unknown health risks. J Nanobiotechnol 2(1):8

17. Eva OR (2004) Manufactured nanomaterials (fullerene, C60) induce oxidative stress in the brain of juvenile largemouth bass. Environ Health Perspect 112(10):1058–1062

18. Panyala NR, Pena-Mendez EM, Havel J (2008) Silver or silver nanoparticles: a hazardous threat to the environment and human health? J Appl Biomed 6(3):117–129

19. Gao J, Zhang X, Yu M, Ren G, Yang Z (2015) Cognitive deficits induced by multi-walled carbon nanotubes via the autophagic pathway. Toxicology 337:21–29

20. Neher E, Sakmann B. (1975) Voltage-dependence of drug-induced conductance in frog neuromuscular junction. Proc Natl Acad Sci 72 (72):2140–2144

21. Liu Z, Zhang T, Ren G, Yang Z (2011) Nano-Ag inhibiting action potential independent glutamatergic synaptic transmission but increasing excitability in rat CA1 pyramidal neurons. Nanotoxicology 6(4):414–423

22. Chen T, Yang J, Zhang H, Ren G, Yang Z, Zhang T (2014) Multi-walled carbon nanotube inhibits CA1 glutamatergic synaptic transmission in rat's hippocampal slices. Toxicol Lett 229(3):423–429

23. Chen T, Yang J, Ren G, Yang Z, Zhang T (2013) Multi-walled carbon nanotube increases the excitability of hippocampal CA1 neurons through inhibition of potassium channels in rat's brain slices. Toxicol Lett 217(2):121–128

24. Liu Z, Liu S, Ren G, Tao Z, Zhuo Y (2011) Nano-CuO inhibited voltage-gated sodium current of hippocampal CA1 neurons via reactive oxygen species but independent from G-proteins pathway. J Appl Toxicol 31(5):439–445

Chapter 13

Comparative Analysis of Neurotoxic Potential of Synthesized, Native, and Physiological Nanoparticles

Arsenii Borysov, Natalia Pozdnyakova, Artem Pastukhov, and Tatiana Borisova

Abstract

The importance of assessing the neurotoxic potential of nanoparticles is underscored by two main factors. From one side, nanoparticles are a perspective matter for use in neurotheranostics, neurosurgery, cancer treatment, and others branches of nanomedicine. From the other side, they are a component of air pollution that is considered to be a potential trigger factor for development of neuropathologies. The novelty of nanoparticle-related research is determined by unexpected physical and chemical properties of nanomaterials that often differ from those in bulk forms. Herein, we performed a comparative analysis of the neuromodulatory effects of synthesized detonation nanodiamonds, carbon dots, nanoparticles from native volcanic ash, and physiological ferritin-based nanoparticles using similar methodological approaches.

Key words Nanoparticles, Carbon dots, Nanodiamonds, Ferum oxide, Environmental-derived particles, Neurotoxicity, Nerve terminals

1 Introduction

Nanoparticles, because of their great biotechnological potential, are envisioned to have a wide set of new applications. Many efforts in research and development are focused on the design of nanomaterials with multiple functions, particularly those that can be used in theranostics, a new branch of nanomedicine, that combines both disease treatment and diagnostic modalities [1–3]. Properties of nanoparticles often differ from those in bulk forms, thereby providing unexpected physical and chemical properties. It is clear that a detailed understanding of principles of influence of nanoparticles on cell functioning is of value for further progress in biotechnology and nanomedicine.

1.1 Nanodiamonds

Nowadays, an increased attention is focused on the investigation of new unusual properties of carbon materials that can be implemented in nanoscale devices and biotechnologies. Nanodiamonds

Fidel Santamaria and Xomalin G. Peralta (eds.), *Use of Nanoparticles in Neuroscience*, Neuromethods, vol. 135,
https://doi.org/10.1007/978-1-4939-7584-6_13, © Springer Science+Business Media, LLC 2018

are one of the most investigated nano-sized particles due to their unique physical and chemical features, e.g., excellent mechanical and optical properties, high surface areas, and tunable surface structures [4–6]. They are mainly composed of carbon sp^3 structures in the core, with sp^2 and disorder/defect carbons on the surface [6]. Nanodiamonds are usually produced by the following methods, using high temperature/high pressure or detonation, laser ablation, and plasma-enhanced chemical vapor deposition [4–9]. Surface properties of nanodiamonds allow modification and conjugation of a variety of biofunctional entities for controlled targeted drug delivery, in particular water-insoluble drugs, and better penetration of the drug complex inside cells. Nanodiamonds have unique thermal properties and they have perspectives among other material regarding a wide range of potential applications in tribology, drug delivery, bioimaging and tissue engineering, and also as a filler material for nanocomposites [4, 6, 10]. Nanodiamonds can be used as a drug delivery system for the treatment of malignant brain gliomas [11]. The physical and chemical properties of nanodiamonds open up possibilities for their use in theranostics. Due to the growth/production procedures, a large number of lattice defects exist in the core of nanodiamonds, which form fluorescent color centers. The centers can be excited with almost any excitation wavelength. Emitted fluorescence is stable and the photobleaching is limited, and the defect centers can be enhanced with high-energy beam treatment followed by thermal annealing [12–16]. In general, nanodiamonds are considered nontoxic which makes them well suited to a wide range of biomedical applications [4]. Nanodiamond-related studies mainly focus on the cellular models or micro-organisms, and so there is a need of research using animal models. The interaction of nanodiamonds with animal organs and tissues, circulation in the organism, and nanodiamonds clearance in the animal body has not been systematically studied [6]. Despite the drug delivery potential of nanodiamonds, fundamental mechanisms of their interaction with cells are still not completely understood [17].

1.2 Carbon Dots

Carbon dots comprise a recently discovered class of strongly fluorescent, emission-color-tuning and non-blinking nanoparticles with great analytical and bioanalytical potential [18, 19]. The most popular methods of obtaining carbon dots are the microwave-assisted pyrolysis [20] of organic matter [21] and even of waste natural products, such as coffee grounds [22] and soy milk [23]. Carbon dots can be seen as a highly defected composition of coexisting aromatic and aliphatic regions, the elementary constituents of which are graphene, graphene oxide, and diamond. They are assembled in proportions and with the variations of surface groups that depend on the original material and the conditions of their synthesis. Regarding optical properties, the absorption and emission spectra of carbon dots obtained by various methods are remarkably similar and strongly resemble that observed for other species of nanocarbon family, graphene, and graphene oxide nanoparticles [24]. Fluorescence emission of carbon dots is typically concentrated in the blue and green ranges of spectra and the positions of their band maxima often depend on the wavelength of excitation, suggesting the evidence for the presence of multiple fluorophores. Carbon dots exhibit nanosecond lifetimes and high two-photonic cross-sections [25]. Their quantum yields were already reported on a relatively high level of 30–40% [20, 24, 26]. Bioimaging sensors for in vivo bioanalytical diagnostics have to be nontoxic and biocompatible. In comparison with carbon dots semiconductor nanomaterials (quantum dots) usually contain toxic metal cadmium. Carbon dots comprise nontoxic elements that make them promising bioanalytical tools [18]. Carbon dots have a very low cytotoxicity and they can be internalized to the cells probably by endocytosis [27–29]. The translocation of carbon dots through cellular membrane was found to be temperature dependent and no internalization was observed at +4 °C. Inside the cells, carbon dots are accumulated preferentially in the membrane and the cytoplasm [18, 30, 31]. At present, the literature data on the assessment of toxic properties of carbon dots is very limited presumably because of their recent discovery. The data of Dorcena et al. [32] using HepG2 liver cell lines demonstrated that carbon dots exhibited cytotoxicity at concentrations greater than 0.2 mg/mL, while their derived nanocomposites (carbon dots in poly(lactic-co-glycolic acid) did not demonstrate cytotoxicity at any concentration tested (i.e., 0.02, 0.1, and 0.2 mg/mL). The uniqueness of carbon dots as fluorescent agents and their prospective usage as possible fluorescent markers for biological objects requires detailed analysis of not only their toxic properties in standard tests but also their possible effects on different cellular processes [24].

1.3 Native Volcanic Ash-Derived Particles

Inorganic native dust particles derived from volcanic air fall ash deposit (JSC-1a, and JSC, Mars-1A) from ORBITEC Orbital Technologies Corporation, Madison, Wisconsin are of interest from the toxicological point of view. We have shown that the average size of the particles in the suspension in a standard salt solution after sonication was equal to 1110 ± 67 nm for JSC-1a and 4449 ± 1030 nm for JSC, Mars-1A, and also minor fractions of nanoparticles with the size of 50–60 nm were found in these preparations [33]. Major elemental composition of JSC-1a (in %) is: SiO_2 (46.67), TiO_2 (1.71), Al_2O_3 (15.79), Fe_2O_3 (12.5), FeO (8.17), MnO (0.19), MgO (9.39), CaO (9.9), Na_2O (2.83), K_2O (0.78), P_2O_5 (0.71). Composition JSC (in %) is: SiO_2 (34.5), TiO_2 (3), Al_2O_3 (18.5), Fe_2O_3 (19), FeO (2.5), MnO (0.2), MgO (2.5), CaO (5), Na_2O (2), K_2O (0.5), P_2O_5 (0.7).

1.4 Physiological Ferritin-Based Nanoparticles

The main function of an iron storage protein ferritin is the sequestration of excess iron ions in an innocuous mineral form [34–36]. Despite considerable differences in the amino acid sequences, the overall structure of ferritins of prokaryotic and eukaryotic organisms is highly conserved [35, 37, 38]. Ferritins are composed of 24 subunits, which form a spherical shell with a large cavity where up to 4500 three-valent iron ions can be deposited as compact mineral crystallites resembling ferrihydrite [34, 35, 37–41]. Ferritin cores exhibit superparamagnetic properties, which are inherent to magnetic nanoparticles [42]. Iron ions enter into the core of ferritin through hydrophilic intersubunit channels [35], the average core diameter varies in different tissues from 3.5 nm to 7.5 nm [42, 43]. Ferritin stores cellular iron in a dynamic manner protecting the cell from potential iron-dependent radical damage and allowing the release of the metal according to demand [40]. Kidane et al. [35] have shown that release of iron from ferritin requires lysosomal activity, and when iron is needed, the metal is released from ferritin by lysosomal proteases. Mammalian ferritins are found intracellularly in the cytosol, in the nucleus, the endo-lysosomal compartment, and the mitochondria. Extracellular ferritins are detected in fluids such as serum, synovial, and cerebrospinal fluid (CSF). Mouse serum ferritin is actively secreted by a non-classical pathway involving lysosomal processing [44]. Experiments with intestinal Caco-2 cells indicate that enterocytes possess a ferritin receptor and absorb ferritin via a receptor-mediated pathway [45]. A ferritin receptor is also present on placental membranes [46]. In insects and worms, ferritin belongs to classically secreted proteins that transport iron. Intracellular and extracellular ferritin may play a role in intra- and intercellular redistribution of iron [44, 47].

Ferritin can be an important player in neurodegeneration [40]. Ferritin was transported across endothelial cells by transcytosis, and the mechanism of ferritin transportation was clathrin-dependent, similar to that previously identified for transferrin [48, 49]. Binding of exogenous ferritin to cell surface receptors has been implicated as an important iron delivery pathway in the brain. Receptor of H ferritin was identified on the cell surface of oligodendrocytes that could take up ferritin via receptor-mediated endocytosis [49–51]. Iron delivered by ferritin is the major source of iron for oligodendrocytes [51]. Disturbances in iron delivery to the brain and its regulation may cause abnormal iron distribution, and thus contribute to a variety of neurological disorders. High plasma and CSF ferritin concentrations within the first 24 h from the onset of ischemic stroke were associated with early neurologic deterioration [52]. In neuroblastoma, an increase in serum ferritin has been directly linked to secretion of ferritin by the tumor. Human ferritins were detected in the sera of nude mice transplanted with human neuroblastoma [51, 53]. Ruddell et al. [54] proposed a new role for extracellular ferritin as a proinflammatory signaling molecule in hepatic stellate cells and this function is independent of the iron content of ferritin molecule suggesting that the role of exogenous ferritin may be entirely independent of its classical assignment as an iron storage protein [51]. The threefold increase of the concentration of ferritin is accompanied by a small enhancement in the total iron concentration in Alzheimer disease. This finding may suggest that ferritin in Alzheimer disease has properties independent of its iron core [40]. Therefore, (1) presence of ferritin in the serum and CSF; (2) ability of ferritin to penetrate the blood-brain barrier; (3) existence of the receptors to ferritin on the cell surface; (4) secretion of ferritin from the cells, and also (5) leakage of ferritin from destroyed cells during insult and brain trauma; make it very important endogenous nanoparticles. Alekseenko et al. [55] considered ferritin as model protein shell-coated nanoparticles, which can serve as good tools to investigate possible toxic properties of synthetic metal nanoparticles coated by polymers [47].

1.5 Glutamate and γ-Aminobutyric Acid Transport in Presynaptic Nerve Terminals

Investigation of interaction of nanoparticles with neurons has shown both negative and positive effects [56, 57]. Nanoparticles can kill the cells by three main pathways, that is, reactive oxygen species formation, mechanical damage of intracellular organelles, and an increase in the cytosolic Ca^{2+} concentration [58]. Detailed mechanisms of nanoparticle interaction with the nerve cells are of critical importance for development of new technologies.

Glutamate and γ-aminobutyric acid (GABA) are key excitatory and inhibitory neurotransmitters in the mammalian central nervous system [59]. These neurotransmitters are implicated in many aspects of normal brain functioning. Abnormal glutamate and GABA homeostasis contributes to neuronal dysfunction and it is involved in the pathogenesis of major neurological disorders. Exocytosis, the fusion of synaptic vesicles containing neurotransmitter with plasma membrane, is the main mechanism of neurotransmitter release from presynaptic nerve terminals. Under normal physiological conditions, extracellular GABA and glutamate between episodes of exocytotic release is maintained at a low level preventing continuing activation of neurotransmitter receptors [60]. The Na^+-coupled neurotransmitter transporters, plasma membrane proteins with several transmembrane domains, are key players in the termination of synaptic neurotransmission and mediate uptake of amino acid neurotransmitters into the cytosol. The ambient level of glutamate and GABA between the episodes of exocytotic release is set at a definite range by permanent transporter-mediated turnover of the neurotransmitters across the plasma membrane [61–63]. Glutamate and GABA transporters belong to the different families. The glutamate transporters belong to the SLC1 family, whereas GABA transporters (as well as carriers for the biogenic monoamines and glycine) belong to the SLC6 family.

The transporters use Na^+/K^+ electrochemical gradients across the plasma membrane as a driving force. Then, neurotransmitters are accumulated in synaptic vesicles, which are the acidic compartments of nerve terminals, which store neurotransmitters and release their contents by exocytosis upon stimulation. Active transport of glutamate, acetylcholine, monoamines, GABA/glycine to synaptic vesicles is mediated by special vesicular transporters, which depend on the proton electrochemical gradient $\Delta\mu H^+$ generated by V-ATPase that electrogenically pumps protons into the vesicle interior. It is clear that the lower the proton gradient in synaptic vesicles, the less the accumulation of neurotransmitter in synaptic vesicles, and also will result in a lower exocytotic release efficiency. Dissipation of the proton gradient by bafilomycin A1, a highly specific inhibitor of V-type ATPase, or by the protonophore FCCP causes a decrease in KCl-evoked exocytotic release of glutamate and in the initial velocity of glutamate uptake by synaptosomes [60, 64, 65]. In stroke, cerebral hypoxia/ischemia, hypoglycemia, traumatic brain injury, etc, the development of neurotoxicity is provoked by an increase in the concentration of ambient glutamate. Excessive extracellular glutamate overstimulates glutamate receptors initiating an excessive entry of calcium through, mainly, N-metyl-D-aspartate ionotropic receptors; this causes excitotoxicity, neuronal injury, and cell death.

Our research is focused on the comparative analysis of neuroactive properties of carbon nanoparticles, native nanoparticles of volcanic ash, and physiological ferritin-based nanoparticles. Neurotoxic effects can be assessed at various levels of nervous system organization. Our research was conducted at the neurochemical level, where an agent might alter the flow of ions across the cellular membranes and block an uptake of the neurotransmitters in nerve terminals (according to *Guidelines for Neurotoxicity Risk Assessment* of US Environmental Protection Agency, 1998, based on paragraph 3. *Hazard Characterization*: 3.1.2. *Animal Studies*, 3.1.2.3. *Neurochemical Endpoints of Neurotoxicity*; 3.1.3.4. In Vitro *Data in Neurotoxicology*).

2   Effects of Different Nanoparticle Types on the Functioning of High-Affinity Na⁺-Dependent Neurotransmitter Transporters in Nerve Terminals

The experiments were carried out in the suspension of nerve terminals isolated from rat brain cerebral hemispheres (synaptosomes). They retain all characteristics of intact nerve terminals, that is, the ability to maintain membrane potential, accomplish uptake and transporter-mediated release of glutamate, exocytosis, endocytosis, etc. Synaptosomes are one of the best systems to explore the relationship between the structure of a protein, its biochemical and cell-biological properties, and physiological role [66]. In the experiments, all nanoparticles were added to synaptosomal suspension 5 min before starting the measurements of transporter-mediated uptake of L-[^{14}C]glutamate and [^{3}H]GABA, or the ambient level of these neurotransmitters (see the next section), and so the acute effects of nanoparticles were analyzed. Concentration ranges within which the nanoparticles were used in the experiments were following: nanodiamonds (0.05–1.0 mg/mL), carbon dots (0.04–0.8 mg/mL), JSC-1a (0.5–2.0 mg/mL), JSC Mars-1A (0.5–2.0 mg/mL), and ferritin (0.008–0.2 mg/mL). Before the experiments with synaptosomes, water suspension of nanoparticles (except ferritin) was subjected to ultrasound treatment at 22 kHz for 1 min.

2.1 Effect of the Nanoparticles on Transporter-Mediated L-[^{14}C] Glutamate Uptake by Nerve Terminals

Influence of the nanoparticles on the initial rate of uptake and accumulation of L-[^{14}C]glutamate by synaptosomes was assessed. The results of comparative analysis were summarized in Table 1 and Fig. 1. As shown in Table 1 and Fig. 1, JSC-1a and JSC Mars-1A were almost inert regarding glutamate transporter functioning. In contrast, significant inhibition of the initial rate of uptake and accumulation of L-[^{14}C]glutamate by nerve terminals was registered in the presence of nanodiamonds, carbon dots, and ferritin. Thus, we observed that the latest nanoparticles with different efficiency were able to inhibit L-[^{14}C]glutamate uptake in synaptosomes.

2.2 Effect of the Nanoparticles on Transporter-Mediated [^{3}H]GABA Uptake by Nerve Terminals

The results of comparative analysis of the effects of the nanoparticles on the initial velocity of uptake and accumulation of [^{3}H]GABA by synaptosomes were summarized in Table 2 and Fig. 2. As shown in Table 2 and Fig. 2, JSC-1a and JSC Mars-1A did not influence GABA transporter functioning and this data is in agreement with glutamate experiments. Significant inhibition of the initial velocity and accumulation of [^{3}H]GABA by nerve terminals

Table 1

Effect of the nanoparticles on transporter-mediated L-[^{14}C]glutamate uptake by nerve terminals

	The initial velocity of L-[^{14}C]glutamate uptake (nmol/min/mg protein)	Accumulation of L-[^{14}C]glutamate for 10 min (nmol/mg protein)
Control	2.8 ± 0.3	9.8 ± 0.5
NDs (1.0 mg/mL)	2.17 ± 0.2	5.48 ± 0.32
CDs (0.4 mg/mL)	1.2 ± 0.1	5.33 ± 0.45
JSC-1a(2.0 mg/mL)	2.76 ± 0.24	9.62 ± 0.48
JSC, Mars-1A (2.0 mg/mL)	2.52 ± 0.14	8.48 ± 0.34
Ferritin(0.08 mg/mL)	1.31 ± 0.3	4.9 ± 0.35

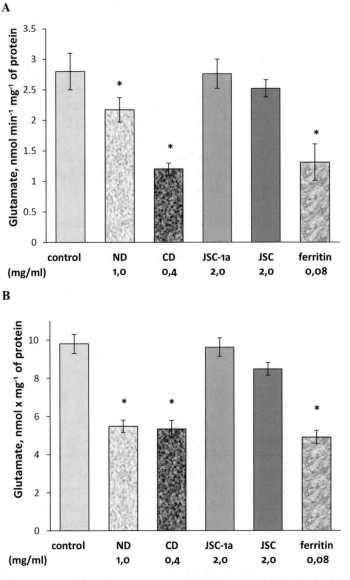

Fig. 1 Comparison of the effects of the nanoparticles on the initial velocity (**a**) and accumulation for 10 min (**b**) of L-[^{14}C]glutamate by synaptosomes. *ND* nanodiamonds, *CD* carbon dots. Data is mean ± SEM of four independent experiments, each of them was performed with different synaptosomal preparations in triplicate. *$p < 0.05$ as compared to control

was shown in the presence of nanodiamonds and carbon dots, which however exhibited different efficiency in this inhibition. Therefore, similarly with the experiments with L-[^{14}C]glutamate, nanodiamonds and carbon dots caused an immediate decrease in the initial velocity of uptake and accumulation of [^{3}H]GABA in synaptosomes.

Table 2
Effect of nanoparticles on transporter-mediated [³H]GABA uptake by nerve terminals

	The initial velocity of [³H]GABA uptake (pmol/min/mg protein)	Accumulation of [³H]GABA for 5 min (pmol/mg protein)
Control	158.6 ± 8.0	557.7 ± 28.1
NDs (1.0 mg/mL)	78.5 ± 8.4	232.5 ± 20.3
CDs (0.4 mg/mL)	65.6 ± 8.4	315.0 ± 22.5
JSC-1a (2.0 mg/mL)	143.3 ± 9.8	470.87 ± 25.3
JSC, Mars-1A (2.0 mg/mL)	149.2 ± 8.3	465.6 ± 33.3

2.3 Influence of Different Nanoparticles on the Ambient Level of L-[¹⁴C]Glutamate and [³H]GABA in the Preparations of Nerve Terminals

Definite level of ambient glutamate and GABA and so proper balance of excitatory/inhibitory signals determines normal synaptic transmission, whereas the changes in this level and misbalance of excitation and inhibition can provoke the development of neurological aberrant conditions. The ambient level of the neurotransmitters is determined mainly by permanent neurotransmitter turnover, that is, balance of transporter-mediated uptake/release and non-transporter tonic release in nerve terminals [61–63]. Nanoparticle-induced decrease in transporter-mediated uptake of glutamate and GABA shown in the previous subsection is expected to result in an increase in the extracellular level of these neurotransmitters in the nerve terminals, similar to that shown by the authors using cholesterol-deficiency models [67, 68]; however, this correlation was not confirmed by a specific centrifuge-induced hypoxia model [69, 70].

Table 3 and Figs. 3 and 4 summarized experimental data concerning the effects of the nanoparticles on the ambient level of L-[¹⁴C]glutamate and [³H]GABA in the preparations of nerve terminals. It was shown that JSC-1a and JSC Mars-1A did not change the ambient level of the neurotransmitters in synaptosomal suspension. A significant increase in this level was registered in the presence of nanodiamonds, carbon dots, and ferritin (was assessed for L-[¹⁴C]glutamate only). The data on ambient neurotransmitters is in agreement with the results from the previous section. It is clear, the lesser uptake there was the higher ambient level was registered.

2.4 Membrane Potential of Nerve Terminals in the Presence of Different Nanoparticles

The key parameter that can significantly alter the functioning of Na^+-dependent transporters of neurotransmitters and their extracellular level is the potential of the plasma membrane of nerve terminals, because Na^+/K^+ electrochemical gradient across the plasma membrane serves as a driving force for glutamate and GABA transporter functioning. Before starting the experiments with

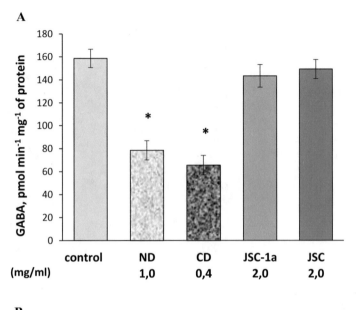

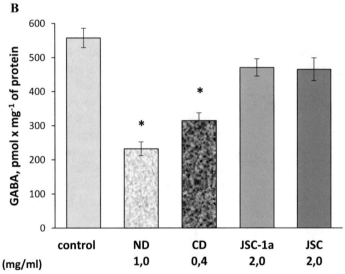

Fig. 2 Comparison of the effects of nanoparticles on the initial velocity (**a**) and accumulation for 5 min (**b**) of [³H]GABA by synaptosomes. *ND* nanodiamonds, *CD* carbon dots. Data is mean ± SEM of four independent experiments, each of them was performed with different synaptosomal preparations in triplicate. *$p < 0.05$ as compared to control

potential-sensitive dye rhodamine 6G, it was assessed whether or not the nanoparticles influenced the fluorescence of rhodamine 6G. No significant changes were found in the emission spectrum of rhodamine 6G in response to the addition of all studied

Table 3

Effects of nanoparticles on the ambient level of L-[¹⁴C]glutamate and [³H]GABA in the preparations of nerve terminals

	The ambient level of L-[^{14}C] glutamate (nmol/mg protein)	The ambient level of [^{3}H]GABA (pmol/mg protein)
Control	0.193 ± 0.013	127.3 ± 5.2
NDs (1.0 mg/mL)	0.383 ± 0.022	167.4 ± 6.24
CDs (0.4 mg/mL)	0.402 ± 0.014	256.9 ± 21.0
JSC-1a (2.0 mg/mL)	0.214 ± 0.029	131.24 ± 11.88
JSC, Mars-1A(2.0 mg/mL)	0.208 ± 0.021	138.5 ± 13.1
Ferritin (0.08 mg/mL)	0.368 ± 0.016	

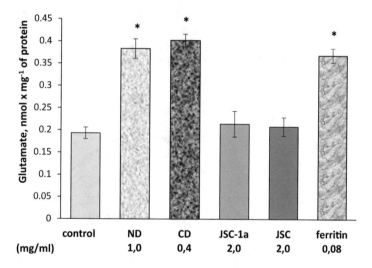

Fig. 3 Comparison of the effects of the nanoparticles on ambient level of L-[^{14}C] glutamate in the preparations of nerve terminals. *ND* nanodiamonds, *CD* carbon dots. Data is mean ± SEM of four independent experiments, each of them was performed with different synaptosomal preparations in triplicate. *$p < 0.05$ as compared to control

nanoparticles at the concentrations within the range 0.05–2.0 mg/mL (Fig. 5). Comparative analysis revealed that among the other studied nanoparticles (at the concentrations used) only nanodiamonds are able to change the membrane potential and cause membrane depolarization (Fig. 6).

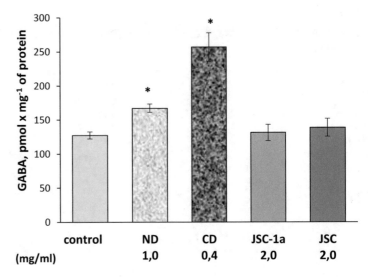

Fig. 4 Comparison of the effects of nanoparticles on ambient level of [³H]GABA in the preparations of nerve terminals. *ND* nanodiamonds, *CD* carbon dots. Data is mean ± SEM of four independent experiments, each of them was performed with different synaptosomal preparations in triplicate. *$p < 0.05$ as compared to control

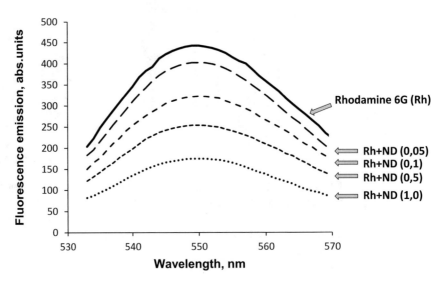

Fig. 5 Fluorescence emission spectra of rhodamine 6G (0.5 μM) in the standard salt solution before and after the application of nanodiamonds (ND) (0.05–1.0 mg/mL). Measurements were performed in dark conditions

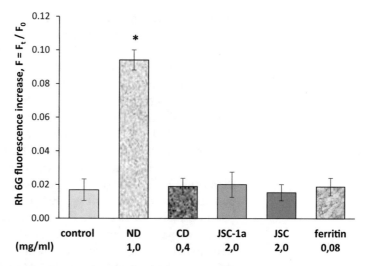

Fig. 6 The fluorescence signal of rhodamine 6G during the application of the nanoparticles. The suspension of synaptosomes was equilibrated with potential-sensitive dye rhodamine 6G (0.5 μM); when the steady level of the dye fluorescence had been reached, the nanoparticles were added to synaptosomes. *ND* nanodiamonds, *CD* carbon dots. Data is mean ± SEM of six independent experiments. *$p < 0.05$ as compared to control (steady level of the dye fluorescence)

3 Comparative Analysis of the Experimental Data on Neuromodulatory Properties of the Nanoparticles

Parameters analyzed in this study, that is, uptake and ambient level of the neurotransmitters are in tight relation with each other [62, 63] consequently, nanoparticle-evoked attenuation of the initial velocity of neurotransmitter uptake is associated with their weak transport to the cytosol of nerve terminals, and so the enhancement of the ambient level. Nanoparticle-induced changes in the above parameters of glutamate and GABA transport can change balance between excitation and inhibition processes [17, 24]. It is so because the maintenance of definite ambient concentration of glutamate and GABA between the episodes of exocytotic release is particularly important for tonic activation of excitatory and inhibitory post- and pre-synaptic receptors of these neurotransmitters. Also, during exocytotic events excess of glutamate and GABA in the synaptic cleft (because of weak neurotransmitter uptake) can be accessible for appropriate receptors for prolonged time intervals. It is well known that glial uptake significantly contributes to the maintenance of appropriate ambient glutamate concentration in the synaptic cleft. It is expected that nanoparticles are able to affect uptake, the ambient level of L-[^{14}C]glutamate, and the plasma membrane potential not only in nerve terminals, but also in glial cells.

We performed experiments with different types of nanoparticles regarding their ability to influence key characteristics of neurotransmission using an analogous methodological approach. The data obtained under similar experimental conditions allow comparing neuromodulatory properties of different nanoparticles. Detonation nanodiamonds, which mainly composed of carbon sp^3 structures in the core with sp^2 and disorder/defect carbons on the surface, revealed the ability to modulate key parameters of glutamate and GABA neurotransmission at concentrations higher than 0.5 mg/mL [17]. Carbon dots consisting of a highly defected composition of coexisting aromatic and aliphatic regions, the elementary constituents of which are graphene, graphene oxide, and diamond [71], were found to possess neuromodulatory properties at concentration 80–800 µg/mL [24]. The next player in comparative analysis was native dust particles derived from volcanic air fall ash deposit. Despite that the average size of the particles in the suspension after sonication was more than three orders larger than those of above-mentioned nanodiamonds and carbon dots, these preparations also possess minor fractions of nanoparticles with the size of 50–60 nm. Interestingly, both JSC-1a and JSC have no effects on uptake and ambient level of the neurotransmitter at a concentration of 2 mg/mL. Only changes in unspecific L-$[^{14}C]$ glutamate binding to synaptosomes in low $[Na^+]$ media and at low temperature in the presence of JSC-1a were found. The fourth compound, neuroactive features of which can be compared to other nanoparticles, is ferritin. Ferritin cores exhibit superparamagnetic properties, the average core diameter varies in different tissues from 3.5 nm to 7.5 nm [34, 42, 43], and this core is coated by 24 specific protein subunits. It was found that ferritin at a concentration of 80 µg/mL (iron content 0.7%) demonstrated neuroactive features.

4 Health and Environmental Issues

The physiological health-related effects of native nanoparticles and carbon dots may be of interest not only from the nanobiotechnological point of view. Accumulating evidence suggests that nano-sized air pollution may have a significant impact on central nervous system in health and disease. This was underlined by the National Institute of Environmental Health Sciences/National Institute of Health convened a panel of research scientists [72]. The critical necessity of the assessment of possible physiological and/or toxic effects of native nanoparticles and carbon dots is underscored by their presence as ubiquitous components of air outdoor pollution components. Also, carbon dots are a component of carbohydrate-contained food caramels, e.g., bread, sugar caramel, corn flakes, and biscuits, where the preparation involves heating of the original material [73, 74]. The facts that carbon nanoparticles are minor nevertheless ubiquitous components of human food chain and the air require considering their health and environmental risks and make this problem multidisciplinary involving medicine, ecology, environmental, and nutrition science [24].

In a mammalian organism, nano-sized particles are efficiently deposited in nasal, tracheobronchial, and alveolar regions due to diffusion and, besides the redistribution between different organs, are transported along sensory axons of the olfactory nerve to the central nervous system [75–79]. Oberdörster et al. [77] showed that intranasally instilled nano-sized particles can target the central nervous system. The most likely mechanism realizes through deposits of these particles on the olfactory mucosa of the nasopharyngeal region of the respiratory tract and their subsequent translocation via the olfactory nerve. In rats, approximately 20% of the nano-sized particles deposited on the olfactory mucosa can move to the olfactory bulb of the brain, which could provide a portal for entry of nano-sized particles into the central nervous system such that they would circumvent the blood-brain barrier [77]. Using TiO_2 nano- and micro-sized particles of the same crystalline structure, Oberdorster et al. [80] demonstrated in a 12-week inhalation experiment on rats resulting in a similar mass deposition of the two particle types in the lower respiratory tract. TiO_2 nano-sized particles were found in the brain of exposed 6-week-old male mice [81]. Kreyling et al. [82] suggested that chronic particle inhalation could trigger or modulate the autonomous nervous system or the release of soluble mediators into circulation and lead to adverse health effects. Besides the brain, it was shown that nano-sized particles can be translocated to, and affect, the liver within 4–24 h post exposure [83, 84]. On the cellular level, nano-sized particles can be transported into the cells through endocytosis [85, 86]. In contrast, Geiser et al. [87] suggested that in vitro uptake of nano-sized particles into the cells does not occur by any of the expected endocytic processes, but rather by diffusion or adhesive interactions. These particles cross cellular membranes by non-phagocytic mechanisms in the lungs and in cultured cells. Within cells they are not membrane bound and have direct access to intracellular proteins, organelles, and DNA that may greatly enhance their toxic potential [87].

As the size of uncoated nanodiamonds and carbon dots varied in different studies from 1 to 10 nm [32, 88], it may be speculated that nanodiamonds and carbon dots because of their small size can enter nerve terminals during synaptic vesicle recycling, i.e., by exo/endocytosis and possibly are accumulated in synaptic vesicles thereby changing their acidification and functional state. However, it cannot be excluded that nanodiamonds and carbon dots somehow interact with the plasma membrane of nerve terminals without penetration through it, and, by changing the membrane potential, affect the functioning of neurotransmitter transporters and normal vesicle recycling, thereby preventing normal exocytotic release of neurotransmitters [17, 24]. It may be speculated that the ability of nanodiamonds and carbon dots to decrease acidification of synaptic vesicles of nerve terminals may also have value for a broad variety of non-neuronal cells, where these nanoparticles can affect proper functioning of acidic compartments.

As stated above, Alekseenko et al. [55] considered ferritin as model nanoparticles with protein shell to study the possible toxicity of metal particles coated by different polymers. Our data indicated that the enlargement of the ambient level of glutamate in ferritin-treated nerve terminals resulted from the inhibition of glutamate transporter functioning. The latest, in turn, was a consequence of a decrease in the proton electrochemical gradient of synaptic vesicles from one side and presumable alterations in the regulation of glutamate transporter activity from the other side. It was shown in additional experiments that apoferritin similarly with ferritin was able to induce changes in transporter-mediated uptake and the ambient level of glutamate in the nerve terminals [47]. This fact should be taken in consideration as the effects of protein shell per se in the ferritin molecule. Taking into consideration (1) the ability of ferritin to pass the brain-blood barrier, (2) its presence in the serum and CSF, (3) the existence of ferritin receptors in the nerve cells, (4) an increase in the concentration of exogenous ferritin in hemorrhagic insult, brain trauma and neuroblastoma cells surroundings, it is suggested that exogenous ferritin can provoke the development of neurologic consequences and pathogenic mechanisms underlying excitotoxicity. Using apoferritin, we found that neurotoxic effect of ferritin on the ambient level and uptake of glutamate was not completely iron-dependent. The current observations showed that ferritin could act in the fields of neuropathology not only as fail of normal iron balance but also as a modulator of glutamate homeostasis in iron-dependent and iron-independent manners [47, 57].

Indirect evidence of neurotoxic action of exogenous ferritin and its ability to increase the ambient glutamate level is a positive correlation between ferritin and glutamate concentrations in the plasma and CSF in stroke patients. Ferritin concentrations >275 ng/mL in the plasma and >11 ng/mL in CSF were significantly related to early neurologic worsening [52]. Serum ferritin levels were higher in amyotrophic lateral sclerosis patients than controls. Patients with a high level serum ferritin had a shorter survival time compared to those with low level serum ferritin [89]. Using a cell culture model of the blood-brain barrier, it was demonstrated that ferritin was transported across endothelial cells by clathrin-dependent endocytosis [48, 49]. It should be noted that ferritin concentration is approximately 30 times higher in the plasma than in CSF. It may be speculated that in neuroferritinopathy excess of synthesized ferritin can disturb ferritin balance between plasma/CSF and exogenous ferritin can provoke neurotoxic consequences through increase in the ambient level of glutamate. The permeability of the blood-brain barrier may be broken down under pathological conditions [90], so the concentration gradient of ferritin between plasma and CSF may be disturbed. Altogether, an increase in ferritin concentration in CSF may consequently enhance the ambient level of glutamate and cause neurological consequences. Also, systemic infections, immunodeficiency virus infection, cancer, liver disease, renal disease are associated with serum ferritin levels greater than 1000 ng/mL. The highest levels occur in patients with sickle cell disease and in the chronically transfused, the extremely high levels of ferritin (mean level of 45,000 ng/mL) suggest reactive hemophagocytic syndrome [51]. During hemorrhagic insult and brain trauma, the intracellular ferritin can freely enter the extracellular space, and so under these pathological conditions it can provoke the development of neurotoxic consequences increasing the ambient level of glutamate. Ferritin may act in a paracrine manner and significantly aggravate the clinical course of these disorders. It can be concluded from our study that the higher intracellular ferritin concentration is, the more significant excitotoxic consequences is after hemorrhagic insult and brain trauma. In neuroblastoma, an increase in serum ferritin has been directly linked to secretion of ferritin by the tumor [51, 53]. It may be speculated that ferritin released from tumor cells can alter glutamate uptake in neighboring nerve terminals and stimulate the development of neurotoxicity [47, 57].

These data indicate that ferritin could be used neither as a model/analogue of polymer-coated magnetic nanoparticles in the assessment of their toxic properties and health risks during diagnostic labeling and manipulation by external magnetic fields, nor as a prototypical nanoparticle for the investigation of possible causes of neurodegeneration associated with exposure to nanoparticles. Still, it could have a biotechnological potential due to its magnetic properties in combination with the ability to change the functional state of nerve terminals [57].

Summarizing, above nanoparticles, that is, carbon dots synthesized from β-alanine [24], detonation nanodiamonds [17], and natural nanocomplex, an iron storage protein ferritin [57] exhibited neuroactivity at concentrations of 0.08 mg/mL, 0.5 mg/mL, and 0.08 mg/mL, respectively.

5 Methods and Materials

5.1 Experiments with Carbon Nanoparticles

Nanodiamonds for our experiments were obtained according to Orel et al. [9] by the method of detonating synthesis using a detonation wave at the explosion of powerful explosive material with negative oxygen balance (trotyl/hexogen, grade TG-40/60). Specific magnetic susceptibility of the preparation consisted of 154.7×10^{-8} m^3/kg. Nanodiamonds synthesized according Orel et al. [9] and their technical characterization were provided by Dr. Olga Leshchenko from the Bakul Institute for Superhard Materials NAS of Ukraine.

Carbon dots were synthesized from β-alanine as described in Borisova et al. [24] and provided by Prof. A. Demchenko and M. Dekaliuk.

5.2 Ethics Statement

Wistar male rats, 100–120 g body weight, were obtained from the vivarium of M.D. Strazhesko Institute of Cardiology, Medical Academy of Sciences of Ukraine. Animals were kept in animal facilities of the Palladin Institute of Biochemistry in accordance with the European guidelines and international laws and policies. They were housed in a quiet, temperature-controlled room (22–23 °C) and were provided with water and dry food pellets ad libitum. Before removing the brain, rats were decapitated. Experimental protocols were approved by the Animal Care and Use Committee of the Palladin Institute of Biochemistry (Protocol from 19/09-2011).

5.3 Isolation of Rat Brain Nerve Terminals (Synaptosomes)

Cerebral hemispheres of decapitated animals were rapidly removed and homogenized in ice-cold 0.32 M sucrose, 5 mM HEPES-NaOH, pH 7.4, and 0.2 mM EDTA. The synaptosomes were prepared by differential and Ficoll-400 density gradient centrifugation of rat brain homogenate according to the method of Cotman [91] with slight modifications [64] (Fig. 7). All manipulations were performed at 4 °C. The synaptosomal suspensions were used in experiments during 2–4 h after isolation. The standard salt solution was oxygenated and contained (in mM): NaCl 126; KCl 5; $MgCl_2$ 2.0; NaH_2PO_4 1.0; $CaCl_2$ 2.0; HEPES 20, pH 7.4; and D-glucose 10. Protein concentration was measured as described by Larson et al. [92].

5.4 L-[^{14}C]glutamate Uptake by Nerve Terminals

Uptake of L-[^{14}C]glutamate by synaptosomes was measured as follows. Synaptosomal suspension (125 µL; of the suspension, 0.2 mg of protein/mL) was pre-incubated in standard salt solution at 37 °C for 10 min, then nanoparticles were added to the synaptosomal suspension and incubated for 5 min. Uptake was initiated by the addition of 10 µM L-glutamate supplemented with 420 nM

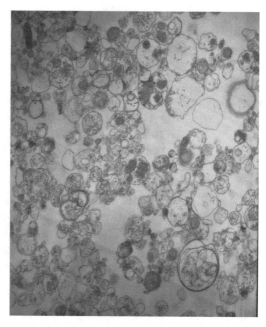

Fig. 7 Electron microscopy of synaptosomes (method of ultrathin sections) [33]

L-[^{14}C]glutamate (0.1 μCi/mL), incubated at 37 °C during different time intervals (1 min and 10 min) and then rapidly sedimented using a microcentrifuge (20 s at 10,000 × g). L-[^{14}C]glutamate uptake was determined as a decrease in radioactivity in aliquots of the supernatant (100 μL) and an increase in radioactivity of the pellet (SDS-treated) measured by liquid scintillation counting with ACS scintillation cocktail (1.5 mL) in a Delta 300 Model 6891(Tracor Analytic, Illinois,USA) scintillation counter [60].

5.5 [^{3}H]GABA Uptake by Nerve Terminals

Synaptosomes were diluted in standard salt solution containing GABA transaminase inhibitor aminooxiacetic acid (100 μM) to minimize the formation of GABA metabolites. Concentration of protein in synaptosomal samples was 200 μg/mL. The samples were pre-incubated at 37 °C for 10 min, then nanoparticles were added to the synaptosomal suspension and incubated for 5 min. Uptake was initiated by the addition of GABA and [^{3}H]GABA (1 μM and 50 nM- 0.1 μCi/mL, respectively). GABA uptake was terminated in different time intervals (1 min and 5 min) by filtering aliquots through a Whatman GF/C filters. After twice washing with 5 mL ice-cold standard saline, the filters were dried, then were drowned in Organic Counting Scintillant (OCS) and counted in a Delta 300 (Tracor Analytic, USA) scintillation counter [93]. Nonspecific binding of the neurotransmitter was evaluated in cooling samples sedimented immediately after the addition of radiolabeled GABA. Each measurement was performed in triplicate.

5.6 Assessment of the Ambient Level of L-[^{14}C]Glutamate in the Preparation of Nerve Terminals

Synaptosomes were diluted in standard saline solution to reach a concentration of 2 mg of protein/mL and after pre-incubation at 37 °C for 10 min they were loaded with L-[^{14}C]glutamate (1 nmol/mg of protein, 238 mCi/mmol) in oxygenated standard saline solution at 37 °C for 10 min. After loading, suspension was washed with 10 volumes of ice-cold oxygenated standard saline solution; the pellet was resuspended in a solution to a final concentration of 1 mg protein/mL and immediately used for release experiments. Synaptosomal suspension (125 μL; 0.5 mg of protein/mL) was pre-incubated for 10 min at 37 °C, then nanoparticles were added and incubated for 5 min and then rapidly sedimented using a microcentrifuge (20 s at 10,000 × g). Release was measured in the aliquots of the supernatants (100 μL) and pellets by liquid scintillation counting with scintillation cocktail ACS (1.5 mL) in a Delta 300 (Tracor Analytic, USA) scintillation counter. The result was expressed as a percentage of total amount of radiolabeled neurotransmitter incorporated [67].

5.7 Assessment of the Ambient Level of [^{3}H]GABA in the Preparation of Nerve Terminals

The synaptosomes were diluted in standard saline solution to 2 mg of protein/mL and after pre-incubation for 10 min at 37 °C were loaded with [^{3}H]GABA (50 nM, 4.7 μCi/mL) in the oxygenated standard saline solution for 10 min. 100 μM aminooxyacetic acid was present throughout all the experiments of [^{3}H]GABA loading and release. After loading, the suspension was washed with 10 volumes of ice-cold oxygenated standard saline solution. The pellet was resuspended in a standard saline solution to obtain protein concentration of 1 mg of protein/mL. Synaptosomes (120 μL of the suspension) were pre-incubated for 5 min with nanoparticles at 37 °C and then rapidly sedimented in a microcentrifuge (10,000 × g, 20 s). [^{3}H]GABA was measured in the aliquots of supernatants (90 μL) by liquid scintillation counting with scintillation cocktail ACS (1.5 mL) in a Delta 300 (Tracor Analytic, USA) scintillation counter and expressed as percentage of a total [^{3}H]GABA accumulated.

5.8 Measurement of Synaptosomal Plasma Membrane Potential (E_m)

Membrane potential was measured using a potentiometric fluorescent dye rhodamine 6G (0.5 μM) based on its potential-modulated binding to the plasma membrane. The suspension of synaptosomes (0.2 mg/mL of final protein concentration) after pre-incubation at 37 °C for 10 min was added to stirred thermostated cuvette. To estimate changes in the plasma membrane potential the ratio (F) as an index of membrane potential was calculated according to Eq. 1:

$$F = F_t / F_0 \qquad (1)$$

where F_0 and F_t are fluorescence intensities of a fluorescent dye in the absence and presence of the synaptosomes, respectively. F_0

was calculated by extrapolation of exponential decay function to $t = 0$.

Rhodamine 6G fluorescence measurements were carried using a Hitachi MPF-4 spectrofluorimeter at 528 nm (excitation) and 551 nm (emission) wavelengths (slit bananodiamonds 5 nm each).

5.9 Statistical Analysis

Results were expressed as mean ± SEM of n independent experiments. Difference between two groups was compared by two-tailed Student's t-test. The differences were considered significant, when $P \leq 0.05$.

5.10 Materials

JSC-1a and JSC were purchased from ORBITEC, Orbital Technologies Corporation, Madison, Wisconsin (U.S.A.). Aminooxyacetic acid, EDTA, HEPES, ferritin, Whatman GF/C filters, analytical grade salts were purchased from Sigma (U.S.A.). Rhodamine 6G were obtained from Molecular Probes (U.S.A). Ficoll 400, L-[^{14}C]glutamate, aqueous counting scintillant (ACS), organic counting scintillant (OCS) were from Amersham (UK). [^{3}H]GABA (γ-[2,3-^{3}H(N)]-aminobutyric acid) was from Perkin Elmer, Waltham, MA, (USA).

Acknowledgments

We would like to thank our colleagues Prof. Alexander Demchenko and Maria Dekaliuk for carbon dots synthesis; Dr. Olga Leshchenko from the Bakul Institute for Superhard Materials NAS of Ukraine for the preparation of nanodiamonds and its technical characterization; Dr. Klaus Slenzka from Jacobs University in Bremen for providing JSC-1a and JSC. This work was supported by Science and Technology Center in Ukraine (#6055); the grants in the frame of Programs of NAS of Ukraine" Molecular and cellular biotechnologies for medicine, industry, and agriculture"; Scientific Space Research; HORIZON 2020, ERA-PLANET. We would like to thank Dr. Sandor Vari for support; Cedars Sinai Medical Center's International Research and Innovation Management Program, the Association for Regional Cooperation in the Fields of Health, Science and Technology (RECOOP HST Association) for their support of our organization as participating Cedars—Sinai Medical Center—RECOOP Research Centers (CRRC).

References

1. Huang P, Lin J, Wang X et al (2012) Light-triggered theranostics based on photosensitizer-conjugated carbon dots for simultaneous enhanced-fluorescence imaging and photodynamic therapy. Adv Mater 24:5104–5110

2. Chen Y-C, Huang X-C, Luo Y-L et al (2013) Non-metallic nanomaterials in cancer theranostics: a review of silica- and carbon-based drug delivery systems. Sci Technol Adv Mater 14:44407

3. Nayak TR, Zhang Y, Cai W (2014) Cancer theranostics with carbon-based nanoplatforms. In: Cancer theranostics. Elsevier, Amsterdam, pp 347–361

4. Mochalin VN, Shenderova O, Ho D, Gogotsi Y (2012) The properties and applications of nanodiamonds. Nat Nanotechnol 7:11–23

5. Man HB, Ho D (2013) Nanodiamonds as platforms for biology and medicine. J Lab Autom 18:12–18

6. Perevedentseva E, Lin Y-C, Jani M, Cheng C-L (2013) Biomedical applications of nanodiamonds in imaging and therapy. Nanomedicine (Lond) 8:2041–2060

7. Butler JE, Sumant AV (2008) The CVD of nanodiamond materials. Chem Vap Depos 14:145–160

8. Dolmatov VY (2001) Detonation synthesis ultradispersed diamonds: properties and applications. Russ Chem Rev 70:607–626

9. Orel VE, Shevchenko AD, Bogatyreva GP et al (2012) Magnetic characteristics and anticancer activity of a nanocomplex consisting of detonation nanodiamond and doxorubicin. J Superhard Mater 34:179–185

10. Chen M, Pierstorff ED, Lam R et al (2009) Nanodiamond-mediated delivery of water-insoluble therapeutics. ACS Nano 3:2016–2022

11. Xi G, Robinson E, Mania-Farnell B et al (2014) Convection-enhanced delivery of nanodiamond drug delivery platforms for intracranial tumor treatment. Nanomedicine 10:381–391

12. Davies G, Hamer MF (1976) Optical studies of the 1.945 eV Vibronic band in diamond. Proc R Soc A Math Phys Eng Sci 348:285–298

13. Davies G, INSPEC (Information service) (1994) Properties and growth of diamond. INSPEC, The Institution of Electrical Engineers, London

14. Gruber A, Dräbenstedt A, Tietz C et al (1997) Scanning confocal optical microscopy and magnetic resonance on single defect centers. Science 276:2012–2014

15. Walker J (1979) Optical absorption and luminescence in diamond. Rep Prog Phys 42:1605–1659

16. Yu S-J, Kang M-W, Chang H-C et al (2005) Bright fluorescent nanodiamonds: no photobleaching and low cytotoxicity. J Am Chem Soc 127:17604–17605

17. Pozdnyakova N, Pastukhov A, Dudarenko M et al (2016) Neuroactivity of detonation nanodiamonds: dose-dependent changes in transporter-mediated uptake and ambient level of excitatory/inhibitory neurotransmitters in brain nerve terminals. J Nanobiotechnology 14:25

18. Esteves da Silva JCG, Gonçalves HMR (2011) Analytical and bioanalytical applications of carbon dots. TrAC Trends Anal Chem 30:1327–1336

19. Li H, Kang Z, Liu Y et al (2012) Carbon nanodots: synthesis, properties and applications. J Mater Chem 22:24230

20. Zhai X, Zhang P, Liu C et al (2012) Highly luminescent carbon nanodots by microwave-assisted pyrolysis. Chem Commun 48:7955

21. Chandra S, Pathan SH, Mitra S et al (2012) Tuning of photoluminescence on different surface functionalized carbon quantum dots. RSC Adv 2:3602

22. Hsu P-C, Shih Z-Y, Lee C-H et al (2012) Synthesis and analytical applications of photoluminescent carbon nanodots. Green Chem 14:917

23. Zhou J, Sheng Z, Han H et al (2012) Facile synthesis of fluorescent carbon dots using watermelon peel as a carbon source. Mater Lett 66:222–224

24. Borisova T, Nazarova A, Dekaliuk M et al (2015) Neuromodulatory properties of fluorescent carbon dots: effect on exocytotic release, uptake and ambient level of glutamate and GABA in brain nerve terminals. Int J Biochem Cell Biol 59:203–215

25. Baker SN, Baker GA (2010) Luminescent carbon nanodots: emergent nanolights. Angew Chem Int Ed Engl 49:6726–6744

26. Dong Y, Wang R, Li H et al (2012) Polyamine-functionalized carbon quantum dots for chemical sensing. Carbon 50:2810–2815

27. Bhunia SK, Saha A, Maity AR et al (2013) Carbon nanoparticle-based fluorescent bioimaging probes. Sci Rep 3:1473

28. Luo PG, Sahu S, Yang S-T et al (2013) Carbon "quantum" dots for optical bioimaging. J Mater Chem B 1:2116

29. Wang X, Qu K, Xu B et al (2011) Microwave assisted one-step green synthesis of cell-

permeable multicolor photoluminescent carbon dots without surface passivation reagents. J Mater Chem 21:2445

30. Cao L, Wang X, Meziani MJ et al (2007) Carbon dots for multiphoton bioimaging. J Am Chem Soc 129:11318–11319

31. Sun Y-P, Zhou B, Lin Y et al (2006) Quantum-sized carbon dots for bright and colorful photoluminescence. J Am Chem Soc 128:7756–7757

32. Dorcéna CJ, Olesik KM, Wetta OG, Winter JO (2013) Characterization and toxicity of carbon dot-poly(lactic-co-glycolic acid) nanocomposites for biomedical imaging. Nano Life 3:1340002

33. Krisanova N, Kasatkina L, Sivko R et al (2013) Neurotoxic potential of lunar and Martian dust: influence on Em, proton gradient, active transport, and binding of glutamate in rat brain nerve terminals. Astrobiology 13:679–692

34. Andrews SC, Arosio P, Bottke W et al (1992) Structure, function, and evolution of ferritins. J Inorg Biochem 47:161–174

35. Kidane TZ, Sauble E, Linder MC (2006) Release of iron from ferritin requires lysosomal activity. Am J Physiol Cell Physiol 291:C445–C455

36. Munro HN, Linder MC (1978) Ferritin: structure, biosynthesis, and role in iron metabolism. Physiol Rev 58:317–396

37. Chasteen ND, Harrison PM (1999) Mineralization in ferritin: an efficient means of iron storage. J Struct Biol 126:182–194

38. Langlois d'Estaintot B, Santambrogio P, Granier T et al (2004) Crystal structure and biochemical properties of the human mitochondrial ferritin and its mutant Ser144Ala. J Mol Biol 340:277–293

39. Ford GC, Harrison PM, Rice DW et al (1984) Ferritin: design and formation of an iron-storage molecule. Philos Trans R Soc Lond Ser B Biol Sci 304:551–565

40. Friedman A, Arosio P, Finazzi D et al (2011) Ferritin as an important player in neurodegeneration. Parkinsonism Relat Disord 17:423–430

41. Linder MC, Kakavandi HR, Miller P et al (1989) Dissociation of ferritins. Arch Biochem Biophys 269:485–496

42. Dubiel SM, Zablotna-Rypien B, Mackey JB (1999) Magnetic properties of human liver and brain ferritin. Eur Biophys J 28:263–267

43. May CA, Grady JK, Laue TM et al (2010) The sedimentation properties of ferritins. New insights and analysis of methods of nanoparticle preparation. Biochim Biophys Acta 1800:858–870

44. Meyron-Holtz EG, Moshe-Belizowski S, Cohen LA (2011) A possible role for secreted ferritin in tissue iron distribution. J Neural Transm 118:337–347

45. Kalgaonkar S, Lönnerdal B (2009) Receptor-mediated uptake of ferritin-bound iron by human intestinal Caco-2 cells. J Nutr Biochem 20:304–311

46. Liao QK, Kong PA, Gao J et al (2001) Expression of ferritin receptor in placental microvilli membrane in pregnant women with different iron status at mid-term gestation. Eur J Clin Nutr 55:651–656

47. Krisanova N, Sivko R, Kasatkina L et al (2014) Excitotoxic potential of exogenous ferritin and apoferritin: changes in ambient level of glutamate and synaptic vesicle acidification in brain nerve terminals. Mol Cell Neurosci 58:95–104

48. Burdo JR, Antonetti DA, Wolpert EB, Connor JR (2003) Mechanisms and regulation of transferrin and iron transport in a model blood-brain barrier system. Neuroscience 121:883–890

49. Fisher J, Devraj K, Ingram J et al (2007) Ferritin: a novel mechanism for delivery of iron to the brain and other organs. Am J Physiol Cell Physiol 293:C641–C649

50. Hulet SW, Heyliger SO, Powers S, Connor JR (2000) Oligodendrocyte progenitor cells internalize ferritin via clathrin-dependent receptor mediated endocytosis. J Neurosci Res 61:52–60

51. Wang W, Knovich MA, Coffman LG et al (2010) Serum ferritin: past, present and future. Biochim Biophys Acta 1800:760–769

52. Dávalos A, Castillo J, Marrugat J et al (2000) Body iron stores and early neurologic deterioration in acute cerebral infarction. Neurology 54:1568–1574

53. Hann HL, Stahlhut MW, Millman I (1984) Human ferritins present in the sera of nude mice transplanted with human neuroblastoma or hepatocellular carcinoma. Cancer Res 44:3898–3901

54. Ruddell RG, Hoang-Le D, Barwood JM et al (2009) Ferritin functions as a proinflammatory cytokine via iron-independent protein kinase C zeta/nuclear factor kappaB-regulated signaling in rat hepatic stellate cells. Hepatology 49:887–900

55. Alekseenko AV, Waseem TV, Fedorovich SV (2008) Ferritin, a protein containing iron nanoparticles, induces reactive oxygen species formation and inhibits glutamate uptake in rat brain synaptosomes. Brain Res 1241:193–200

56. Yang Z, Liu ZW, Allaker RP et al (2010) A review of nanoparticle functionality and toxic-

ity on the central nervous system. J R Soc Interface 7(Suppl 4):S411–S422

57. Borysov A, Krisanova N, Chunihin O et al (2014) A comparative study of neurotoxic potential of synthesized polysaccharide-coated and native ferritin-based magnetic nanoparticles. Croat Med J 55:195–205

58. Fedorovich SV, Alekseenko AV, Waseem TV (2010) Are synapses targets of nanoparticles? Biochem Soc Trans 38:536–538

59. Danbolt NC (2001) Glutamate uptake. Prog Neurobiol 65:1–105

60. Borisova T, Krisanova N, Sivko R, Borysov A (2010) Cholesterol depletion attenuates tonic release but increases the ambient level of glutamate in rat brain synaptosomes. Neurochem Int 56:466–478

61. Borisova T, Borysov A, Pastukhov A, Krisanova N (2016) Dynamic gradient of glutamate across the membrane: glutamate/aspartate-induced changes in the ambient level of L-[(14)C]glutamate and D-[(3)H]aspartate in rat brain nerve terminals. Cell Mol Neurobiol 36:1229–1240

62. Borisova T (2016) Permanent dynamic transporter-mediated turnover of glutamate across the plasma membrane of presynaptic nerve terminals: arguments in favor and against. Rev Neurosci 27:71–81

63. Borisova T, Borysov A (2016) Putative duality of presynaptic events. Rev Neurosci 27:377–383

64. Borisova TA, Krisanova NV (2008) Presynaptic transporter-mediated release of glutamate evoked by the protonophore FCCP increases under altered gravity conditions. Adv Space Res 42:1971–1979

65. Borisova T (2014) The neurotoxic effects of heavy metals: alterations in acidification of synaptic vesicles and glutamate transport in brain nerve terminals. Horizons Neurosci Res 14:89–112

66. Sudhof TC (2004) The synaptic vesicle cycle. Annu Rev Neurosci 27:509–547

67. Borisova T (2013) Cholesterol and presynaptic glutamate transport in the brain. Springer, New York, NY. https://doi.org/10.1007/978-1-4614-7759-4

68. Borisova T, Sivko R, Borysov A, Krisanova N (2010) Diverse presynaptic mechanisms underlying methyl-β-cyclodextrin-mediated changes in glutamate transport. Cell Mol Neurobiol 30:1013–1023

69. Borisova T, Krisanova N, Himmelreich N (2004) Exposure of animals to artificial gravity conditions leads to the alteration of the glutamate release from rat cerebral hemispheres nerve terminals. Adv Space Res 33:1362–1367

70. Borisova TA, Himmelreich NH (2005) Centrifuge-induced hypergravity: [3H]GABA and L-[14C]glutamate uptake, exocytosis and efflux mediated by high-affinity, sodium-dependent transporters. Adv Space Res 36:1340–1345

71. Georgakilas V, Perman JA, Tucek J, Zboril R (2015) Broad family of carbon nanoallotropes: classification, chemistry, and applications of fullerenes, carbon dots, nanotubes, graphene, nanodiamonds, and combined superstructures. Chem Rev 115:4744–4822

72. Block ML, Elder A, Auten RL et al (2012) The outdoor air pollution and brain health workshop. Neurotoxicology 33:972–984

73. Wang J, Sahu S, Sonkar SK et al (2013) Versatility with carbon dots—from overcooked BBQ to brightly fluorescent agents and photocatalysts. RSC Adv 3:15604

74. Sk MP, Jaiswal A, Paul A et al (2012) Presence of amorphous carbon nanoparticles in food caramels. Sci Rep 2:383

75. Kao Y-Y, Cheng T-J, Yang D-M et al (2012) Demonstration of an olfactory bulb-brain translocation pathway for ZnO nanoparticles in rodent cells in vitro and in vivo. J Mol Neurosci 48:464–471

76. Mikawa M, Kato H, Okumura M et al (2001) Paramagnetic water-soluble metallofullerenes having the highest relaxivity for MRI contrast agents. Bioconjug Chem 12:510–514

77. Oberdörster G, Sharp Z, Atudorei V et al (2004) Translocation of inhaled ultrafine particles to the brain. Inhal Toxicol 16:437–445

78. Qingnuan L, Yan X, Xiaodong Z et al (2002) Preparation of 99mTc-C60(OH)x and its biodistribution studies. Nucl Med Biol 29:707–710

79. Wang H, Wang J, Deng X et al (2004) Biodistribution of carbon single-wall carbon nanotubes in mice. J Nanosci Nanotechnol 4:1019–1024

80. Oberdörster G, Ferin J, Lehnert BE (1994) Correlation between particle size, in vivo particle persistence, and lung injury. Environ Health Perspect 102(Suppl 5):173–179

81. Takeda K, Suzuki K, Ishihara A et al (2009) Nanoparticles transferred from pregnant mice to their offspring can damage the genital and cranial nerve systems. J Health Sci 55:95–102

82. Kreyling WG, Semmler-Behnke M, Takenaka S, Möller W (2013) Differences in the bioki-

netics of inhaled nano- versus micrometer-sized particles. Acc Chem Res 46:714–722

83. Bourdon JA, Saber AT, Jacobsen NR et al (2012) Carbon black nanoparticle instillation induces sustained inflammation and genotoxicity in mouse lung and liver. Part Fibre Toxicol 9:5

84. Oberdörster G, Sharp Z, Atudorei V et al (2002) Extrapulmonary translocation of ultrafine carbon particles following whole-body inhalation exposure of rats. J Toxicol Environ Health A 65:1531–1543

85. Garred Ø, Rodal SK, van Deurs B, Sandvig K (2001) Reconstitution of clathrin-independent endocytosis at the apical domain of permeabilized MDCK II cells: requirement for a Rho-family GTPase. Traffic 2:26–36

86. Xia T, Kovochich M, Liong M et al (2008) Cationic polystyrene nanosphere toxicity depends on cell-specific endocytic and mitochondrial injury pathways. ACS Nano 2:85–96

87. Geiser M, Rothen-Rutishauser B, Kapp N et al (2005) Ultrafine particles cross cellular membranes by nonphagocytic mechanisms in lungs and in cultured cells. Environ Health Perspect 113:1555–1560

88. Tao H, Yang K, Ma Z et al (2012) In vivo NIR fluorescence imaging, biodistribution, and toxicology of photoluminescent carbon dots produced from carbon nanotubes and graphite. Small 8:281–290

89. Nadjar Y, Gordon P, Corcia P et al (2012) Elevated serum ferritin is associated with reduced survival in amyotrophic lateral sclerosis. PLoS One 7:e45034

90. Lossinsky AS, Shivers RR (2004) Structural pathways for macromolecular and cellular transport across the blood-brain barrier during inflammatory conditions. Review. Histol Histopathol 19:535–564

91. Cotman CW (1974) Isolation of synaptosomal and synaptic plasma membrane fractions. Methods Enzymol 31:445–452

92. Larson E, Howlett B, Jagendorf A (1986) Artificial reductant enhancement of the Lowry method for protein determination. Anal Biochem 155:243–248

93. Pozdnyakova N, Dudarenko M, Borisova T (2015) New effects of GABAB receptor allosteric modulator rac-BHFF on ambient GABA, uptake/release, Em and synaptic vesicle acidification in nerve terminals. Neuroscience 304:60–70

Chapter 14

A Stoichiometrically Defined Neural Coculture Model to Screen Nanoparticles for Neurological Applications

Stuart I. Jenkins and Divya M. Chari

Abstract

In neuronanotherapeutics, regenerative goals could be achieved by designing therapeutic nanoparticles to target, or evade, specific neural cell types. However, effective screening of candidate particles is hampered by the limited neuromimetic capacity of available biological models. Central nervous system (CNS) tissue is composed of multiple specialized cell types, with dramatically differing particle uptake profiles, dominated by microglia, the ubiquitous immune component of the CNS, resulting in competition for particle uptake. Such dynamics are difficult to monitor in vivo, while in vitro monocultures lack competitive uptake and so predictive value. Available coculture systems are frequently oversimplistic, lack reproducible composition and/or fail to include the immune component. Further, cell-specific culture media are often employed for each neural cell type, leading to differences in protein corona formation around particles, potentially confounding cross-cellular analyses. We describe a novel coculture system that can overcome these limitations, and discuss its utility for assessing uptake, toxicity, and functional efficacy of nanoparticles intended for neurological applications.

Key words Corona, Astrocytes, Microglia, Oligodendrocytes, Oligodendrocyte precursor cells, Glia, Central nervous system, Stealth, Magnetic particles, Time-lapse microscopy

1 Introduction

Nanoparticles, especially magnetic particles, are demonstrating great potential as multifunctional tools for numerous clinical applications, including both in vitro and in vivo engineering (e.g., gene/drug delivery, cell labeling, and hyperthermic ablation of tumors [1–3]). Experimental research models intended to screen novel particles must take account of the unique complexity of central nervous system (CNS) tissue, which is composed of neurons and various specialized glial cell types. Importantly, particle uptake has been shown to vary dramatically in both rate and extent between these CNS cell types [4], particularly the various glia [microglia (immune cells) > astrocytes (homeostatic support cells) > oligodendroglia (cells forming insulating sheaths around nerve fibers)]. The dominance of microglial uptake is of particular

Fidel Santamaria and Xomalin G. Peralta (eds.), *Use of Nanoparticles in Neuroscience*, Neuromethods, vol. 135,
https://doi.org/10.1007/978-1-4939-7584-6_14, © Springer Science+Business Media, LLC 2018

note, as uptake by these cells is likely to result in intracellular trafficking through the lysosomal pathway, and so degradation of the particles and probably any cargo [5–7]. These major differences in particle uptake/processing between the key neural cell populations implies that particle fate in mixed cell populations (e.g., in vivo) will depend on intercellular "competitive uptake dynamics."

Given the differences in uptake profiles and the fact that glia vastly outnumber neurons, glial cell types are likely to dominate particle uptake in vivo. Indeed, neurons typically exhibit very limited levels of particle uptake, with one study showing that glia accounted for 99.5% of particle uptake in vivo [8]. Therefore, glial competitive particle uptake dynamics must be studied in order to predict the overall in vivo fate of different particle designs. As microglia exhibit the greatest avidity for particle uptake, it is essential for models to incorporate this neural immune component. This is true both for cell-targeting strategies, where particles are intended to be taken up by specific cell types (e.g., gene therapy, tumor targeting), and also for uptake evasion strategies, for example where "stealth" particles are intended to release their payload extracellularly. Despite this, many nanoparticle studies rely on the use of purified monocultures which cannot be predictive of particle handling in the CNS.

It is also now widely acknowledged that when introduced into biological media (e.g., culture media, blood) nanoparticles develop a corona of (largely proteinaceous) biomolecules, with corona composition influenced by particle size, nanotopography, and surface chemistry [9–11]. Cellular interactions and particle fate are therefore dictated by the properties of this particle–corona composite, rather than features of the bare particle. Further, corona formation is dependent on both particle- and media-specific properties: the same particle may develop physicochemically different coronas within different media, with potential consequences for cellular interactions. As different neural cells are routinely cultured in cell-specific media, this can confound cross-cellular comparisons: the reliability of comparative data will depend on cells being cultured in the same medium, so standardizing corona formation.

A further point to note is that attempts have been made to engineer various CNS cell types using physicochemically diverse nanoparticles, but there is considerable inconsistency within the literature in respect of experimental methodologies, particle design/characterization, and outcome measures—leaving uncertainty regarding the particle properties required for successful outcomes [2]. Interpreting published data to inform the design of nanoparticles for applications in the CNS will require both thorough nanoparticle characterization, ideally in appropriate experimental conditions, in conjunction with detailed assessment of particle interactions with major neural cell types. For example, as factors such as temperature [12], viscosity [13], and pH can influence corona formation and particle behaviors, it is clear that the utility of particle characterization data depends on studies being performed in biologically-relevant fluids such as pre-warmed culture media, rather than room temperature, unbuffered water or saline [14].

As particles can be synthesized with near-limitless permutations of physicochemical characteristics, **simple, standardized biomimetic screening systems will be required to assess cell–particle interactions**. However, there is a critical lack of biological models that (1) mimic CNS cellular complexity, (2) incorporate the CNS immune component, and (3) can standardize corona formation whilst (4) offering ease of analysis for high throughput screening of therapeutic particles. We describe one such novel coculture model developed by ourselves, with defined proportions of each neural cell type in a standardized culture medium (and so standardized corona), and discuss how it can be used to assess uptake, toxicity and functional efficacy of nanoparticles intended for neurological applications [15]. We also describe associated methods of pre-characterization of nanoparticles used. We focus here on magnetic nanoparticles, with some methods applicable exclusively to iron-based and/or magnetic particles, although many of these methods are applicable to nanoparticles of *any* composition.

2 Materials

The care and use of animals was in accordance with the Animals (Scientific Procedures) Act of 1986 (UK) with approval by the local ethics committee.

2.1 Cell Culture Reagents and Media

1. TrypLE trypsin replacement (Sigma-Aldrich, UK).

2. Sterile syringe filters, 0.22 μm pore size (VWR, UK).

3. Heated rotary shaker (e.g., MaxQ 4000, Thermo Scientific).

4. Primary mixed glial ("parent") cultures were maintained in D10 medium [Dulbecco's modified Eagle's medium (DMEM) supplemented with 10% fetal bovine serum, 2 mM GlutaMAX-I, 1 mM sodium pyruvate, 50 U/mL penicillin, and 50 μg/mL streptomycin].

5. Oligodendrocyte precursor cell (OPC)-specific maintenance medium (OPC-MM): DMEM supplemented with 2 mM GlutaMAX-I, 1 mM sodium pyruvate, 10 nM biotin, 10 nM hydrocortisone, 30 nM sodium selenite, 50 μg/mL transferrin, 5 μg/mL insulin, 0.1% bovine serum albumin, 50 U/mL penicillin, 50 μg/mL streptomycin, 10 ng/mL PDGF-AA, and 10 ng/mL FGF2.

6. Oligodendrocyte-specific maturation medium (Sato): DMEM supplemented with 2 mM GlutaMAX-I, 1 mM sodium pyruvate, 1× N2 supplement (insulin, human transferrin, progesterone, putrescine, selenite [16]), 30 nM thyroxine, 30 nM triiodothyronine, 50 U/mL penicillin, and 50 μg/mL streptomycin.

2.2 Microscopy, Imaging, and Analysis

1. Sphero magnetic particles (mean diameter 360 nm, range 200–390 nm, 15–20% Fe w/v; Spherotech Inc., Illinois, USA), previously characterized in detail [4].

2. Mounting medium with and without DAPI (Vector Laboratories, Peterborough, UK).

3. Monoclonal anti-biotin-FITC (fluorescein isothiocyanate) secondary antibody (clone BN-34; Sigma-Aldrich, Poole, UK).

4. All other secondary antibodies were from Jackson ImmunoResearch Laboratories Inc. (West Grove, PA, USA).

5. Normal goat and donkey sera for blocking (Stratech Scientific Ltd., UK).

6. Confocal laser scanning microscope: Bio-Rad MRC-1024.

7. Time-lapse experiments were performed using an Axiozoom V16 microscope (AxioCam ICm1 camera; Carl Zeiss, Germany) with ZEN software (Blue Ed., v.1.1.1.0, Carl Zeiss, Germany).

8. ImageJ image processing and analysis software was developed by the US National Institutes of Health (NIH) and is freely available [17, 18].

2.3 Particle Characterization

1. Formvar in 1,2-dichloroethane (09823, Sigma-Aldrich).

2. Dynamic light scattering (hydrodynamic diameter) and surface charge (zeta potential) were determined using a Zetasizer Nano ZS (Malvern, UK).

3. Magnetic holder for 1.5 mL tubes (e.g., DynaMag, Life Technologies).

4. Fourier transform infrared spectroscopy (FTIR) was performed by Dr. Paul Roach (University of Loughborough, UK) using a Bruker Alpha system using a DRIFT attachment, with 512 scans being averaged at a resolution of 4 cm^{-1} [15]. Amide I band component peak fitting was performed using previously defined parameters [19], and an in-house program built using Omnic Macros Basic (Thermofisher Scientific). Eigen Vector Solo was used for PCA analysis, with all data being mean-centered.

3 Methods

3.1 Staggered Cell Culture System for Stoichiometrically Defined Cocultures

We have recently demonstrated major differences between glial cell types in terms of nanoparticle uptake/fate, including demonstration of competitive uptake dynamics in cocultures [15]. This was achieved using a stoichiometrically defined coculture system, consisting of defined ratios of microglia, astrocytes, and oligodendroglial cells, derived from the same primary culture. This system offers numerous advantages for assessing intercellular uptake dynamics, including definition of cellular ratios to model specific CNS regions/disease states, standardization of protein coronas, ease of nanoparticle delivery, and amenability to various (including real-time) imaging techniques.

3.1.1 Standardization of Corona Formation Through Universal Medium for All Cell Types

As each cell type is typically grown in cell-specific culture media, it was recognized that differences in corona formation could confound cross-cellular comparisons: each cell type would be interacting with a different particle–corona composite. Further, the development of cocultures would require media supportive of all glial cell types (*see* **Notes 1** and **2**). Therefore, a single gliosupportive medium was developed that could support all of the glial cell types, individually and in coculture, and so standardize corona formation for each test particle [15]. This medium (D10-CM) consists of standard D10 medium supplemented (20%) with conditioned medium derived from mixed glial parent cultures (collected 48 h after last medium change, filter-sterilized to remove cells/debris, stored at 4 °C <1 week).

3.1.2 Mixed Glial Culture Preparation and Maintenance

Primary mixed glial cultures were prepared from dissociated cerebral cortices of Sprague-Dawley rats (postnatal day 1–3 (*see* **Note 3**)). This technique is based on the McCarthy and de Vellis mixed glial culture method [20], with modifications by Chen et al. [21] Terminally anesthetized animals were decapitated, and whole brains were extracted into dissection medium on ice, then olfactory bulbs and mid/hindbrain (rostral to the superior colliculi) removed, leaving the regions covered by the cerebral hemispheres. These were then bisected and inverted (rolled sideways) to expose the ventral tissue. Forceps were used to gently spread the outer margins of the cortex, and scoop out the tissue underlying the cortex. The remaining cortical tissue was then gently rolled along an autoclaved hand towel to remove meninges and blood vessels (*see* **Note 4**), before being collected in fresh, chilled dissection medium. All cortices were then pooled in a clean, dry Petri dish and minced by scalpel (*see* **Note 5**). Fresh D10 was added (~0.5 mL, room temperature) and the minced tissue collected by Pasteur pipette into a fresh universal tube. A further 2 mL of D10 was then used to rinse the Petri dish and collect remaining tissue fragments, before all of the tissue was triturated ~30 times by Pasteur (repeat until lumps are <3 mm (*see* **Note 6**)). Further trituration was performed using a 23 gauge, then 21 gauge, needle and syringe (×3 each) to dissociate the cells, before all of the tissue was sieved into 50 mL tubes (70 μm then 40 μm filters) to remove debris/clumps (*see* **Note 7**). Tubes were centrifuged (1200 rpm, 5 min) and pellets resuspended in D10 (*see* **Note 8**), before performing a trypan blue viability assay/cell count. Cells were seeded in PDL-coated flasks (*see* **Note 9**) at different cellular densities to stagger the onset of confluence and synchronize the time-point at which each cell type is harvested (*see* Fig. 1 and the following section). All cultures were maintained at 37 °C in 5% CO_2/95% humidified air with 50% medium changes every 2–3 days (*see* **Note 10**).

3.1.3 "Staggered Culture" System for Synchronized Derivation of Neural Cell Types

The flasks seeded at the highest density reach confluency (the entire substrate being covered by cells) first (~6 days). Approximately 2 days post-confluency, astrocytes dominate the base layer, with microglia and OPCs being present above the astrocytes. Once this stratification is complete, the high density flasks (denoted as MG1 in Fig. 1) can be shaken (16–20 h, 37 °C, 200 rpm, caps sealed with Parafilm to preserve a 5% CO_2 headspace (*see* **Note 11**)) yielding microglia and OPCs in the medium with negligible numbers of astrocytes, which can be centrifuged, resuspended, and transferred to medium density flasks (MG2). MG1 should then receive fresh D10 and be returned to the incubator to regas (>3 h; 5% CO_2, restores pH). After shaking for two nights, high density flasks (with no microglia and few OPCs) can be washed with PBS (×2), trypsinized and replated as high purity astrocyte cultures (PDL-coated flasks (*see* **Note 12**)). Meanwhile, MG2 can be shaken for 2 h to remove most microglia, which can be transferred to low density flasks (MG3). MG2 should then be regassed before shaking (16–18 h) to yield an OPC fraction for coculture. One day later, MG3 should be confluent, enriched with microglia, but with limited OPCs. Shake MG3 (90 min) to yield a high purity microglial fraction for coculture. In parallel, trypsinize the astrocyte flasks to yield a high purity astrocyte fraction for coculture.

3.1.4 Plating High Purity Cellular Fractions for Stoichiometrically Defined Cocultures

High purity cell fractions were plated as monocultures (6×10^4 cells per cm^2) and 50:50 cocultures (3×10^4 cells per cm^2, each cell type) on PDL-coated coverslips (*see* **Note 13**) in 24-well plates, and maintained in D10-CM [15]. At derivation, OPC fractions were transferred to non-tissue-culture-grade Petri dishes to which microglia (unlike OPCs) readily attach, reducing contamination (30 min, incubated (*see* **Note 14**)). Unattached OPCs were then plated, 24 h ahead of astrocytes and microglia, as they typically required at least 8 h to adhere and regrow processes.

For initial assessment of competitive uptake dynamics, 50:50 cocultures were used to ensure comparable cell numbers were

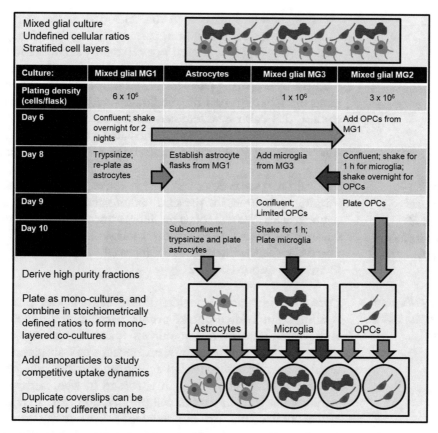

Fig. 1 Schematic illustrating derivation of high purity glial fractions from primary mixed glial cultures, and establishment of stoichiometrically defined cocultures. Plating density is for a 75 cm² flask. A simplified version is available [15]

present for head-to-head analyses [15]. Plating densities have been determined such that sufficient survival factors are available to support OPCs intended for coculture until microglia or astrocytes were added (*see* **Note 15**), while avoiding confluency over the experimental time course. It is possible to plate these cell types in alternative stoichiometries, to model specific CNS regions, developmental stages or disease/injury states [15, 22, 23], but minimum cellular densities necessary for survival, the avoidance of confluency, and differences in proliferation rates must be considered.

3.1.5 Competitive Nanoparticle Uptake Studies

Monocultures and cocultures should be established as described above. To assess competitive uptake dynamics in a neural coculture system, it is important to have control monocultures, e.g., separate high purity astrocyte and microglial cultures treated identically to the test astrocyte–microglia coculture. These high purity cultures will reveal "intrinsic" levels of particle uptake, which can then be compared to the uptake observed in coculture.

Test particles can then be introduced to each culture. Immediately prior to addition, particles should be mixed/dissociated thoroughly (e.g., sonication, vortexing) in culture medium. It is good practice to minimize disturbance to the culture milieu, for example by preparing particles at 10× final concentration and then replacing 10% of the culture medium. In addition to particle uptake studies, this culture system can be employed to assess functional outcomes such as nanoparticle-mediated gene delivery (transfection) efficiency, for example using fluorescent reporter genes.

3.2 Imaging and Analytical Techniques for Cell–Nanoparticle Interactions

The culture system described here is amenable to a wide variety of imaging techniques for the study of interactions between cells and nanoparticles. We will focus on fluorescence microscopy, but these cultures are amenable to both transmission and scanning electron microscopy, allowing analysis of intracellular particle fate and cell membrane responses [4, 24].

3.2.1 Live Cell Dynamic Time-Lapse Fluorescence Microscopy

Dynamic, time-lapse imaging of live cultures can provide data relating to individual uptake events, particle retention over time, and particle fate following mitosis (particle inheritance). These data are important, as long-term tracking of transplanted cells is highly dependent upon label retention, but dilution of particle labeling is known to occur, in vitro and in vivo, with this being attributed at least in part to cell proliferation [25, 26]. However, analysis of fixed cultures can only hint at the outcome of mitosis in a particle-labeled cell. Live imaging can reveal the entire sequence of events, including changes in intracellular organization of particles and separation of particles into the daughter cells.

For time-lapse studies, we use a light and fluorescence microscope with incubation system and dynamic imaging software (*see* Sect. 2). Conditions such as temperature, pH and medium volume must remain within normal ranges for cell culture. We perform these experiments in a chamber that has a heated base and lid, drip-fed with 5% CO_2/95% air (*see* **Note 16**) (to maintain pH), which first passes through an air-stone in a water bottle to be humidified, reducing evaporation. The microscope must be operated in an environment with very limited vibration and changes in temperature or air flow, so that the sample will remain in focus over prolonged periods (days). However, some "drift" of the focal plane often occurs, whether due to physical movement of the sample, or change in light path due to evaporation. Incorporating z stack imaging (multiple focal planes per time-point) can compensate for this drift.

3.2.2 Cell Fixation

Fixation protocols should be selected based on intended imaging applications. Paraformaldehyde (PFA; 4% in PBS) is favored for fluorescence microscopy to limit autofluorescence, while glutaraldehyde is preferred when preparing samples for electron microscopy. We wash cells twice with PBS, then fix at room temperature for 15–20 min, to limit loss of antigens such as NG2 (OPC marker, sensitive to fixation). Samples are then washed twice with PBS, being stored in PBS (4 °C) until further processing.

3.2.3 Perls' Prussian Blue Iron Staining

Perls' stain is of value for assessing iron oxide particle uptake, as dye/fluorophores could leach from particles in culture conditions and be taken up by cells, potentially being misinterpreted as cellular uptake of particles. Cells should be cultured on coverslips, incubated with fluorescent/dyed iron oxide particles, then fixed. Fixed cells can be incubated with 2% potassium ferricyanide in 2% HCl for 10 min (*see* **Note 17**), washed three times with distilled water and then mounted in glycerol-based mounting medium without DAPI. Brightfield and fluorescence micrographs can then be merged to assess the coincidence of fluorescence and iron oxide within cells (Fig. 2).

3.2.4 Immunocytochemistry

Cell-specific markers can be used to confirm the cellular stoichiometries of cocultures. We use double-staining, coincubating coverslips with two primary antibodies then two secondary antibodies. For microscopes with additional filter sets, extra antibodies can be included. Fixed cells should be incubated with blocking solution (room temperature; 30 min), then primary antibody or lectin in blocking solution (Table 1; 4 °C; overnight), washed with PBS (×3), incubated with an appropriate fluorophore-conjugated secondary antibody (room temperature; 2 h; typically 1:200; raised in same species as source of blocking serum, e.g., 5% donkey serum with donkey anti-rabbit secondary) and mounted with mounting medium containing the nuclear stain DAPI.

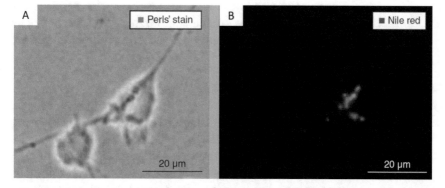

Fig. 2 Perls' staining can be used to confirm the coincidence of iron with other particle labels such as dyes or fluorophores. Here, intracellular iron staining (**A**, blue) is clearly colocalized with Nile Red (**B**), a dye used in the core of Sphero magnetic particles. This demonstrates that cells have taken up intact particles, rather than fluorophores that have leached from the particles. Reproduced with permission from Nanomedicine as agreed by Future Medicine Ltd. [4]

Table 1
Antibodies and protocols for immunocytochemistry

Antibody/marker	Epitopes (marker for cell type)	Product code and supplier	Blocking solution, in PBS	Antibody dilution
Mouse anti-A2B5	Cell surface ganglioside (OPC)	A8229; Sigma-Aldrich, UK	5% serum	1:200
Rabbit anti-GFAP	Glial fibrillary acidic protein (astrocyte)	Z0334; DakoCytomation, UK	5% serum, 0.3% Triton	1:500
Lectin (*Lycopersicon esculentum*, biotin-conjugated)	Specific cell surface sugar groups (microglia)	L0651; Sigma-Aldrich, UK	5% serum	1:200
Rat anti-MBP	Myelin basic protein (oligodendrocyte)	MCA409S; BioRad, formerly AbD Serotec, UK	5% serum, 0.3% Triton	1:200
Rabbit anti-NG2	Chondroitin sulfate proteoglycan (OPC)	AB5320; Millipore, UK	5% serum	1:150
OX42 (mouse anti-CD11b)	Integrin alpha chain membrane protein (microglia)	MCA275; BioRad, formerly AbD Serotec Ltd., UK	5% serum[a]	1:500

OPC oligodendrocyte precursor cell, *PBS* phosphate buffered saline, *Triton* Triton X-100
[a]Pre-permeabilized cells with 1% Triton in PBS, 20 min, room temperature

3.2.5 Z Stack and Confocal Fluorescence Microscopy

Conventional microscopy can show seemingly cell-associated nanoparticles, but leave doubt as to whether these particles are intracellular. By employing z stack or confocal microscopy, the presence of intracellular particles can be confirmed. For example, the series of confocal micrographs in Fig. 3 demonstrates that red fluorescent particles are not visible in focal (z) planes below or above cells, and not in focal planes that include the uppermost and lowermost regions of cell membrane. However, they are clearly identifiable in focal planes between these upper and lower regions of membrane.

3.2.6 Fluorescence Microscopy for Toxicity and Uptake Analyses

We recommend that a minimum of three micrographs and 100 nuclei per culture are assessed for all conditions. Toxicity can be assessed by morphological observations and by comparing proportions of pyknotic nuclei (number of pyknotic nuclei divided by healthy plus pyknotic nuclei), identified as small, intensely stained and often fragmenting. Culture purity and stoichiometry can be determined by assessing the percentage of cells expressing cell-specific markers. The percentage of cells exhibiting particle uptake should be determined by examining micrographs. Control cultures, without particles, should be examined to determine the minimum exposure time that does not detect autofluorescence (for

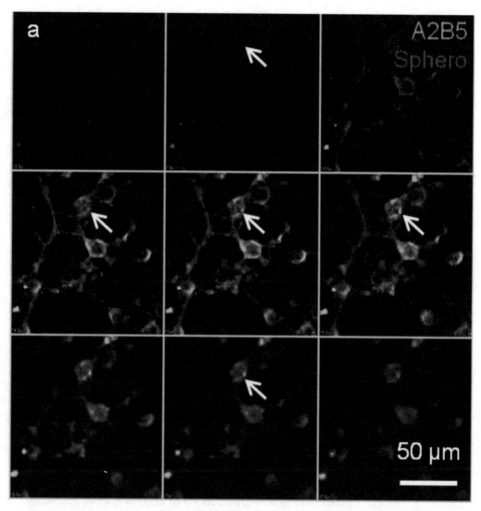

Fig. 3 Confocal fluorescence microscopy can demonstrate intracellular localization of fluorescent nanoparticles. Here, Sphero magnetic particles are shown to be coincident with A2B5 staining (a marker for oligodendrocyte precursors), but not present in focal planes (z axis) above or below the labeled cell (arrow). Adapted from [27].

each filter), and this exposure time should be used for all micrographs. However, the mere demonstration of uptake having occurred does not guarantee that an effective quantity of particles will have been taken up, and attempts should be made to quantify levels of cellular uptake. This is not always a straightforward procedure, and we will now describe a microscopic technique for assessment of relative extent of particle uptake per cell.

3.2.7 Semiquantitative Assessment of Extent of Nanoparticle Uptake

It is useful to attempt to measure the quantity of particles taken up by each cell, as this may indicate levels of drug delivery, or likelihood of achieving sufficient levels for imaging applications. To achieve this, researchers often report measurements of "intracellular" iron content per cell (using colorimetric absorbance assays).

However, these include substantial proportions of extracellular (membrane-bound) particles: reportedly 20% of the iron per cell value for microglia [28], and up to 50% for astrocytes [29]. Such techniques also assume an even distribution between cells and we have shown that considerable heterogeneity exists within glial sub-types in terms of extent of uptake [4], which will mask any hetero-geneity of particle accumulation within a cell population. This is particularly relevant to primary populations (the most likely cell source for transplantation therapies, and the more relevant model of in vivo cellular behaviors) which show considerable heterogene-ity in behavior including particle uptake [4]. This is in contrast to cell lines, which behave in a relatively clonal manner.

So, we have developed a system whereby extent of particle-loading can be assessed semiquantitatively by comparing the area occupied by particles with the average cross-sectional area of a cell nucleus [27]. ImageJ software (NIH) was employed for the follow-ing area measurements and analyses. The areas of 30 nuclei were measured, and an average derived. A circle with this area was then drawn, along with circles occupying 50% and 10% of this area. Each cell in a micrograph was then analyzed by comparing the intracel-lular area occupied by particles (e.g., particle-associated fluores-cence) with these nucleus size guides (either printed or kept on-screen). When numerous disparate particles/clusters were pres-ent, the judgment of the scorer was applied. When particles were gathered in few clusters or a single cluster, they could be rapidly and objectively analyzed by delineating particles/clusters using ImageJ. Uptake for each cell was therefore scored as low (<10% of the area of an average nucleus), medium (10–50%) or high (>50%). Figure 4e, f illustrates data derived using this system to compare uptake of magnetic particles by neural cells, demonstrating that the presence of microglia limits particle uptake by astrocytes and OPCs.

We have since developed a more sophisticated system, reduc-ing observer bias, whereby ImageJ is used to derive *numerical* semiquantitative values for particle accumulation by individual cells [30, 31]. Micrographs of the nanoparticle-relevant color channel (identical exposure times) are converted to grayscale as a

---

Fig. 4 (continued) (D) OPC–microglia coculture showing extensive loading in microglia—note lack of OPC labeling (DAPI+/lectin-unreactive; phase contrast counterpart inset). (E) Bar graph showing proportions of MP-labeled cells/extent of loading in microglia–astrocyte cultures. Proportions of MP-labeled astrocytes were significantly reduced versus monocultures ($^{+++}P < 0.001$) with more astrocytes exhibiting "medium" ($^{*}P < 0.05$) or "high" ($^{*}P < 0.05$) loading, and fewer exhibiting "low" loading ($^{***}P < 0.001$); $n = 3$. (F) Bar graph showing proportions of MP-labeled cells/extent of loading in microglia–OPC cultures. When cocultured with microglia, proportions of MP-labeled OPCs were significantly reduced ($^{+++}P < 0.001$). More OPCs exhibited "medium" ($^{***}P < 0.001$) loading in monocultures than in cocultures ($n = 4$). Reprinted from Nanomedicine: NBM, Jenkins SI, Roach P and Chari DM, Development of a nanomaterial bio-screening platform for neuro-logical applications, 11:77–87, Copyright (2015), with permission from Elsevier [15]

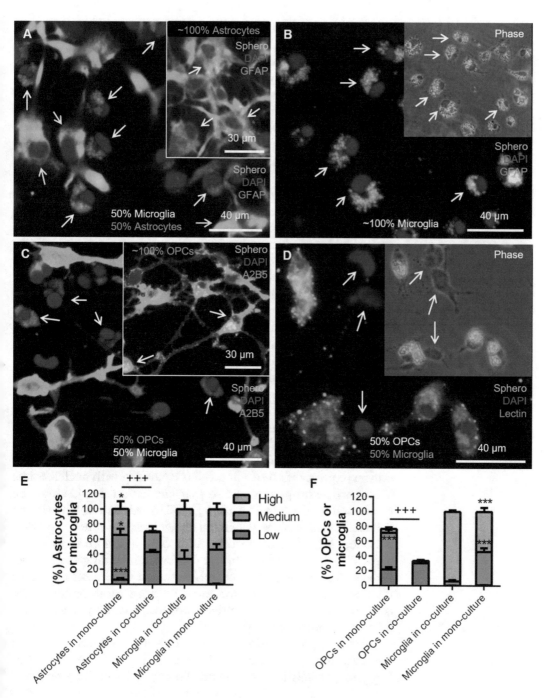

Fig. 4 Astrocytes and oligodendrocyte precursor cells (OPCs) show marked reduction in proportions of Sphero magnetic particle (MP)-labeled cells and extent of loading in coculture with microglia. Fluorescence micrographs of (**A**) astrocyte–microglia coculture showing extensive microglial loading (white arrows). Contrast unlabeled GFAP+ astrocytes (yellow arrows) with several labeled astrocytes in monocultures (inset; arrows show "high" loading). (**B**) Microglial monoculture exhibiting extensive loading (arrows; GFAP−; phase contrast counterpart inset). (**C**) OPC–microglia coculture showing extensive microglial loading (arrows; DAPI+/A2B5−)—note lack of labeled OPCs. Inset, OPC monoculture with multiple labeled OPCs, arrows show "high" loading.

batch for each experiment (using ImageJ software, National Institutes of Health, USA [32]). ImageJ is then used to "stack" a micrograph (fluorescent or phase "cell image," revealing cell membrane/morphology) with the counterpart grayscale particle image. An image with a scale bar is used to set a global pixels/micron scale for all images. Individual cells are delineated free-hand in the cell image, before switching to the grayscale image for measurement, deriving values for area, mean gray value and integrated density for each cell. These numerical values provide an assessment of relative accumulation of intracellular particles per cell, based on fluorescence intensity. For each micrograph, the mean gray value from five cell/particle-free regions ("blanks") is averaged to derive the background intensity. For each cell in that micrograph, cell area is multiplied by this background intensity to generate a micrograph- and cell-specific correction value. For each cell, the correction value is subtracted from that cell's integrated density measurement, giving corrected integrated density values which can be compared. This is summarized here in equation form:

Intracellular particle-associated fluorescence = cellular integrated density − (cellular area × mean gray value of blanks)

3.3 Physicochemical Characterization of Nanoparticles and Coronas

Whether synthesized in-house or obtained commercially, all nanoparticles should be characterized prior to use. This applies to new batches of previously characterized particles, and at least some measurements should be repeated if particles are to be used after prolonged periods of storage, to ensure properties are unchanged. It is also important to acknowledge that modification/functionalization (e.g., addition of a fluorophore, coating with nucleic acids) alters particle properties, and so characterization should always be performed on the "final" version [2, 33]. Such data will be vital to determine which physicochemical characteristics influence particle behavior, including interactions with cells, and so inform future particle design. Physicochemical properties including size, shape, surface charge, and surface chemistry should all be assessed, but as multiple techniques exist for some of these measurements, the context of each should be understood, such that they can be compared and contrasted [14].

Further, corona formation can influence properties such as relaxivity and MRI contrast efficiency [34], and drying processes (e.g., for transmission electron microscopy, TEM) will cause aggregation that may not be representative of aggregation in suspension. So, although some particle characterization techniques should be performed on dried samples, or particles in water/saline, it is important to gather data on particle properties and behaviors following incubation in biological media.

We will describe a small selection of methods for nanoparticle characterization, including the application of TEM to whole particles. We also briefly describe iron staining to confirm retention of fluorophores/dyes. Dynamic light scattering and zeta potential measurement are widely used techniques, but we argue that these should be applied to particles in the presence of biological media.

3.3.1 Formvar TEM for Determination of Intact Particle Size and Shape

Particles can be embedded in resin and sectioned for TEM. However, the act of sectioning can result in artifacts (e.g., stretching/distortion), and such samples will not reliably reveal the particle diameter, as individual particles could be sectioned through any plane (not exclusively the midline). TEM relies on differences in electron density, and many particle components (e.g., polymers) will not be visible without staining, which introduces electron-dense materials which are potentially confounding when imaging at the nanoscale. However, metallic (e.g., iron oxide) particles with an overall diameter less than approximately 300 nm (comparable to microtome sections) can be subjected to TEM whole, without sectioning or staining, and so more accurately revealing particle morphology and size.

Sonicate copper TEM grids in 95% ethanol, rinse with distilled water, and air-dry on filter paper (*see* **Note 18**). Pour formvar (polyvinyl formaldehyde) solution into a clean beaker of water, forming a surface film. Lay grids face-down in the film using inverse forceps, then gently insert a clean glass slide into the liquid and use to retrieve the grids. The grids (with formvar support films) can then be transferred to a clean slide until use. Nanoparticles should be heavily diluted in distilled water (*see* **Note 19**), and 30 μL droplets placed on Parafilm. Formvar grids should be inserted horizontally into the center of these droplets using inverse forceps, and released. After 30 s, retrieve grids and wick away excess liquid by inserting filter paper between the forceps arms. Air-dry grids then image using a transmission electron microscope. Multiple particles can be identified and measured to determine the range and uniformity of particle/core diameter, as well as indicating electron-dense elements of morphology.

3.3.2 Perls' Iron Staining to Determine Particle Integrity

Particle stability/label retention is an important parameter, especially when different components may be used to track those particles. For example, many iron oxide nanoparticles are fluorescently labeled or dyed, providing multiple ways to detect them. However, dyes/fluorophores could leach from the particles, and so it should be established that fluorescent signals correlate exclusively with the presence of particles. One method for this is to check the coincidence of iron oxide staining with fluorescence using microscopy, as described for cells in Sect. 3.2.3.

Incubate a small quantity of particles (e.g., 0.1–5 μL of stock) in 2% potassium ferricyanide, 2% HCl (100 μL; 10 min; room temperature; fume hood). Apply a drop (~50 μL) of this solution to a glass slide and take counterpart light and fluorescence (use long exposure times) micrographs. Merge these images and check for coincidence of fluorescence with iron staining. If fluorescence is detectable away from iron (a blue stain), then the fluorophore/dye may be leaching, potentially confounding uptake and toxicity studies (*see* **Note 20**).

3.3.3 *Hydrodynamic Diameter and Surface Charge*

Measuring the hydrodynamic diameter of nanoparticles provides an estimate of the size of particles (singly or in realistic aggregates) which includes attendant corona formation, and the solvation layer. By performing these measurements in biological media, alterations to surface charge or aggregation tendencies due to temperature, pH and media components can be discovered, which would not be possible from measurements in water/saline at room temperature (the dominant paradigm (*see* **Note 21**)). Here, we describe the use of a Zetasizer (Malvern, UK) to determine both hydrodynamic diameter and zeta potential.

Nanoparticles should be prepared identically to experimental conditions [e.g., vortexed, sonicated, suspended in medium (37 °C, experimental pH) at the experimental concentration of particles] and aliquots placed in a cuvette and a culture plate. The cuvette should then be analyzed three times using the DLS/zeta potential equipment, for an early/immediate assessment (*see* **Note 22**). The culture plate should be incubated alongside cells for the duration of the experiment (e.g., 24 h, 5% CO_2), then similarly analyzed. Culture media typically contain carbonate buffer to maintain a pH of ~7.4 when incubated at 37 °C in 5% CO_2. Therefore, during preparation, samples should have their headspace sealed to preserve the CO_2 levels/pH. Both diameter and zeta potential values should be reported with standard deviation (SD) or standard error of the mean (SEM), to indicate uniformity/variation. Example data are shown in Table 2, which also includes analysis of the amide I component following FTIR spectroscopy. This provides data relating to the proportions of particular protein features in the corona. Further corona characterization is described in the following section.

3.3.4 *Characterization of the Protein Corona*

By characterizing protein coronas for different particles in different media, an understanding can be gained of how particle properties influence corona formation, and also how corona properties influence cellular interactions.

Particles should be pre-incubated in media to facilitate corona formation (37 °C, experimental pH). Magnetic particles with coronas can be retrieved by applying an external magnetic field to the tube, then removing the media by pipette (*see* **Note 23**). Samples should be washed with saline to remove unattached proteins, then air-dried onto aluminum discs. FTIR analyses can then be carried out. Amide I band (1600–1700 cm^{-1}; reflecting protein secondary structure [19, 35]) component peak fitting and principal component analysis (PCA) can be performed to assess characteristics of the particle-associated proteins (Fig. 5). PCA is a

Table 2
Properties of Sphero magnetic particles in various neural cell culture media

Medium	Relevant cell type	ζ-potential (mV ± SEM)		d_{DLS} (nm ± SEM)		Amide I component band (% ± SD; 3 h)		
		5 min	24 h	5 min	24 h	α-helix	β-sheet/turn	Extended chain
OPC-MM	OPCs	−13.3 ± 0.8	−11.8 ± 0.6	875 ± 154	1041 ± 140	14.9 ± 0.0	63.6 ± 0.2	21.5 ± 0.1
Sato	Oligodendrocytes	−11.9 ± 0.5	−13.0 ± 1.3	609 ± 21	1536 ± 53	13.0 ± 0.2	66.2 ± 0.8	20.9 ± 0.4
D10	Astrocytes, microglia	−12.4 ± 0.6	−11.8 ± 1.1	1117 ± 467	1051 ± 128	16.4 ± 0.1	63.2 ± 0.4	20.4 ± 0.1
D10-CM	All glia	−12.9 ± 0.6	−12.2 ± 1.0	890 ± 94	1274 ± 89	17.0 ± 0.2	65.8 ± 0.4	17.2 ± 0.2

Reprinted from Nanomedicine: NBM, Jenkins SI, Roach P and Chari DM, Development of a nanomaterial bio-screening platform for neurological applications, 11:77–87, Copyright (2015), with permission from Elsevier [15]

See Sect. 2 for composition of media

d_{dls} hydrodynamic diameter, determined by dynamic light scattering; *OPC* oligodendrocyte precursor cell; *OPC-MM* OPC maintenance medium; *D10-CM* D10 medium containing 20% conditioned D10 medium from parent mixed glial cultures; *SEM* standard error of the mean; *n* = 3 for ζ-potential, dDLS; *n* = 5 for amide I bands

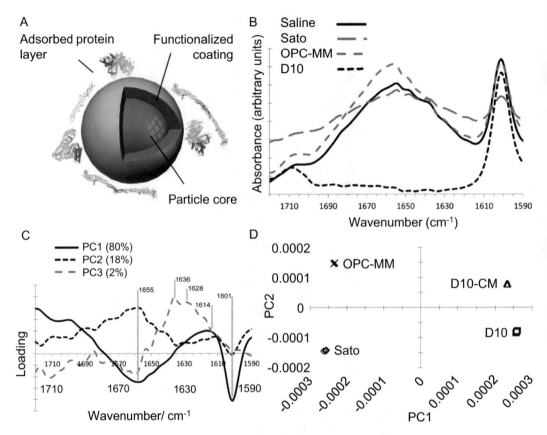

Fig. 5 Media-specific protein coronas form around nanoparticles in biological fluids. (**A**) Schematic of a nanoparticle-associated protein corona. (**B**) FTIR spectra for the protein-relevant Amide I region for Sphero nanoparticles in saline and three neural culture media (Sato, OPC-MM and D10; *see* Sect. 2 for details). The nanoparticle signal is common to all, but the protein corona-related signal is apparent when compared to particles in saline. Different profiles are evident for each medium, as also shown in principal component analyses (PCA) of FTIR spectra (**C**). (**D**) There is excellent discrimination between particle–corona combinations using only two principal components (PC1 and PC2). Reprinted from Nanomedicine: NBM, Jenkins SI, Roach P and Chari DM, Development of a nanomaterial bio-screening platform for neurological applications, 11:77–87, Copyright (2015), with permission from Elsevier [15]

multivariate technique, making it useful for analyzing complex datasets. It identifies sources of variability (parameters, e.g., subsets of FTIR peaks that describe corona features) that best describe differences between samples, and groups these parameters into principal components (PC1, PC2, etc.) [36]. In Fig. 5c, PC1 accounts for 80% of all variability (Fig. 5d, *x* axis, shows that PC1 distinguishes D10/D10-CM from OPC-MM/Sato). This demonstrates considerable similarities in coronal profile between the two media containing serum, and the two serum-free media. However, these pairs of media can be discriminated by including PC2. The distances between datapoints in the PCA plot (Fig. 5d) indicate how distinct the sample measurements are, and ideally will show that groupings of datapoints do not overlap. In summary, FTIR with PCA can be used to determine relative similarities in features of the nanoparticle corona.

4   Notes

1. Attempts were made to culture each cell type in various neural culture media, but all resulted in deleterious effects on survival, proliferation or differentiation. For different combinations of neural cell types, established media may prove suitable, or different media may need to be developed.

2. An alternative neural coculture system can be derived from neural stem cells (NSCs), containing neurons, astrocytes and oligodendroglia in a single medium. However, the cell types are not present in defined ratios, and astrocytes are numerically dominant [37].

3. Mixed glial cultures can also be derived from mice. Transgenic and knockout animals could be used as the source of one or all cell types.

4. When rolling cortices to remove meninges, use the "heel" of curved forceps to gently nudge the tissue without puncturing it. Do not allow the tissue to dry out, as it will then be difficult to retrieve from the towel. Paper hand towels can be wrapped in tin foil and autoclaved to provide a sterile surface for dissection.

5. When mincing tissue, use a chopping motion. Do not cut, as this will scratch the plastic and tissue will be lost in the grooves. Also, use the base of a Petri dish, not the lid, as the lid often has grooves/markings which can hamper the process.

6. Do not introduce excessive bubbles into the medium, as frothing will result in cell loss. Eject cells from the syringe onto the tube wall above the medium. This will reduce the chance of introducing bubbles.

7. Once cells have been transferred to the sieve, improve yield by rinsing the tube with saline and then using this to wash cells through the sieves.

8. When resuspending pelleted cells, start with ~800 µL medium using a 1 mL pipette. Triturate to dissociate cells.

9. Flasks can be preprepared with medium (e.g., 75% final volume) and allowed to gas in the incubator before adding cells (25% final volume), improving cell recovery.

10. If greater microglial yields are required, whenever medium is refreshed, "waste" medium can be centrifuged and pelleted cells (almost exclusively microglia) can be returned to flasks.

11. To seal the headspace in flasks (5% CO_2), Parafilm can be used to wrap the cap. However, this can tear due to the force of shaking, with subsequent changes in pH leading to compromised cell survival. Two pieces of Parafilm should be used, one to wrap the neck and one to form a "bubble" over the cap filter. This bubble is formed by pressing opposite edges of a Parafilm rectangle onto the sides of the cap, leaving a loose dome of Parafilm. As this is not taut, it will not tear due to shaking.

12. When seeding high purity astrocyte flasks/cultures, medium should be replenished once cells show signs of adherence (~30–45 min), as this will reduce OPC contamination (OPCs are slower to adhere).

13. Coverslips can be nitric acid-washed prior to use, but we find that it is usually sufficient to simply sterilize in 24-well plates with 70% ethanol, prior to PDL coating.

14. Microbiology-grade Petri dishes work well as non-tissue-culture plates for removing contaminating microglia. This step is necessary, as even 5% microglial contamination can affect particle uptake by OPCs.

15. Many cell types secrete survival factors, and if plated at too low a density, these survival factors fall below the threshold for culture survival.

16. A CO_2 controller can be used with a 100% CO_2 supply. We use 5% CO_2/95% air as it is more economical, and safer should the cylinder empty into the room.

17. Prepare stocks of both HCl and potassium ferricyanide at 4%, then combine immediately prior to use.

18. Grids for TEM must be clean to ensure no imaging artifacts or nanoscale contaminants.

19. As drying will result in particle aggregation, we recommend testing increasingly dilute samples until single particles are readily apparent in TEM.

20. Alternatively, this fluorophore/dye may be a residue from synthesis. Trap magnetic particles in a tube using a magnet, dispose of supernatant (note this volume, in order to resuspend at original concentration), then wash with water to remove free fluorophore/dye and resuspend.

21. Do not use pure deionized water for DLS measurements, as the lack of ions will result in inaccurate readings.

22. Many DLS devices can be set to maintain samples at 37 °C.

23. Nonmagnetic particles of sufficient density can be collected by high speed centrifugation and disposal of the supernatant.

Acknowledgments

We thank the following for advice on protocol development, equipment operation and analysis techniques: Chris Adams, James Beardmore, Alinda Fernandes, David Furness, Mark Pickard, Paul Roach, Emma Shardlow, Jacqueline Tickle, Karen Walker, Alan Weightman (all Keele University), and Humphrey Yiu (Heriot-Watt University). These protocols were developed while SJ was supported by an Engineering and Physical Sciences Research Council (EPSRC; UK) Engineering Tissue Engineering and Regenerative Medicine (E-TERM) Landscape Fellowship (EP/I017801/1).

References

1. van Landeghem FKH et al (2009) Post-mortem studies in glioblastoma patients treated with thermotherapy using magnetic nanoparticles. Biomaterials 30:52–57

2. Jenkins SI, Yiu HHP, Rosseinsky MJ, Chari DM (2014) Magnetic nanoparticles for oligodendrocyte precursor cell transplantation therapies: progress and challenges. Mol Cell Ther 2:23

3. Knežević NŽ, Lin VSY (2013) A magnetic mesoporous silica nanoparticle-based drug delivery system for photosensitive cooperative treatment of cancer with a mesopore-capping agent and mesopore-loaded drug. Nanoscale 5:1544–1551

4. Jenkins SI, Pickard MR, Furness DN, Yiu HHP, Chari DM (2013) Differences in magnetic particle uptake by CNS neuroglial subclasses: implications for neural tissue engineering. Nanomedicine (Lond) 8:951–968

5. Soenen SJH et al (2010) Intracellular nanoparticle coating stability determines nanoparticle diagnostics efficacy and cell functionality. Small 6:2136–2145

6. Petters C, Thiel K, Dringen R (2016) Lysosomal iron liberation is responsible for the vulnerability of brain microglial cells to iron oxide nanoparticles: comparison with neurons and astrocytes. Nanotoxicology 10:332–342

7. Wu H-Y et al (2013) Iron oxide nanoparticles suppress the production of IL-1beta via the secretory lysosomal pathway in murine microglial cells. Part Fibre Toxicol 10:46

8. Maysinger D, Behrendt M, Lalancette-Hébert M, Kriz J (2007) Real-time imaging of astrocyte response to quantum dots: in vivo screening model system for biocompatibility of nanoparticles. Nano Lett 7:2513–2520

9. Walczyk D, Bombelli FB, Monopoli MP, Lynch I, Dawson KA (2010) What the cell 'sees' in bionanoscience. J Am Chem Soc 132:5761–5768

10. Mahmoudi M, Serpooshan V (2011) Large protein absorptions from small changes on the surface of nanoparticles. J Phys Chem C 115:18275–18283

11. Ghavami M et al (2013) Plasma concentration gradient influences the protein corona decoration on nanoparticles. RSC Adv 3:1119

12. Mahmoudi M et al (2013) Temperature: the 'ignored' factor at the NanoBio interface. ACS Nano 7:6555–6562

13. Soukup D, Moise S, Céspedes E, Dobson J, Telling ND (2015) In situ measurement of magnetization relaxation of internalized nanoparticles in live cells. ACS Nano 9(1):231–240

14. Brun E, Sicard-Roselli C (2014) Could nanoparticle corona characterization help for biological consequence prediction? Cancer Nanotechnol 5:7

15. Jenkins SI, Roach P, Chari DM (2015) Development of a nanomaterial bio-screening platform for neurological applications. Nanomedicine 11:77–87

16. Rubio N, Rodriguez R, Arevalo MA (2004) In vitro myelination by oligodendrocyte precursor cells transfected with the neurotrophin-3 gene. Glia 47:78–87

17. Rasband WS. U.S. National Institutes of Health, Bethesda, Maryland, USA. http://imagej.nih.gov/ij/

18. Abràmoff MD, Magalhães PJ, Ram SJ (2004) Image processing with ImageJ. Biophoton Int 11:36–42

19. Roach P, Farrar D, Perry CC (2006) Surface tailoring for controlled protein adsorption:

effect of topography at the nanometer scale and chemistry. J Am Chem Soc 128:3939–3945

20. McCarthy KD, de Vellis J (1980) Preparation of separate astroglial and oligodendroglial cell cultures from rat cerebral tissue. J Cell Biol 85:890–902

21. Chen Y et al (2007) Isolation and culture of rat and mouse oligodendrocyte precursor cells. Nat Protoc 2:1044–1051

22. Occhetta P, Glass N, Otte E, Rasponi M, Cooper-White JJ (2016) Stoichiometric control of live cell mixing to enable fluidically-encoded co-culture models in perfused microbioreactor arrays. Integr Biol 8:194–204

23. Herculano-Houzel S (2014) The glia/neuron ratio: how it varies uniformly across brain structures and species and what that means for brain physiology and evolution. Glia 62:1377–1391

24. Fernandes AR, Adams CF, Furness DN, Chari DM (2015) Early membrane responses to magnetic particles are predictors of particle uptake in neural stem cells. Part Part Syst Charact 32:661–667

25. Bulte JWM et al (1999) Neurotransplantation of magnetically labeled oligodendrocyte progenitors: magnetic resonance tracking of cell migration and myelination. Proc Natl Acad Sci U S A 96:15256–15261

26. Cianciaruso C et al (2014) Cellular magnetic resonance with iron oxide nanoparticles: long-term persistence of SPIO signal in the CNS after transplanted cell death. Nanomedicine (Lond) 9:1457–1474

27. Jenkins SI (2013) Applications of magnetic particles for oligodendrocyte precursor cell transplantation strategies. Doctoral thesis, Keele University. http://eprints.keele.ac.uk/3821/

28. Luther EM et al (2013) Endocytotic uptake of iron oxide nanoparticles by cultured brain microglial cells. Acta Biomater 9:8454–8465

29. Geppert M et al (2011) Uptake of dimercaptosuccinate-coated magnetic iron oxide nanoparticles by cultured brain astrocytes. Nanotechnology 22:145101

30. McCloy RA et al (2014) Partial inhibition of Cdk1 in G2 phase overrides the SAC and decouples mitotic events. Cell Cycle 13:1400–1412

31. Gavet O, Pines J (2010) Progressive activation of CyclinB1-Cdk1 coordinates entry to mitosis. Dev Cell 18:533–543

32. Jenkins SI et al (2014) Identifying the cellular targets of drug action in the central nervous system following corticosteroid therapy. ACS Chem Neurosci 5:51–63

33. Yiu HHP et al (2012) Fe3O4-PEI-RITC magnetic nanoparticles with imaging and gene transfer capability: development of a tool for neural cell transplantation therapies. Pharm Res 29:1328–1343

34. Amiri H et al (2013) Protein corona affects the relaxivity and MRI contrast efficiency of magnetic nanoparticles. Nanoscale 5:8656–8665

35. Roach P, Farrar D, Perry CC (2005) Interpretation of protein adsorption: surface-induced conformational changes. J Am Chem Soc 127:8168–8173

36. Abdi H, Williams LJ (2010) Principal component analysis. Wiley Interdiscip Rev Comput Stat 2:433–459

37. Fernandes AR, Chari DM (2014) A multicellular, neuro-mimetic model to study nanoparticle uptake in cells of the central nervous system. Integr Biol 6:855–861

Chapter 15

Long-Term Organism Distribution of Microwave Hydrothermally Synthesized ZrO₂:Pr Nanoparticles

Jarosław Kaszewski, Paula Kiełbik, Anna Słońska-Zielonka, Izabela Serafińska, Jakub Nojszewski, Marek Godlewski, Zdzisław Gajewski, and Michał M. Godlewski

Abstract

Studies over ZrO_2:Pr nanoparticles distribution kinetics were performed. It was found that excitation of material may occur at 488 nm as it directly populates the $^3P_0$ sublevel of the Pr^{3+} ions. Also, multiphoton excitation allows using the near infrared region tissue transparency window. The suspensions made with ZrO_2:Pr nanoparticles of various sizes appeared stable by means of zeta potential. Nanoparticles administrated to mice were distributed to the majority of tissues and organs, including rapid transfer through the blood-brain barrier. It was observed tendency for accumulation of nanoparticles in the cells of the aqueduct of midbrain. Two possible pathways of nanoparticle elimination from the brain were postulated.

Key words Nanoparticles, Luminescence, Zirconium dioxide, Biomarkers, Alimentary application, Long-term organ kinetics

1 Introduction

Lanthanides-doped nanoparticles are promising luminescent contrast agent in biological imaging. Luminescence is originating mainly from the trivalent ions of the 4f shell elements [1]. Characteristic emission results from the intrashell 4f transitions and comes out as sharp spectral lines located from ultraviolet to infrared region [2]. These spectral features in first approximation are not sensitive to the symmetry of the chemical environment of the lanthanide ion, do not age with time or temperature [3]. Pr^{3+} is an ion with electronic structure $[Xe]4f^2$ and ground state $^3H_4$. It offers a variety of emission lines in regions dependent on the host lattice. Fluorides doped with Pr^{3+} exhibit violet emission due to $^1S_0 \rightarrow {}^1I_6$ transitions [4]. YAG:Pr shows greenish 487 nm line due to $^3P_0 \rightarrow {}^3H_4$ transition [5]. Y_2O_3:Pr exhibits red luminescence at ca. 625 nm mainly due to $^3P_0 \rightarrow {}^3H_6$ and $^1D_2 \rightarrow {}^3H_4$ transitions [6]. Versatility of praseodymium ions allows applying oxide matrices

Fidel Santamaria and Xomalin G. Peralta (eds.), *Use of Nanoparticles in Neuroscience*, Neuromethods, vol. 135, https://doi.org/10.1007/978-1-4939-7584-6_15, © Springer Science+Business Media, LLC 2018

doped with it in biological imaging. Additional advantage of Pr^{3+} application is that it exhibits two photon absorption resulting in visible emission [7], which increases tissue penetration depth, reduces the background noise and scattering in samples [8]. One of the best host lattices for lanthanide ions is zirconium dioxide, ZrO_2. It possesses low phonon energy (470 cm^{-1}), high host absorption coefficient, and high refractive index [9]. Additionally, it is chemically resistant, which is important for biocompatible material design [10]. Technology of luminescent biomarkers crystallization also should be impurity free. In this work, we have used the microwave-driven hydrothermal technique of nanoparticle synthesis [11]. The use of microwaves eliminates the necessity of reaction mixture contact with heating elements [12]. As it is known, hydrothermal conditions are very corrosive and any subject introduced into the reaction environment must be treated as the source of impurities. Microwave heating also improves reaction rates in relation to the classic hydrothermal process [13]. One can control the properties of product by adjustment of reaction conditions [14]. Most importantly, nanocrystals grow in the aqueous environment, therefore no further purification is needed, as the crystals are terminated by hydroxyl and water species [15]. The attempts to use non-functionalized oxide nanoparticles as imaging agents in living organisms were conducted earlier [16–20].

There is a great need in biomedical sciences for stable, non-bleaching, and nontoxic labels. Traditional chemical dyes, although widely available, have one major disadvantage—their photostability is extremely limited, especially in the biological media. The first generation of fluorescent nanoparticles provided stable and high-yield signal [21–23]; however, their core based on heavy metal ions (i.e., Cd) proved unstable and toxic [24]. Second limitation for the wide medical applicability of nanoparticles is the route of entry to the organism. Hitherto, majority of the experiments utilized intravenous injection of the nanoparticle suspension [21–23]. The great advantages of our system are the oral route of administration [18] which greatly simplifies the preparation of suspension, high stability in biological media, low toxicity of dopants, and high fluorescent yield following excitation in the visible range [16–20].

2 Experiment

2.1 Nanoparticles Preparation

Nanoparticles were prepared using the microwave hydrothermal technique [25]. The solution of zirconyl, yttrium, and praseodymium ions was prepared by dissolving $ZrO(NO_3)_2 \cdot xH_2O$ ($x \approx 6$, 99%, Sigma-Aldrich), $Y(NO_3)_3 \cdot 6H_2O$ (99.9%, Sigma-Aldrich), and $Pr(NO_3)_3 \cdot 6H_2O$ (99.9%, Sigma-Aldrich) compounds in distilled water. The solution was then alkalized with the aqueous ammonia solution (25%, J.T.Baker) and the residue was moved to the funnel and suction filtered. Resulting precipitate was triply washed with distilled water to remove nitrate(V) and ammonium ions. Residue was placed in the Ertec microwave hydrothermal reactor with teflon lining and average power of 7 W/ml of reaction mixture. The hydrothermal reaction was conducted at 6 MPa by 20 min. The product was dried at 40 °C for 24 h and divided into four parts. Three were calcined at 400, 800, and 1200 °C and one left uncalcined. The final ZrO_2 product contained 0.5 molar % of praseodymium and 6 molar % of yttrium as a stabilization agent, in the text it is referred to as ZrO_2:Pr.

2.2 Biological Experiments

2.2.1 Cell Culture

The human Caco-2 cells were seeded in 6-well plates and propagated to full confluence in Dulbecco's Modification of Eagles Medium (DMEM) with essential aminoacids (1%), streptomycin and penicillin (1%), sodium bicarbonate (0.2%) and supplemented with 10% fetal bovine serum (FBS). The cells are incubated at 37 °C in a 5% CO_2 incubator over 9 days. When the cells reached the 100% of coverage the growth medium was replaced with a medium containing nanoparticles at various concentrations (1, 0.1, 0.01, and 0.001 mg/ml) and incubated for 24 h at 37 °C. After the incubation cells were collected (0.05% tripsin and 0.2% EDTA for 10 min), cell suspension was centrifuged and the pellet was resuspended in 1 ml growth medium. 100 μl of sterile 0.4% trypan blue solution was mixed with 100 μl cell suspension. After 10 min incubation a hemocytometer chamber was filled with 10 μl of cell suspension. Cells were viewed under an inverted phase contrast microscope (20× lens). The viable (bright) and dead (blue) cells were counted. The concentration of viable and nonviable cells and the percentage of viable cells were calculated using the following equations:

$$\text{Viable cell count} = \frac{\text{number live cells}}{\text{number of large corner squares counted}} \times \text{dilution} \times 10^4.$$

$$\text{Non viable cell count} = \frac{\text{number dead cells}}{\text{number of large corner squares counted}} \times \text{dilution} \times 10^4.$$

$$\text{Viability} = \text{number viable cells} / \text{total number of cells} \times 100\%.$$

2.2.2 Mouse Model

All the procedures were approved by the Local Ethical Committee (agreement No 44/2012) and conducted according to the EU and local directives. Adult Balb-c mice (3–6 months, $n = 24$) were kept in the standard and controlled conditions (UniProtect Air

Flow Cabinet, Merazet SA, Poland) with 12/12 day/night cycle and feed and water provided ad libitum. Following 7 days of acclimatization, suspension of NPs in RO water (10 mg/ml, 0.3 ml per animal) was administrated via gastric gavage (IG) to mice (Herberholz HAUPTNER 31011, Merazet). Mice were sacrificed 1, 2, 3, 4, 8, and 16 weeks following IG (CO_2 Box, Bioscape, Merazet). Afterward the following internal organs were collected: duodenum, liver, kidney, spleen, and brain. Specimens were dehydrated in the tissue processor (Leica-TP1020, KAWA.SKA Sp. z o.o., Poland), then embedded in paraffin (Leica-EG1150H, KAWA.SKA), cut into 5 μm-thin sections (Leica-RM2255, KAWA. SKA), and mounted on the microscope slides (Equimed Sp. J. Warsaw, Poland). For scanning cytometry analyses cell nuclei of the samples were counterstained by HOECHST 33342 (Sigma-Aldrich) and mounted with Fluoromount (Sigma-Aldrich) under the coverslips.

For the identification of ZrO_2:Pr NPs in the studied organs and tissues SCAN^R scanning cytometry system (Olympus Polska Sp. z o.o., Warsaw, Poland) was used. Identification of cells within tissues was performed based on the HOECHST 33342-related fluorescence of cell nuclei. The identification of NPs within the cells was based on Pr-related emission in the green (505IF) range following 488 nm excitation. Final result was presented as the index of cells positive for the ZrO_2:Pr fluorescence in the whole tissue sample. In the control samples (where NPs were omitted) the false-positive signals did not exceed 2% (not shown). Additionally, the distribution maps of the brain tissue showing localization of the ZrO_2:Pr-positive cells were collected.

2.3 Methods of Analysis

Photoluminescence (PL) and photoluminescence excitation (PLE) spectra were measured at the room temperature with the Horiba/Jobin-Yvon Fluorolog-3 spectrofluorimeter. Zeta potential and hydrodynamic radii of the nanoparticles were measured using Beckman Coulter DelsaMax Pro light scatter analyzer equipped with DelsaMax Assist cell pressurization system. Light scattering versus luminescence experiments were performed using Apogee A50 Micro Plus Flow Cytometer. Fluorescence measurements were performed with Red and DRed detectors operating at 662–698 and >740 nm regions respectively. The samples were excited with blue 488 nm, 50 mW laser. Fluorescence images of nanoparticles were taken with Olympus Fluoview FV1000MPE Multiphoton laser scanning microscope. Excitation source was Mai Tai HP DS Ti:Sapphire laser operating at 690–1040 nm region, with average 2.1 W power. Spinning disc confocal microscopy was performed with Leica MM AF (Leica Microsystems, Wetzlar).

3 Results: Properties of Nanoparticles

3.1 Luminescence

After excitation in ultraviolet (260 nm) one can observe activation of Pr^{3+} ions as well as matrix luminescence (Fig. 1). Broad band with maximum at ca. 540 nm is likely related to the presence of Pr^{4+} ions [26] or monoclinic ZrO_2 phase [27], and it gradually disappears with the increase of calcination temperature. The appearance of Pr^{3+} ions is demonstrated by a number of sharp lines, related to intrashell transitions (Fig. 1). The Pr^{3+} emission does not change after calcination, relative intensities of particular lines are preserved. The most intense line is located in red region (613 nm) due to $^3P_0 \rightarrow {}^3H_6$ transition. Spectral response of this line for various excitation wavelengths was investigated in the photoluminescence excitation spectra (Fig. 2). All the spectra contain sharp lines related to 4f ← 4f electronic transitions and 4f ← 5d band associated with promotion of the 4f electron to 5d orbital. Intensity relationship between both does not change significantly with increase of calcination temperature, hence slightly fluctuates. 4f ← 4f transitions are much less intense and mainly display as four lines peaking at: 451, 457, 474, and 487 nm (Fig. 2). Interconfigurational band is located in ultraviolet region, peaking at 299 nm and with high integral intensity. As one may notice a certain excitation line is very close to the wavelength of 488 nm argon laser, often used in the optical laboratory setups. Next the samples were optically excited at 487 nm, which resulted with spectra shown in Fig. 3. To the measurements sequence, it was added sample before microwave hydrothermal treatment, to notice its influence on the luminescence properties. The sample is most likely amorphous or only short ranged ordering of the ions in the crystal lattice is present. This is confirmed by photoluminescence, as the emission lines related to the transitions from $^3P_1$ and $^3P_0$ sublevels are present, but are very broadened in this sample. Changes in the relative intensities of the lines indicate that Pr^{3+} ions change its crystallographic position compared to the ZrO_2 lattice and broadening shows that luminescence activator ions are in many different crystallographic positions, varying slightly from one to another. Calcined samples' spectra do not differ qualitatively from each other. However, if one compare it to the sample as crystallized in the microwave hydrothermal process, the raised background or broadened $^3P_1$ level related emission is noticeable. It is dominating the spectrum in the amorphous sample; therefore, one can conclude that not reacted leftovers of amorphous matter are present in the as grown sample. The 1200 °C calcined sample was excited with near infrared radiation (800 nm) and observed in a multiphoton microscope (Fig. 4). The first picture was taken with transmitted light to show the shapes of the ZrO_2 grains (Fig. 4a). Then luminescence images were taken in green and red channels

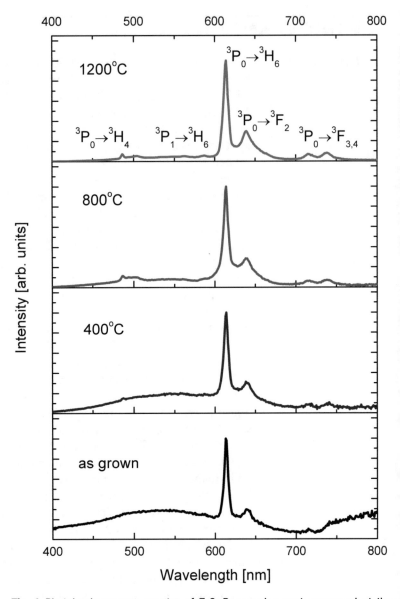

Fig. 1 Photoluminescence spectra of ZrO$_2$:Pr samples post-processed at the various temperatures. Excitation wavelength is 260 nm

(Fig. 4b, c) to show if the luminescence overlaps with the nanoparticles shape. In the last picture all the images are combined.

3.2 Suspension Properties

Zeta potential and hydrodynamic radius measurements are shown in Fig. 5. The samples were prepared by suspending ca. 5.5 mg of nanopowder into the 5 ml of miliporous water. Then the samples were shaken out and diluted in the rate 1:25 with pure water. The suspension was injected rapidly using a syringe into the flow cell to avoid air bubbles in the cell's quartz windows. Between measurements the cell was washed with ca. 5 ml of pure water. The measurements

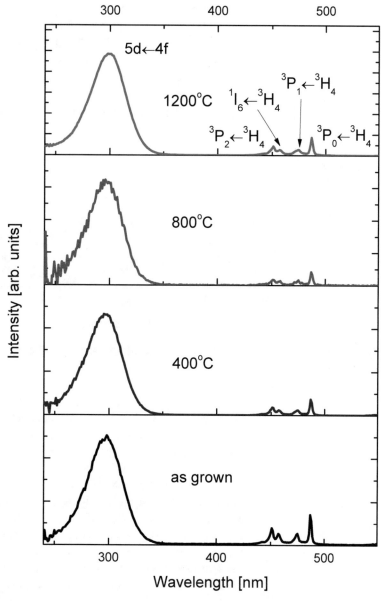

Fig. 2 Photoluminescence excitation spectra of ZrO₂:Pr nanopowders prepared at the various temperatures. Emission wavelength is 613 nm

were conducted with 532 nm laser operating in normal mode (45 mW) and using electric field frequency 10 Hz. Each measurement was repeated five times, then the results were averaged. All the measurements were conducted at 25 °C. In all the samples it was detected one fraction in population of nanoparticles (Fig. 5). Calcination caused the increase of hydrodynamic radius from ca. 200 nm for as grown and 400 °C heated sample to over 400 nm for the 1200 °C heated one. However, the difference between primary nanoparticles' sizes in pure ZrO₂ is larger [25]. It may be the

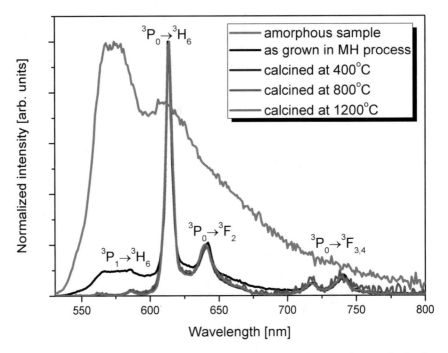

Fig. 3 Effect of microwave hydrothermal (MH) treatment and thermal processing on the luminescence of ZrO$_2$:Pr excited at 487 nm

result of suspension preparation, as no special techniques were applied to disassociate single nanoparticles from the agglomerates. The existence of the agglomerates was additionally proven using spinning disc confocal microscopy, where 290 and 1120 nm sized objects were found in the same sample calcined at 1200 °C. Depending on the calcination temperature zeta potential does not change significantly. Agglomerates are negatively charged and value around −30 mV indicates that aqueous suspensions of ZrO$_2$:Pr nanoparticles are very stable. The absence of dependence on the calcination temperature (and thus primary nanoparticles sizes) indicates that the creation of a double electric layer strongly relies on the crystallites' chemical composition.

3.3 Cytometry

The cytometry technique was used to determine nanoparticles' suspension properties and the results are shown in Fig. 6. ZrO$_2$:Pr nanoparticles' suspension was prepared by dispersing 1 mg of nanoparticles in 1 ml of deionized water. Then the samples were mechanically shaken out for white opaque liquid. Initial suspension was diluted 1:25 with deionized water to form a clear product. The magnitude of dilution was tuned to not exceed 3000 events per second during the measurement. 130 μl sample volume was taken on each cycle with sheath fluid set to 150 mbar. In all the photomultipliers (PMTs) noise levels were around or below 0.1. Gain was set to 1 in all the detectors. All the suspensions contained one

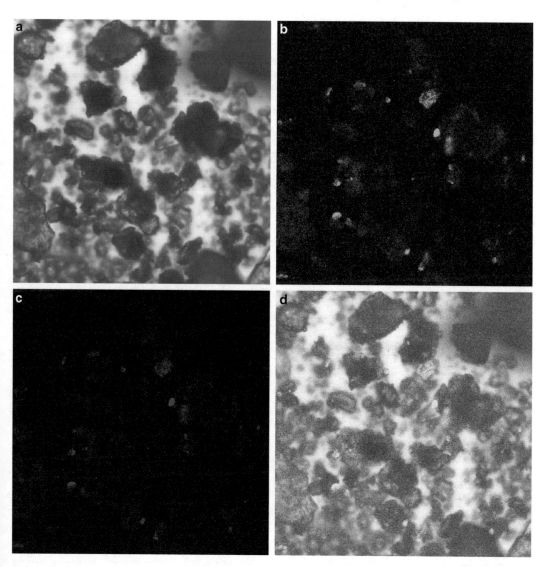

Fig. 4 Pure ZrO$_2$:Pr nanopowders observed in multiphoton scanning microscope with excitation wavelength 800 nm. (**a**) Transmitted light, (**b**) green channel luminescence, (**c**) red channel luminescence, (**d**) combined transmitted light and luminescence channels

population of particles (Fig. 6, SALS vs. Red). The distribution was monomodal, but dispersion of sizes is very large. The samples also indicated large uniformity of light scattering using various angles of scattering detection. SALS (short angle light scattering) is in the range 0.5–5°, while MALS in 5–15°. SALS plotted against MALS (Fig. 6) shows graph close to the $f(x) = x$ function, meaning various angle light scattering results in the same output and likely round shaped and uniform objects. Different is sample calcined at 1200 °C, where one can observe, except the plot present in all the samples, also second function. It probably indicates that the shapes of objects in the suspension are not uniform.

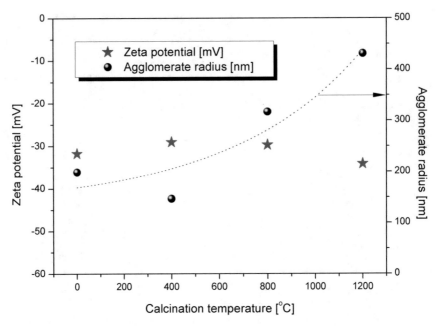

Fig. 5 Zeta potential (red stars) and hydrodynamic radius of ZrO$_2$:Pr nanoparticles (black balls) in water suspension

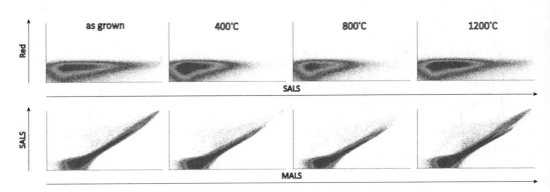

Fig. 6 Cytometric results for ZrO$_2$:Pr nanoparticles

Since excitation of $^3P_0$ sublevel in Pr^{3+} ions results in radiative emission, as we have used 488 nm laser, we have expected luminescence to occur. The Red detector in Apogee A50 analyzes photons in the wavelengths region of 662–698 nm; therefore, the slope of the $^3P_0 \rightarrow {}^3F_2$ transition was expected to give a response. However, population of nanoparticles does not exhibit detectable luminescence. Signal from detectors in other regions, i.e., DRed (>740 nm), Orange (560–590 nm), and Green (500–550 nm) was also analyzed to indicate complete lack of luminescence in the samples. Proposed explanation assumes that nanoparticles spend less time in the measurement chamber than optical phenomenon needs to occur. This is not the case in routinely used fluorophores, as their luminescence decays very fast, usually in the nanosecond region. For instance, Alexa Fluor 488 fluorescence lifetime is 4.1 ns [28]. On the other hand, Pr^{3+} ions related luminescence lifetimes in the yttria stabilized zirconia matrix, depending on the coordination number of activator ions which is up to 310 µs [29], five orders of magnitude longer than organic fluorophores. The nanoparticles are excited, but the emission occurs beyond the measurement zone.

4 Results: Toxicology

4.1 CD50

Trypan blue labeling showed that cell density and percentage of viable cells in culture significantly decreases only after incubation with high dose of nanoparticles (Tables 1 and 2, Fig. 7). Concentration nanoparticles at physiological levels (0.01 and 0.001 mg/ml) caused only a slight decrease of cells viability (Table 2, Fig. 7). For further experiments, the concentration of 0.001 mg/ml of ZrO_2:Pr or its equivalent (for experiments on living organisms) has been selected.

4.2 The Morphology of Caco-2 Cell Culture

Following the CD50 assay, Caco-2 cultures were incubated with working (0.001 mg/ml) and high (0.1 mg/ml) (Fig. 8). After 4 days at the lower concentration (Fig. 8a, b), neither the morphology nor confluence of the Caco-2 monolayer showed any changes and the number of unattached cells did not exceed the control (not shown). However, when a high (0.1 mg/ml) concentration was used (Fig. 8c, d), a clear toxic effect can be observed in the form of lesions in the monolayer and increased number of floating, unattached cells. Also groups of detaching cells can be observed (Fig. 8d, upper left corner).

Table 1
Cell density per ml

	CTRL	0.001 mg/ml	0.01 mg/ml	0.1 mg/ml	1 mg/ml
Viable cells	6.38E + 06	4.68E + 06	4.12E + 06	3.52E + 06	1.68E + 06
Dead cells	2.52E + 06	2.42E + 06	2.04E + 06	2.92E + 06	2.84E + 06
Total	8.90E + 06	7.10E + 06	6.16E + 06	6.44E + 06	4.52E + 06

Table 2
The percentage of viable cells in culture

	CTRL	0.001 mg/ml	0.01 mg/ml	0.1 mg/ml	1 mg/ml
Viable cells	72%	66%	66%	56%	37%

a

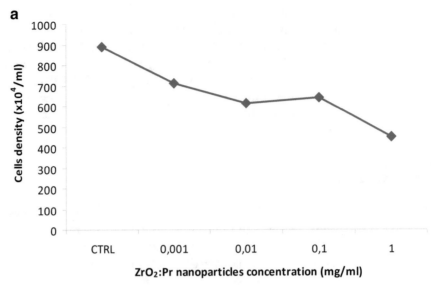

b

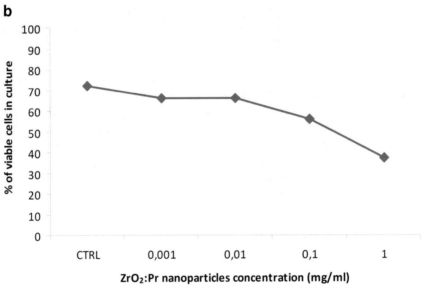

Fig. 7 (**a**) Cells density and (**b**) percent of viable cells, after 24 h incubation with increasing concentrations of ZrO_2:Pr

5   Distribution of ZrO_2:Pr Nanoparticles in Mice

Quantitative results obtained with scanning cytometry analyses are presented as an average ± SD index of cells positive for ZrO_2:Pr NPs-related fluorescence in analyzed tissues, following different times after the administration of ZrO_2:Pr NPs (Table 3). In duodenum a highly statistically significant drop of the index of ZrO_2:Pr—positive cells was observed between at second week after administration of NPs (Table 3). In subsequent time points

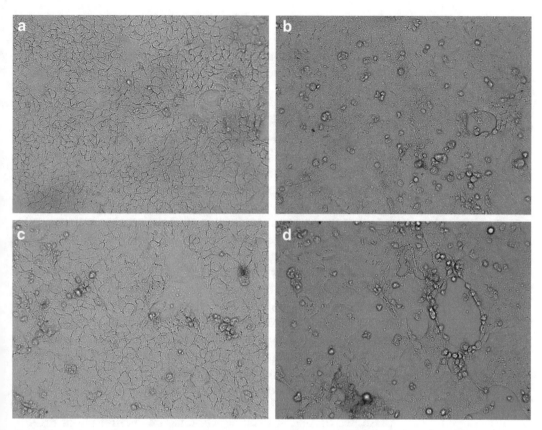

Fig. 8 Caco-2 cells after 10 min and 4 days incubation with nanoparticles ZrO_2:Pr in two concentrations—0.001 mg/ml (**a**, **b**) and 0.1 mg/ml (**c**, **d**). Pictures from inverted phase contrast microscope (20× magnification)

Table 3
Percent of ZrO_2:Pr-positive cells in the analyzed organs evaluated by scanning cytometry

	Week 1	Week 2	Week 3	Week 4	Week 8	Week 16
Duodenum	11.05 ± 1.62	5.77 ± 1.2[a]	6.65 ± 0.73[a]	5.35 ± 0.96[a]	6.66 ± 1.3[a]	4.42 ± 0.83[a]
Liver	7.90 ± 1.31	9.54 ± 1.46[b]	7.08 ± 0.88	10.43 ± 3[b]	9.33 ± 2.46[b,c]	4.63 ± 1.49
Kidney	7.04 ± 0.99	8.81 ± 1.67	7.53 ± 1.35	5.72 ± 0.84	6.39 ± 1.07	8.50 ± 1.03
Spleen	5.97 ± 0.35	6.41 ± 1.12	8.89 ± 1.29	7.50 ± 1.09	8.93 ± 1.07	6.39 ± 0.97
Brain	9.57 ± 2.15	10.97 ± 2.59	12.16 ± 1.83	9.69 ± 1.4	8.83 ± 1.71[c]	8.07 ± 0.96[d]

Data presented as average ± SD from 4 mice per time-point. Statistical significance: [a]$p \leq 0.001$ vs. week 1, [b]$p \leq 0.001$ vs. week 16, [c]$p \leq 0.05$ vs. week 3, [d]$p \leq 0.001$ vs. week 3

no significant changes were observed. In the liver, a significant increase of ZrO_2:Pr-positive cells was detected between the third and fourth weeks after administration of NPs. Afterward, the highly significant decrease of the index was observed between the

4th and 16th weeks as well as between the 8th and 16th weeks (Table 3). In the kidneys all the observed changes in the index of ZrO_2:Pr-positive cells were not statistically significant. However, an increase in the index was noticeable 2 weeks after the administration of NPs with a subsequent decrease until the fourth week post IG. Following, a gradual increase in the index between the 4th and 16th weeks was observed (Table 3). Similarly, in the spleen there were no statistically significant changes in the index of ZrO_2:Pr-positive cells in all time points. An increase in the index was observed 3 weeks after the administration of NPs and remained elevated up to the eighth week. In the brain nonsignificant increase of percent of ZrO_2:Pr-positive cells was detected between the first and third weeks after the administration of NPs, with subsequent decrease up to the 16th week (Table 3). Afterward, the statistically significant decrease of the index was detected in the latter time points (8th and 16th weeks) (Table 3). Results suggest that unlike ionic form, the nanoparticles are characterized by their long-term biological half-life, probably related to the processes of reabsorption of nanoparticles eliminated with the bile by duodenal enterocytes. Crucially, the 16th week after oral exposure seems to be at an onset of final elimination phase for non-biodegradable nanoparticles.

Representative maps of the distribution of cells positive for ZrO_2:Pr NPs-related fluorescence in analyzed brain sections are presented on Fig. 9. In all time points, the distribution of ZrO_2:Pr NPs in the brain tissue was relatively uniform. However, NPs showed the tendency to accumulate in the cells of the aqueduct of midbrain (Fig. 9b, c, e, arrows), suggesting the role of cerebrospinal fluid in the recirculation of NPs.

6   Conclusions

Concluding, presented results confirm the high potential of proposed ZrO_2:Pr NPs for the applications as fluorescent markers for biology and medicine. Nanoparticles showed a great bioavailability after alimentary administration, effective tissue distribution and following redistribution processes. NPs used in these experiments underwent quick distribution to majority of tissues and organs, including rapid transfer through the blood-brain barrier. The high fluorescent signal after excitation in the visible spectrum enabled easy visualization and distinction from the tissue autofluorescence. Based on this and previous studies, two possible pathways of nanoparticle elimination from the brain were postulated: through the blood-brain barrier into the bloodstream and along neuronal projections. Finally, the long-term accumulation of nanoparticle-positive cells around the aqueduct of midbrain

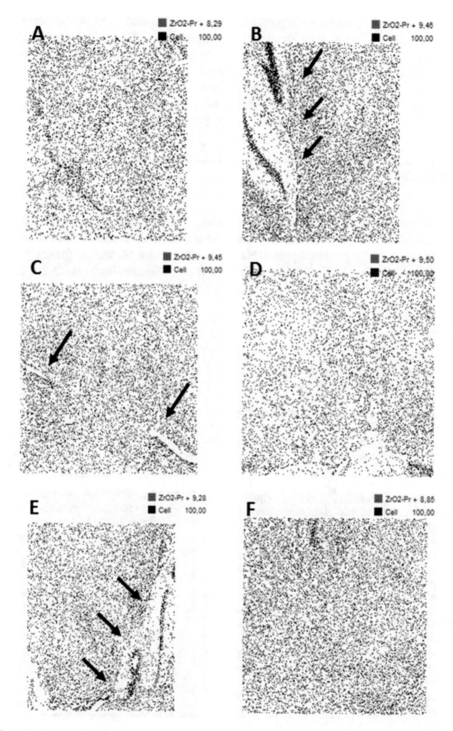

Fig. 9 Scanning cytometry maps of the distribution of cells positive for ZrO₂:Pr fluorescence (red) in the analyzed brain tissue. Time after application of ZrO₂:Pr—(**a**) 1 week, (**b**) 2 weeks (**c**) 3 weeks, (**d**) 4 weeks, (**e**) 8 weeks, (**f**) 16 weeks. Arrows indicate areas of accumulation of NPs in the cells of the aqueduct of midbrain

suggests involvement of cerebrospinal fluid in the accumulation, recirculation, and/or elimination of NPs. However, neither acute nor chronic behavioral or pathological changes were observed in the organism and studied tissues following ZrO_2:Pr NPs application, their biostability and redistribution/reabsorption processes enhance their half-life in the organism, suggesting caution in the cyclic application to the living organism.

Acknowledgments

The authors wish to thank Olympus Polska for multiphoton confocal microscopy measurements. The research was partially supported by the National Centre for Research grants "Maestro" 2012/06/A/ST7/00398 and "Sonata-Bis" UMO 2012/05/E/NZ4/02994.

References

1. Ronda CR, Justel T, Nikol H (1998) Rare earth phosphors: fundamentals and applications. J Alloys Compd 275–277:669–676

2. Bunzli J-CG, Piguet C (2005) Taking advantage of luminescent lanthanide ions. Chem Soc Rev 34:1048–1077

3. Hemmil I, Laitala V (2005) Progress in lanthanides as luminescent probes. J Fluoresc 15(4)

4. Gusowski MA, Swart HC, Karlsson LS, Trzebiatowska-Gusowska M (2012) $NaYF_4$:Pr^{3+} nanocrystals displaying photon cascade emission. Nanoscale 4:541

5. Song H, Chunhua L, Liu X, Zhongzi X (2016) Optical temperature sensing based on the luminescence from YAG:Pr transparent ceramics. Opt Mater 60:394–397

6. Mishra K, Dwivedi Y, Rai A, Rai SB (2012) Spectral characteristics of intense red luminescence in Pr:Y_2O_3 nanophosphor on UV excitation. Appl Phys B 109(4):663–669

7. Bloembergen N (1984) The solved puzzle of two-photon rare earth spectra in solids. J Lumin 31–32:23–28

8. Pospichalova V, Svoboda J, Dave Z, Kotrbova A, Kaiser K, Klemova D, Ilkovics L, Hampl A, Crha I, Jandakova E, Minar L, Weinberger V, Bryja V (2015) Simplified protocol for flow cytometry analysis of fluorescently labeled exosomes and microvesicles using dedicated flow cytometer. J Extracell Vesicles 4:25530

9. Isasi-Marın J, Perez-Estebanez M, Dıaz-Guerra C, Castillo JF, Correcher V, Cuervo-Rodrıguez MR (2009) Structural, magnetic and luminescent characteristics of Pr3+-doped $ZrO2$ powders synthesized by a sol–gel method. J Phys D Appl Phys 42:7 pp. 075418

10. Wulfman C, Sadoun M, Lamy de la Chapelle M (2010) Interest of Raman spectroscopy for the study of dental material: the zirconia material example. IRBM 31:257–262

11. DMP M (1994) The applications of microwaves in chemical syntheses. Res Chem Intermediates 20:85–91

12. Whittaker AG, DMP M (1994) Application of microwave heating to chemical syntheses. J Microw Power Electromagn Energy 29:195–219

13. Mingos DMP, Baghurst DR (1991) Applications of microwave dielectric heating effects to synthetic problems in chemistry. Chem Soc Rev 20:1–47

14. Opalińska A, Leonelli C, Łojkowski W, Pielaszek R, Grzanka E, Chudoba T, Matysiak H, Wejrzanowski T, Kurzydłowski KJ (2006) Effect of pressure on synthesis of Pr-doped zirconia powders produced by microwave driven hydrothermal reaction. J Nanomater 1–8

15. Bondioli F, Ferrari AM, Leonelli C, Siligardi C, Pellacani GC (2001) Microwave-hydrothermal synthesis of nanocrystalline zirconia powders. J Am Ceram Soc 84(11):2728–2730

16. Kielbik P, Kaszewski J, Witkowski BS, Gajewski Z, Gralak MA, Godlewski M, Godlewski MM. Cytometric analysis of Zn-based nanoparticles for biomedical applications, chapter in "Microscopy and imaging science: practical approaches to applied research and education," Microscopy Book Series – 2017 Edition, ISBN-13: 978-84-942134-9-6

17. Kielbik P, Kaszewski J, Rosowska J, Wolska E, Witkowski BS, Gralak MA, Gajewski Z, Godlewski M, Michal MG (2017) Biodegradation of the ZnO:Eu nanoparticles in the tissues of adult mouse after alimentary application. Nanomedicine 13(3):843–852

18. Słońska A, Kaszewski J, Wolska-Kornio E, Witkowski B, Wachnicki Ł, Mijowska E, Karakitsou V, Gajewski Z, Godlewski M, Godlewski MM (2016) Luminescent properties of ZrO_2:Tb nanoparticles for applications in neuroscience. Opt Mater 59:96–102

19. Kaszewski J, Godlewski MM, Witkowski BS, Słońska A, Wolska-Kornio E, Wachnicki Ł, Przybylińska H, Kozankiewicz B, Szal A, Domino MA, Mijowska E, Godlewski M (2016) Y_2O_3:Eu nanocrystals as biomarkers prepared by a microwave hydrothermal method. Opt Mater 59:157–164

20. Godlewski MM, Kaszewski J, Szal A, Slonska A, Domino MA, Godlewski M (2014) Size of nanocrystals affects their alimentary absorption in adult mice. Med Water 70(9)

21. Akerman ME, Chan WC, Laakkonen P, Bhatia SN, Ruoslahti E (2002) Nanocrystal targeting in vivo. Proc Natl Acad Sci U S A 99(20):12617–12621

22. Jackson H, Muhammad O, Daneshvar H, Nelms J, Popescu A, Vogelbaum MA, Bruchez M, Toms SA (2007) Quantum dots are phagocytized by macrophages and colocalize with experimental gliomas. Neurosurgery 60(3): 524–530

23. Luo G, Long J, Zhang B, Liu C, Ji S, Xu J, Yu X, Ni Q (2012) Quantum dots in cancer therapy. Expert Opin Drug Deliv 9(1):47–58

24. Cho SJ, Maysinger D, Jain M, Roder B, Hackbarth S, Winnik FM (2007) Long-term exposure to CdTe quantum dots causes functional impairment in live cells. Langmuir 23(4):1974–1980

25. Kaszewski J, Yatsunenko S, Pełech I, Mijowska E, Narkiewicz U, Godlewski M (2014) High pressure synthesis versus calcination—different approaches to crystallization of zirconium dioxide. Pol J Chem Technol 16(2):99–105

26. Hoefdraad HE (1975) Charge-transfer spectra of tetravalent lanthanide ions in oxides. J Inorg Nucl Chem 37:1917–1921

27. Kiisk V, Puust L, Utt K, Maaroos A, Mändar H, Viviani E, Piccinelli F, Saar R, Joost, U IlmoSildos (2016) Photo-, thermo- and optically stimulated luminescence of monoclinic zirconia. J Lumin 174:49–55

28. Fluorescence quantum yields (QY) and lifetimes (τ) for Alexa Fluor dyes—Table 1.5. www.thermofisher.com

29. Munoz-Santiuste JE et al (2001) Pr3+ Centers in YSZ single crystals. J Alloys Compd 323–324:768–772

Chapter 16

Gold Nanoparticles as Nucleation Centers for Amyloid Fibrillation

Yanina D. Álvarez, Jesica V. Pellegrotti, and Fernando D. Stefani

Abstract

The aggregation of proteins into amyloid fibrils is related to more than 30 diseases, including the most common neurodegenerative conditions. Amyloid fibrillation is a nucleation-dependent polymerization reaction where monomeric protein first assembles into oligomers that in turn serve as nuclei for fibril formation. Recently, nanoparticles of various compositions and sizes have been investigated as nucleation centers for amyloid fibrillation. The interaction of nanoparticles with amyloid proteins can generate intermediate structures able to accelerate or inhibit fibrillation, and therefore, they constitute a tool to control and manipulate amyloid fibrillation which may be the key to elucidate molecular mechanisms or to devise therapies. In this chapter, we first give a general overview about the use of nanoparticles as artificial nucleation centers for amyloid aggregation, and then we focus on gold nanoparticles providing detailed protocols for their functionalization and use in amyloid fibrillation assays.

1. Amyloid fibrillation as a nucleation and growth polymerization.

2. Nanoparticles as nucleation centers.

3. Unique properties of gold nanoparticles.

4. Fabrication and surface modification of gold nanoparticles.

5. Amyloid aggregation assays with gold nanoparticles.

6. Protocols.

Key words Amyloid aggregation, Nucleation and growth, Protein self-assembly, Surface functionalization, Plasmonic nanoparticles

1 Amyloid Fibrillation as a Nucleation and Growth Polymerization

Amyloid fibrillation refers to the misfolding and subsequent aggregation into fibrils of peptides and proteins that are normally soluble. At a molecular level, the mechanistic details of amyloid fibrillation are complex and not yet fully understood. Overall, amyloid fibril formation occurs as a nucleation and growth polymerization reaction (Fig. 1a). The first step consists of the formation of various oligomers and intermediate supramolecular assemblies,

Fidel Santamaria and Xomalin G. Peralta (eds.), *Use of Nanoparticles in Neuroscience*, Neuromethods, vol. 135, https://doi.org/10.1007/978-1-4939-7584-6_16, © Springer Science+Business Media, LLC 2018

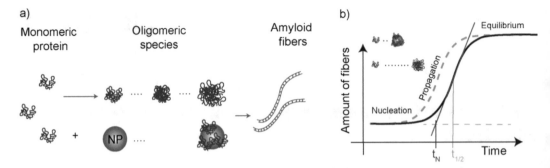

Fig. 1 (**b**) Schematic representation of the steps of amyloid fibrillation. Nanoparticles (NPs) act as artificial nuclei. (**a**) Typical time evolution of amyloid formation in the absence (solid) and presence (dashed) of NPs that accelerate nucleation

which in turn function as nuclei for subsequent growth into amyloid fibrils by monomer protein addition [1, 2].

Accumulation of amyloid fibrils in the affected cells, tissues, or organs is the hallmark of more than 30 diseases, including the most common neurodegenerative disorders like Alzheimer's, Parkinson's, and spongiform encephalopathies, among others. Pathogenic aggregates are associated with a loss of function of the proteins involved and with the generation of oligomeric intermediates which are often toxic [3–6]. There is solid experimental data about the supramolecular organization of amyloid fibrils indicating their core is composed of stacked β-sheets [7]. However, there is still a severe lack of knowledge about the structural features of the oligomeric fibril precursors.

The in vitro fibrillization of amyloid proteins and peptides follows a kinetic profile typical of a nucleation-dependent polymerization process (Fig. 1b). Three distinct phases can be identified: nucleation or lag-phase, propagation, and equilibration [8–11]. The durations of the phases have been found to depend on several factors such as protein concentration, temperature, pH, mechanical agitation, ionic strength, point mutations in the amino acid sequence, and the presence of chaperone proteins [12]. Typical parameters used to describe the fibrillation kinetics are the duration of the lag-phase or nucleation time (t_N), the time needed to reach half the total amount of fibrils ($t_{1/2}$), and the propagation rate (V).

The manipulation and understanding of amyloid fibrillation is of great importance for several reasons. First, understanding how fiber formation occurs and the mechanisms by which they induce pathogenic behaviors is relevant for the understanding of the pathogeny, and in turn for the design of therapies. Second, the regulation of the concentrations of the different states of a protein, as well as the rates of conversion between them, is essential to the maintenance of protein homeostasis under vital conditions. Finally, and related to the two points above, it may be particularly important for therapies to reduce the population of toxic oligomeric species [13], either by disrupting the processes of their formation or by promoting pathways for their removal by incorporating them into fibers. For example, different small molecules [14–17] have been shown to inhibit the amyloid fibrillation. Catalyzers for fibril formation [18] seem at first sight as a more efficient strategy, since smaller concentrations would be able to remove larger amounts of oligomers from solution.

2   Nanoparticles as Nucleation Centers

An attractive approach to achieving control over the aggregation process of amyloid proteins is the use of nanoparticles (NPs) as artificial chaperones to promote and regulate the formation of amyloidogenic as well as inhibitory structures. Recently, NPs of different compositions, sizes, and surface modifications have been investigated as nucleation centers for the amyloid fibrillation of a variety of proteins and peptides. The following table summarizes the work done on this direction, describing for each protein or peptide the NPs investigated and their influence on the fibrillation process. The table describes the type (T), size (S), and concentration (C) of NPs used in each case, as well as a brief description of the observations.

Monellin		
T: copolymeric (NiPAM:BAM) S: 40 nm C: 8.1–270 pM	Acceleration or inhibition of amyloid fibrillation depending on stability and intrinsic aggregation rate of different protein mutants	[19]
T: Au NPs, AuNRs, PS, SiO$_2$ coated with proteins (protein corona) S: 20, 30, 100, 12 × 50, 20 × 400 nm C: 5–80 µg/mL	All bare NPs showed inhibitory behavior for the fibrillation, which became weaker with the protein corona	[20, 21]

Human serum albumin		
T: AuNPs coated with HAS S: 5–100 nm C: 66–0.008 nM	Inhibition of fibrillation, strongest effect as NP sizes reduce	[22]

α-Synuclein		
Type of nanoparticle, size, concentration	**Effect**	**Ref.**
T: CdSe/ZnS QDs surface modified with α-synuclein S: approximately 25 nm C: 0.1–5 nM	Up to 15-fold shortening of $t_{1/2}$	[23]
T: citrate capped AuNPs S: 10, 14 and 22 nm C: 0.3–32 nM	Size-dependent reduction of t_N and increase of V. Strongest effect for smaller NPs: fivefold reduction of t_N 2.5-fold increase of V	[18]

(continued)

α-Synuclein

Type of nanoparticle, size, concentration	Effect	Ref.
T: small unilamellar vesicles S: 20–100 nm C: –	Reduction of t_N and increase of V. No size dependency	[24]
T: human serum albumin S: 35–280 nm C: 25–100 µg/mL	No significant effect	[25]
T: human serum albumin coated with poly-ethyleneimine S: 35–180 nm C: 25–100 µg/mL	Up to fivefold reduction of t_N. No change of V. No size dependency	[25]
T: iron oxide—bare and coated with L-lysine S: C: 1 mg/mL	Acceleration of amyloid fibrillation	[26]
T: iron oxide coated with L-lysine S: C: 1 mg/mL	Strong inhibition of fibrillation	[26]

Insulin

T: citrate capped AuNPs S: 10 nm C: –	Formation of insulin fibrils was significantly delayed, although no detailed kinetic analysis was made	[27]
T: Au and Ag NPs, coated with tyrosine and tryptophan S: 5–30 nm C:	Inhibition of amyloid fibrillation, both spontaneous and in the presence of seeds for nucleation	[28]

Prion

T: AuNPs layer-by-layer coated with PSS and PAH S: 20 nm C: 15–30 nM	Both positively and negatively charged NPs were found to reduce t_N in vitro. Interestingly, in mice, positively charged NPs showed therapeutic effects, whereas negatively charged particles were toxic	[29]

Human β₂-microglobulin

T: cerium oxide, CdSe/ZnS quantum dots coated with hydrophilic polymer, carbon nanotubes, and copolymeric NPs (NiPAM:BAM) S: 6–200 nm C: 10 µg/mL/100 nM	Two- to fivefold reduction of t_N	[30]

IAPP

T: NiPAM:BAM copolymeric S: 40 nm C: 90 pM	Increase of t_N and reduction of V. Strongers effect for 100% NiPAM NPs	[31]

Amyloid-β peptide

T: PS with surface amine groups S: 57, 120, and 180 nm C: 1–1100 µg/mL	Reduction of t_N at low particle surface area. Inhibition of fibrillation at high particle surface area	[32]
T: copolymeric (NiPAM:BAM) with ratios from 100:0 to 50:50 S: 40 nm C: 0.1–10 µg/mL (0.9–90 pM)	Up to sevenfold increase of t_N. The strongest inhibitory effect was found for the most hydrophobic NPs (NiPAM:BAM = 100:0)	[33]
T: AuNPs coated with polyoxometalates (POMD) and LPFFD peptides S: 21 nm C: 40 nM	Formation of fibrils was significantly delayed, although no detailed kinetic analysis was made	[34]
T: AuNPs coated with citrate ions, or polymers with amino or carboxyl groups S: 30 nm C: 0.02–1.36 nM	The negatively charged NPs (citrate and carboxyl) inhibited fibrillation and led to spherical oligomers and fragmented fibrils. The positively charged (amine) NPs showed almost negligible influence	[35]
T: PEGylated phospholipid micelles S: 14 nm C: –	Attenuated aggregation	[36]
T: CdTe quantum dots coated with thioglycolic acid (TGA) S: 3.5 nm C: 25–1250 nM	Up to 2.5-fold increase of t_N	[37]
T: TiO$_2$ S: 20 nm C: 0.1–100 µg/mL	Reduction of t_N and increase of V. Also SiO$_2$, ZrO2, CeO$_2$ NPs were tested, as well as C60, and C70. The latter showed no influence on the aggregation kinetics	[38]
T: CdTe quantum dots capped with N-acetyl-L-cysteine (NAC) S: 3–5 nm C: 10 nM–1 µM	Inhibition of fibrillogenesis at any reaction time	[39]
T: NPs composed of cholesterol-bearing pullulan (CHP), polysaccharide hydrogel S: 20–30 nm C: –	Inhibition of aggregation and reduced cytotoxicity	[40]
T: iron oxide coated with polyethylene glycol (PEG), bare, as well as terminated with amine and carboxyl groups S: 20 nm C: 5 mg/mL	Acceleration or inhibition of amyloid fibrillation depending on concentration, surface charge (−NH$_2$ or −COOH modification), and magnetic field	[41]

The overview given above clearly shows that several different peptide or protein properties (such as sequence, stability, and intrinsic aggregation tendency), as well as NP composition, size, and surface modification influence the way in which proteins react in the presence of the NPs leading to divergent behaviors.

3 Unique Properties of Gold Nanoparticles

Metallic NPs exhibit unique physical, chemical, and electronic properties which make them useful in multiple fields of science and technology [42]. Gold nanoparticles (AuNPs) in particular outstand in comparison to other nanomaterials when it comes to biological assays and applications because of their chemically stability under a wide range of conditions and their demonstrated biocompatibility [43–47]. Tuning the size and shape of metallic NPs gives control not only over the specific surface area but also over optoelectronic properties [42], both of which are exploited in biological assays and therapies [48–53].

AuNPs can be functionalized with great flexibility. A large number of protocols are available by which all kinds of (bio-) molecules and polymers can be attached to their surface [45, 47]. The most common strategy consists of using molecules containing thiol moieties that bind with high affinity to gold surface atoms. Typically, a thiolated molecule is used to directly bring the desired chemical function or to provide the first step for subsequent reactions.

Also, and highly relevant to potential neurological applications, AuNPs with various surface functionalizations are able to cross the blood brain barrier and reach different regions of the brain in useful concentration [34, 47, 54–56].

AuNPs have distinctive opto-electronic properties too [42, 57]. The conduction electrons of AuNPs can sustain collective oscillations resonant at optical frequencies called localized surface plasmons resonances (LSPR) [42, 57, 58]. Due to the LSPRs, AuNPs interact strongly with light. Specifically, the absorption and scattering cross sections of AuNPs at resonance take values that are several times the physical cross section. The frequency of the LSPRs depends on the size and shape of the AuNP, as well as on the optical properties of the surrounding medium [58]. For example, the LSPR of small (~30 nm) AuNPs in water shows up as a peak in the extinction spectra at around 530 nm (Fig. 2a). This absorption gives the typical red color to AuNPs colloids. For larger AuNPs the extinction peak shifts toward longer wavelengths. Elongated NPs such as Au nano-rods (AuNRs) present two extinction peaks, corresponding to two different LSPRs: one corresponding to electrons oscillating along the transverse direction, and the other along the longitudinal dimension of the rod (Fig. 2b). The contribution of scattering to the extinction becomes more important as the NP

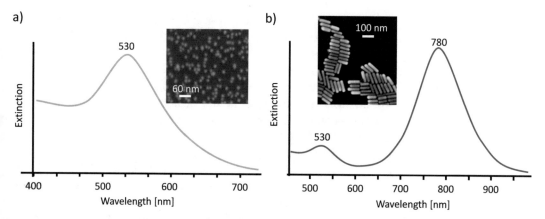

Fig. 2 Extinction spectra of gold nanoparticles: (**a**) spherical AuNPs of 15 nm, (**b**) gold nano-rods of 100 × 25 nm

size increases. For smaller AuNPs, the extinction is dominated by absorption [42, 57].

AuNPs interact electromagnetically with fluorophores too [42]. Illumination at the LSPR frequency generates a near field distribution around the AuNPs with regions of enhanced electric field. A fluorophore located in such a region experiences an increased rate of fluorescence excitation. On the other hand, fluorescence emission is strongly quenched when fluorophores are closer than ca. 20 nm to the AuNP surface. At longer separation distances, fluorescence emission may be enhanced or suppressed by modulations of the local density of photonic states [42, 59]. These three effects are not only dependent on the distance to the AuNP but also on spectrum and mode (dipolar, cuadrupolar, etc.) of the LSPRs involved and the relative orientation of the fluorophore to the NP surface [42]. Therefore, performing quantitative fluorescence measurements that involve AuNPs is not trivial, and a proper interpretation of results requires calculations and/or careful control measurements.

This combination of properties makes AuNPs rather unique, and has enabled their application in various areas of biotechnology and diagnostics [52, 60, 61], as for example gene and drug delivery [62–66], biolabeling and biosensing [67–71], photothermal therapy [72], and in other areas such as plasmonics [73], optical nanoantennas [74–76], and fluorescence enhancement [77].

4   Fabrication and Surface Modification of Gold Nanoparticles

4.1   Synthesis

The synthesis of gold nanoparticles with diameters ranging from a few to several hundreds of nanometers is well established in an aqueous solution as well as in organic solvents. In typical syntheses, gold salts such as $AuCl_3$ are reduced by the addition of a reducing agent which leads to the nucleation of Au ions and nanoparticle growth. The simple reduction of metal salt in a controlled fashion generally produces spherical nanoparticles because it is the lowest energy shape.

In addition, specific molecules or ions are added to the reaction mixture in order to stop the growth at a given size and to stabilize the nanoparticles in suspension. These stabilizing agents either adsorb or chemically bind to the surface of the AuNPs, and are typically charged, so that the NPs are stabilized in colloidal suspension by electrostatic repulsion. If the capping agent binds preferentially to different crystallographic planes of Au, or forms micellar structures, NPs of non-spherical shapes can be obtained, for example nano-rods [78].

A number of compounds, both aqueous and organic, including sodium borohydride ($NaBH_4$) and trisodium citrate ($C_6H_5Na_3O_7$) had been used as common reducing agents. The most widely used method to prepare spherical AuNPs is the one introduced by Turkevich, which uses citrate buffer at boiling point as both reducing and stabilizing agent producing citrate-capped AuNPs of 15 nm diameter [79]. Smaller diameter NPs can be obtained by using borohydrure as a reduction agent [80]. There are numerous modifications of the Turkevich method as well as different alternative methods to produce a large variety of AuNPs.

4.2 Size and Shape Characterization

There are typically two ways to obtain information about the size and shape of a set of AuNPs. The most direct way is the observation of dispersed AuNPs by electron microscopy techniques (EM). AuNPs offer great contrast in basically all modalities of EM. For Scanning EM (SEM) one should take into account that a conductive substrate (e.g., silicon or ITO) is recommended in order to avoid surface charging effects and reach the maximum resolution possible. Typically, a resolution of a few nm is achieved by SEM (Fig. 2). Higher, even atomic resolution is reached with Transmission EM (TEM), which enables the examination of very small clusters as well as crystalline and morphological details of NPs.

An alternative, simpler, and widely used technique for the characterization of AuNPs is UV-VIS spectrophotometry. A careful examination of UV-VIS extinction spectra provides quantitative information about the size distribution and shape of AuNPs. This is possible because (1) the scattering and absorption of metallic NPs strongly depends on size and shape, and (2) the electromagnetic response of AuNPs can be modeled precisely. For spherical AuNPs, there is the analytical solution provided by Mie [81]. For NPs of arbitrary shapes there are currently several numerical tools to calculate accurately the absorption and scattering spectra of NPs, such as finite difference time domain (FDTD) or the boundary elements method (BEM) [82]. Figure 2 show example SEM images and extinction spectra of spherical Au NPs and Au nanorods.

Scattering spectra of single AuNPs with sizes above 40 nm are readily obtained by dark-field microscopy. This methodology provides size and shape information of individual NPs [83, 84], and gives therefore direct insight into the distribution of shapes and sizes of an ensemble of NPs.

4.3 Surface Functionalization of Au Nanoparticles

Surface functionalization is essential to adjust the surface properties of the nanoparticles and thereby to modify and control their interaction with (bio) molecules, other particles, or substrates. After the synthesis of the NPs the stabilizing molecules can be replaced in a ligand exchange reaction. The most frequently used strategy consists of using thiol-modified ligands, taking advantage of the high affinity between thiol moieties and gold surface atoms.

Conjugation of macromolecules, e.g., proteins, with AuNPs may involve a number of interactions, which are often not totally under control. In order to prepare a stable complex, it is necessary to know the surface net charge of the particle, which will determine the attraction or repulsion between the particle and the protein. Also, possible adsorption involving hydrophobic pockets on the protein, as well as the covalent binding possible via sulfhydryl groups present in the protein [85], must be taken into account.

There are basically two options to create the protein-AuNP conjugate: (1) adsorption or (2) covalent binding via sulfhydryl groups on the protein (natural or synthetically added). In both the cases it is important to know the isoelectric point of the protein and to control the pH of the adsorption process, in order to adjust the surface charge of the nanoparticles and proteins to minimize electrostatic repulsion between them.

In a stable colloidal suspension exists a balance between electrostatic repulsion and attractive forces that cause coagulation. This equilibrium can be distorted by the addition of electrolytes to the solution which screen electrostatic repulsion. At a certain concentration of electrolytes, the colloid will begin to collapse as the AuNPs adsorb into one another, forming large aggregates which ultimately precipitate out of suspension. Electrolyte-mediated coagulation forms the basis for creating all gold conjugates with other molecules. If proteins are present in excess, as the electrolyte concentration is raised to surpass the repulsion effects, proteins will be able to contact the NP surface and either adsorb or covalently bind to them, stabilizing the NPs back in suspension. Thus, sufficient protein amounts are necessary in order to produce a stable suspension of protein-conjugated AuNPs. Otherwise, the colloidal suspension precipitates (Fig. 3a). In **Protocol 1** we provide detailed procedures to check the concentration of protein needed to produce a stable suspension of protein-conjugated AuNPs. Briefly, the protocol consists of a titration with electrolyte (the most commonly used are NaCl or buffer salts) in the presence of different

a)   b)

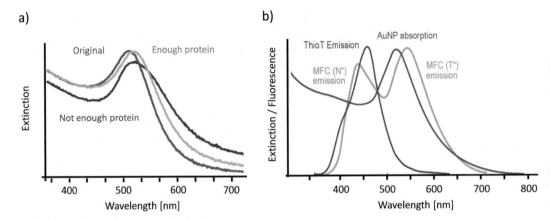

Fig. 3 (a) Typical UV-VIS extinction spectra of a 20 nm AuNPs colloid during the process of α-synuclein adsorption. If the protein is not enough, nanoparticle clusters form and the colloid precipitates as evidenced in the changes of the UV-VIS extinction spectrum (purple curve) **(b)** Emission spectra of Thio-T and the ESIPT fluorophore MFC [86], together with the extinction spectra of 30 nm AuNPs

concentrations of protein. If coagulation occurs, more protein should be added in order to stabilize the colloid.

Conjugation of proteins to AuNPs by adsorption is rather straightforward (**Protocol 2**), although it has the following shortcomings:

(a) Unspecific binding: there is no control over the orientation of the protein after binding to the surface. This may be critical in some cases as specific binding sites of the protein could be blocked after the conjugation. It should be noted, however, that not all protocols for covalent binding provide control over the protein orientation on the NP surface either.

(b) The amount of protein needed to stabilize the solution is usually one or two orders of magnitude larger than for specific covalent binding.

Covalent conjugation of proteins to AuNPs can be achieved in several ways, depending on the sequence of the protein:

1. If the wild-type protein already has exposed cysteines on its structure, it is possible to use those sulfhydryl groups to covalently bind the protein to the AuNP (**Protocol 4**).

2. If the wild-type protein does not have any cysteine it is possible to add sulfhydryl groups by:

    (a) Modifying the protein sequence before the synthesis [23] having as a result a mutant with a cysteine in a specific place of the protein. This procedure is of special interest for studying the functionality of specific regions of the protein.

    (b) Modifying the protein with cystamine (**Protocol 3**). Here, the sulfhydryls are attached to exposed amine groups of the protein. This procedure provides less control over the final orientation of the protein after conjugation to the NP.

In all the cases, an accurate determination of the amount of protein conjugated is important to compare results obtained under different experimental conditions. In **Protocol 5** and **Protocol 6** we detail two methods for the quantification of protein attached to the AuNPs.

5   Amyloid Aggregation Assays with Gold Nanoparticles

In-vitro aggregation assays consist of submitting a solution of monomeric protein to, shaking at a constant temperature, while the amount of amyloid fibrils formed is measured as a function of time. A general problem of these measurements is the reproducibility, partly due to the high sensitivity of the nucleation process to initial conditions. Performing a high number of replicate aggregation assays on a plate reader is essential to obtain reliable results [87].

Most practical methods for the in-vitro monitoring of amyloid protein aggregation are based on optical readout, either by UV-VIS photometry or fluorescence emission [88].

The simplest method consists of monitoring the light transmittance of the sample at 320–350 nm. As the amyloid fibrils form, more light is scattered and therefore the transmittance reduces. This assay is simple and requires no labeling, but is rather insensitive to detect the first stages of aggregation and is not specific for amyloid structures. This method is not suitable when AuNPs are included in the assay, because of the strong extinction of the AuNPs at those wavelengths.

Alternatively, there are several assays based on fluorescence readout. There exist various fluorophores that report specifically different protein aggregation states. The most widely used is Thioflavin T (Thio-T) which presents a fluorogenic signal upon binding to the β-sheet structures of amyloids [89].

Other assays have been developed aiming to obtain more reliable information. For example, the assay based on fluorophores presenting excited state intramolecular proton transfer (ESIPT) [86]. These fluorophores exhibit dual emissions states: the "normal" excited state (N^*) and a phototautomeric excited state (T^*) product of ESIPT. Since the ESIPT process is highly sensitive to the polarity of the molecular environment, its efficiency changes dramatically when the fluorophore intercalates into amyloid fibrils. The aggregation process can be monitored by a ratiometric fluorescence measurement of the T^*/N^* bands. The ratiometric measurement provides additional reliability in comparison to single intensity measurements, such as the Thio-T assay, because it is not affected by concentration differences nor by the eventual (photo-)

degradation of the fluorescent probe during the extended time of the aggregation experiment.

When AuNPs are used, special care must be taken at the moment of interpreting the fluorescence signals. The AuNPs may influence the fluorescence signal in two ways. First, the strong extinction in the visible region may produce an internal filter effect leading to a lower detected fluorescence signal (see Fig. 3b). Second, near field interaction with the fluorophores may affect the excitation and decay rates, as well as the directionality of fluorescence [42]. The latter effects can usually be neglected if the concentration of AuNPs used is 2–3 orders of magnitude smaller than the concentration of fluorophore. In ratiometric fluorescence measurements, the AuNPs may affect differently the involved fluorescence bands.

In addition, the influence of the AuNPs on the measured signal may vary during the aggregation process. AuNPs aggregates may form which have different extinction spectra, or fluorophores may get in close contact with the AuNPs and thereby change their fluorescence performance. For this reason it is important to take into account the potential influence of the AuNPs on the kinetics recorded.

In summary, a correct interpretation of aggregation assays with AuNP requires control measurements such as the ones described in **Protocol 8** and **Protocol 9**.

6   Protocols Step-by-Step

In this section, we describe the step-by-step protocols for conjugation of proteins to AuNPs, for the quantification of the amount of protein attached to NPs, and for carrying out amyloid aggregation assays in the presence of AuNPs.

6.1   Protocol 1: Finding the Amount of Protein Needed for Conjugation

Depending on the size, shape, and concentration of the AuNPs, as well as on the protein nature, different amounts of protein will be needed to form a stable colloid of protein-conjugated AuNPs. This protocol is a guideline to determine the minimum amount of protein necessary.

1. Prepare the same volume (10 µL for example) of four different concentrations of the protein to be studied. Add the protein to four tubes with 200 µL of the AuNP suspension in order to have several ratios NP: protein. The correct amount will strongly depend on the size and charge of the particle and protein. Typical values 1:150, 1:300, 1: 600, 1:1000. Mix well.

   Note: If the particles are positively charged, add 1 µL of NaOH 0.5 M to each tube.

2. After 30–60 min at room temperature centrifuge the sample. The relative centrifugal force (RCF) must be adjusted depending on the size of the NPs. For spherical AuNPs of 20 nm, 30 min at 2000 RCF is enough, whereas for Au nanorods of 10 × 40 nm 10 min would be adequate. Discard the supernatant and resuspend the pellet in PBS 1× (or NaCl 10%).

3. Measure the extinction spectrum of each suspension and compare it to the spectrum of the AuNP suspension. For low concentrations of protein aggregation occurs, evidenced in the spectra as a strong red-shift or as a new peak appearing at longer wavelengths. When the concentration of the protein added is enough to stabilize the colloid, the extinction spectrum should have the same shape as the original one, with possibly a small red-shift of less than 3 nm due to the protein layer formed around the NPs.

6.2 Protocol 2: Protein Conjugation to AuNPs by Adsorption

1. Define the amount of protein to be used with **Protocol 1**. From the minimum amount found, add a 10–20 % excess. Use a pH near to the isoelectric point of the proteins, if needed, adjust the pH by adding NaOH or HCl. A pH 6–7 may work for most proteins. If you are uncertain about the isoelectric point, it is recommended to use a higher pH.

2. Add the protein dropwise to the NP suspension. Try to keep the colloid solution concentrated. The more diluted the solution, the more time will need each protein to collide with a nanoparticle. If necessary use centrifugal filter concentrators [like Amicon Millipore, select a Molecular Weight Cutoff that is 2× smaller than the Molecular Weight (MW) of the protein] to concentrate the protein. Stir well.

3. Let the mixture react 2 h at room temperature or overnight at 4 °C while gently agitated.

4. To remove the excess of protein centrifuge the preparation. For small AuNPs (e.g., 10–15 nm) 13 kRCF for 30 min is sufficient. For larger AuNPs 6 kRCF is typically enough. Discard the supernatant and resuspend the AuNP-protein pellet in the buffer needed (compatible with your sample). Repeat three times.

5. **Note**: For Au nanorods or NPs with diameter larger than 80 nm, it is suggested to add 1% PEG or 0.25% BSA, after the centrifugation in order to stabilize the solution against aggregation making the colloid stable during longer periods of time.

6. Take a UV-VIS extinction spectrum and compare to the spectrum of the starting solution in order to check that there is no aggregation. Store the solution at 4 °C.

Note: The washing step by centrifugation is a critical step that could lead to irreversible agglomeration. If you are uncertain about the needed RCF, centrifuge first only an aliquot at low values of RCF, e.g., 1.5 kRCF for 15 min, and check the supernatant. If it is still rose-red in color you need to increase the RCF. The right settings depend not only on the size and grade of labeling of the NPs, but also on the dimensions and form of the tube used.

Note: Protein-conjugated AuNP colloids are usually stable for long periods of time (months). Nevertheless, before use it is recommended to take a UV-VIS extinction spectrum and compare it to the original spectrum of that batch. Any wavelength shift or broadening of peaks, even if the colloid looks stable, is indicative of the formation of small aggregates that make protein functionality uncertain. In such cases, it is advisable to prepare a new conjugate.

6.3 Protocol 3: Thiolating Proteins: Addition of Sulfhydryl Groups via Cystamine

1. Prepare a protein solution at a concentration of 2 mg/mL in a pH between 4.7 and 7.5. For physiological conditions, use PBS 1×, pH 7.2. Use concentrators if needed. Avoid several centrifugations with the concentrators as some proteins tend to lose their structure and functionality upon such treatments. Do not use buffer containing carboxylates or amines (e.g., Tris–HCl). A small amount of azide may not interfere.

2. Prepare a stock solution of Cystamine (MW 152.28) 10 mM in the same buffer. Add 10- to 20-fold molar excess of the stock solution to the protein. For example, for a small protein like α-synuclein of ca. 14 kDa at a concentration of 2 mg/mL, add 100 μL of stock solution to each mL of protein. Mix well.

3. Prepare a stock solution of EDC [N-Ethyl-N'-(3-dimethylaminopropyl)carbodiimide hydrochloride] in the reaction buffer. Add EDC solution to the Cystamine solution so as to have fivefold molar excess (molar ratio EDC:Cystamine = 5:1). React for 2 h at room temperature.

4. Wash the excess of EDC and Cystamine from the solution by gel filtration or dialysis. For this step, it is highly recommended to use desalting spin columns as they can be used with reduced volumes and they do not dilute the solution much, being the less aggressive method, preserving the protein structure.

5. In order to reduce disulfide groups, add TCEP (tris(2-carboxyethyl)phosphine) or DTT (Dithiothreitol) at a final concentration of 0.5 mg per mg of modified protein. Mix and let it react for 30 min.

6. Purify the thiolated protein from excess of DTT by dialysis or gel filtration. If TCEP was used as a reductant, this step is not necessary because this agent does not contain any thioles that could interfere.

6.4 Protocol 4: Protein Conjugation to AuNPs Using Thiolated Proteins

1. Define the amount of protein to be used with **Protocol 1**. From the minimum amount found, add a 10–20% excess. Use a pH near to the isoelectric point of the protein, if needed, adjust the pH by adding NaOH or HCl. A pH between 6 and 7 is suitable for most proteins.

2. Prepare 1 mL of AuNP suspension. AuNP colloids with OD (optical density) between 1 and 2 have typically an adequate concentration. Add Tween-20 to final concentration 0.02%. Stir.

3. Add the protein to the AuNP suspension dropwise.

4. Add 100 μL of PBS 10× pH 7.2.

5. Let the mixture react for 1 h at room temperature and then overnight at 4 °C.

6. To remove the excess of protein, centrifuge the preparation. For small AuNPs (e.g., 10–15 nm) 13 kRCF for 30 min is sufficient. For larger NPs 6 kRCF is typically enough. Discard the supernatant and resuspend the pellet of protein-conjugated AuNPs in the buffer of choice (compatible with your sample). Repeat three times.

Note: If you are working with NPs stabilized in CTAB as AuNRs, it is highly recommended to remove the CTAB first by adding methoxy-PEG-thiol (mPEG-SH). To this end use a short chain PEG (MW 2 kDa) in a concentration of 0.5 mM in water. Add 100 μL per mL of AuNR suspension (with O.D. $\cong$3) and sonicate for 30 min. After sonication free CTAB and free mPEG-SH must be removed by centrifugation. Then proceed with the protocol above.

6.5 Protocol 5: Quantification of the Amount of Proteins per Nanoparticle

Quantifying accurately the amount of protein conjugated to the AuNPs is a challenging task. There exist only few protocols in the literature and all of them have some shortcomings. Here, we present the two ways we consider more effective.

In the first approach protein absorption is interrogated. The problem here is that AuNPs also absorb at the usual protein interrogation wavelength of 280 nm. For this reason the absorption of the supernatant after centrifugations is measured, and the amount of protein attached to the NPs is obtained by subtraction.

As an alternative we present the second protocol based on fluorescence. Any influence of the AuNPs can be minimized by properly choosing the fluorophore far away from any absorption band of the NPs. Naturally, extra steps for the fluorescent labeling of the protein are required. It should be taken into account that the yield of protein conjugation to AuNPs may be different for bare proteins and fluorescently labeled proteins.

For the measurement of protein absorption in the supernatant, it is highly recommended to use a microplate reader or a spectrophotometer of small volumes (e.g., Nanodrop).

1. Prepare at least four tubes with 200 µL of AuNP suspensions of different known concentrations (for example 3, 2, 1, and 0.5 nM) and one more tube with the same volume of buffer.

2. Follow the protein conjugation protocol until the washing step. To the fifth tube (control without NPs) also add protein so that the final concentration is the same in all the tubes.

3. Centrifuge the samples until the supernatant looks colorless (e.g., typical conditions for AuNPs with diameters of 10–15 nm are: 13 kRCF for 1 h. For larger AuNPs: 6 kRCF for 30 min to 1 h).

4. Extract a sample of the supernatant of each sample and analyze the absorption at 280 nm of all the samples. The difference between each sample and the control (no AuNPs) corresponds to the protein attached to the AuNPs. Check that the absorption in the control tube did not change. In case protein was lost in the control tube upon centrifugation, that value must be subtracted from the measurements of all the samples.

5. Calibration. To convert absorption signal to concentration, a calibration curve is needed. Prepare (at least) four tubes with a known concentration of protein in the range of interest and fit absorption vs. concentration curve.

Note: It is very important that the supernatants are free of NPs. Use a small micro pipette and uptake the solution as slow as you can.

Note: Obtaining the amount of protein attached to the AuNPs from the difference in absorbance of a given sample before and after centrifugation is not possible, because the absorbance of AuNPs at 280 nm is too strong.

This method requires the conjugation of fluorophores to the proteins (see **Protocol** 7 below).

Bulk Measurement

1. Calibration. Prepare four samples of AuNP colloids at an optical density of 0.1, each one including a different and known concentration of the fluorescently labeled protein. The calibration measurements are carried out in the presence of AuNPs in order to account for any influence on the fluorescence readout.

   Note: It is important to note that even if the measurements are done in the presence of AuNPs, the calibration may not be accurate as the quenching or enhancement of the fluorescence intensity produced by the AuNPs is highly dependent on the distance of the probe to the NP; i.e., it may be different for proteins free in the solution and proteins conjugated to the AuNPs.

2. Conjugate the nanoparticles with the labeled protein following the protocol of your convenience.

3. After washing the conjugate thoroughly adjust the concentration to an optical density of 0.1 to match the calibration measurements. The number of proteins per NP can be determined from the comparison of the conjugate's fluorescence to the calibration curve.

Note: It is important to choose a fluorophore whose absorption and emission bands do not overlap to the LSPRs of the AuNPs. In this way all electromagnetic interactions are minimized and the determination of the amount of protein is most accurate.

Single nanoparticle measurement

1. Deposit protein-conjugated NPs on a microscope glass coverslip. The surface density should be adjusted until individual protein-conjugated AuNPs can be studied in a far-field fluorescence microscope; i.e., approximately 0.5 NP/μm^2. Substrates must be free of any fluorescence background. Typical cleaning procedures involve washing with detergent, sonication, and subsequent rinsing with water, ethanol, and acetone. After that, they should be treated in a plasma cleaner or in an oven at 250 °C for 2 h.

2. On a fluorescence microscope with single-molecule sensitivity, follow the fluorescence emission of single particles in time until all fluorophores are bleached.

3. The number of proteins conjugated to a NP can be obtained in two ways: (1) counting the bleaching steps, or (2) comparing the fluorescence intensity of the NPs to the average fluorescence intensity of a single fluorophore. The latter is obtained from the emission level before the last single-step photobleaching event of each NP.

Note: an accurate determination of the exact number of proteins conjugated to a AuNP requires knowledge of the average number of fluorophores attached to a protein, for example obtained as described in step 6 of **Protocol 6** below.

6.6   Protocol 6: Protein Fluorescent Labeling

First, it should be noted that some fluorescently labeled proteins are nowadays commercially available. That said, there are several protocols to attach fluorophores to proteins well described in the literature [90–94]. Here we present one example, for fluorophores with succinimidyl-ester groups.

1. Prepare a protein solution at a concentration of 2–10 mg/mL in a 0.1 M sodium bicarbonate buffer pH 8.2. Concentrations lower than 2 mg/mL are not recommended since the effi-

ciency of the reaction is highly concentration dependent. Avoid amino-containing compounds in the solution.

2. Dissolve the succinimidyl-ester fluorophore in DMSO at 10 mg/mL. This step must be done immediately before starting the reaction in order to ensure the stability of the probe. Vortex or sonicate briefly if necessary.

3. Under stirring, slowly add 50–100 μL of the reactive dye to the protein solution.

4. Incubate the reaction for 1 h at room temperature with continuous stirring.

5. Equilibrate a desalting column (10×300 mm may be enough) with PBS 1×. Follow the protocol recommended by the column manufacturer. The first fluorescent band to elute corresponds to the conjugate.

6. To determine the degree of labeling (DOL), calculate the concentration of the protein with the absorption at 280 nm and measure the absorption at the maximum of the dye (A_{max}). Calculate:

$$DOL = A_{max} \times MW \left[protein\right] \times \varepsilon_{dye}$$

where MW = the molecular weight of the protein, ε_{dye} = the extinction coefficient of the dye at its absorbance maximum, and the protein concentration is in mg/mL.

6.7 Protocol 7: Amyloid Protein Aggregation Assay with Fluorescence Readout

It is advisable to perform the assays on plate reader, making as many replicate measurements as possible, with a minimum of 6 for each condition.

1. In order to minimize the presence of protein at any stage of aggregation (e.g., dimers or higher order oligomers) at the starting point of the assay, ultracentrifugate the protein solution for 1 h at $100,000 \times g$ before each experiment and use the supernatant. The buffer should also be submitted to ultracentrifugation in order to minimize the presence of small particles that could act as external nucleation centers.

2. Determine the concentration of protein solutions by measuring the UV absorbance at 275 nm after centrifugation.

3. Prepare the samples containing the protein, the fluorescent probe, and sodium azide at 0.2% (to avoid microbial contamination). All plastic material should be perfectly clean so as to avoid the presence of particles that could act as external nuclei. Prepare control samples containing only the buffers and the fluorescent probe.

4. If the assay involves AuNPs, prepare samples as in point 3 including AuNPs at different concentrations. Follow **Protocols 8** and **9**.

5. Set up the measurement conditions as temperature, shaking, and detection parameters.

6.8   Protocol 8: Quantifying the Effect of AuNPs on the Amount of Fibrils

1. Incubate the protein long enough to assure amyloid fibrillation has reached equilibrium.

2. Take two samples of fibrils and add the fluorescent probe (e.g., Thio-T). Mix well before extracting the samples to ensure homogeneous solution and equivalent samples.

3. Add the same concentration of AuNPs that was used in the experiment into one of the samples, and the same volume of buffer to the other.

4. Perform the fluorescence readout on the two samples.

5. Compare the ratio between the measurements with the ratio between the final values of the aggregations with the AuNP. If the ratios are not significantly different, no effect on the amount of fiber can be ascribed to the presence of the AuNPs.

Note: the same protocol must be done for all (or at least extreme) concentrations and types of AuNPs used.

Note: Example for Thio-T measurement. Take 10 μL of each aggregation sample and dilute them in 10 mL of a solution 5 μM of Thio-T in 50 mM glycine buffer pH 8.2. After 15 min of shaking, measure the fluorescence at 460 ± 10 nm by exciting at 420 ± 10 nm.

6.9   Protocol 9: Quantifying the Effect of AuNPs on the Kinetics of the Aggregation

For this purpose it is necessary to take aliquots at different time points during the amyloid aggregation. Enough samples should be tested in order to evaluate the effect of AuNPs on the different stages of the aggregation process. For each point in time:

1. Extract two aliquots from an aggregation assay in the same experimental conditions, but without AuNPs.

2. Add the concentration of AuNPs used in the experiment to one set of samples and the same volume of buffer to the other set of samples.

3. Perform the fluorescence readout for the two samples.

4. Compare the kinetic curves obtained. If the normalized curves are identical, it can be concluded that the AuNPs do not affect the readout of the kinetics of aggregation.

References

1. Knowles TPJ, Buehler MJ (2011) Nanomechanics of functional and pathological amyloid materials. Nat Nanotechnol 6:469–479

2. Arosio P, Knowles TPJ, Linse S (2015) On the lag phase in amyloid fibril formation. Phys Chem Chem Phys 17:7606–7618

3. Lansbury PT, Lashuel HA (2006) A century-old debate on protein aggregation and neurodegeneration enters the clinic. Nature 443:774–779

4. Chiti F, Dobson CM (2006) Protein misfolding, functional amyloid, and human disease. Annu Rev Biochem 75:333–366

5. Winner B et al (2011) In vivo demonstration that alpha-synuclein oligomers are toxic. Proc Natl Acad Sci 108:4194–4199

6. Lashuel HA, Overk CR, Oueslati A, Masliah E (2013) The many faces of α-synuclein: from structure and toxicity to therapeutic target. Nat Rev Neurosci 14:38–48

7. Otzen DE (ed.) (2013) Amyloid fibrils and prefibrillar aggregates: molecular and biological properties. Wiley-VCH Verlag GmbH & Co. KGaA: Weinheim, Germany ISBN: 978-3-527-33200-7

8. Morris AM, Watzky MA, Finke RG (2009) Protein aggregation kinetics, mechanism, and curve-fitting: a review of the literature. Biochim Biophys Acta 1794:375–397

9. Morris AM, Finke RG (2009) Alpha-synuclein aggregation variable temperature and variable pH kinetic data: a re-analysis using the Finke-Watzky 2-step model of nucleation and autocatalytic growth. Biophys Chem 140:9–15

10. Buell AK, Dobson CM, Knowles TPJ (2014) The physical chemistry of the amyloid phenomenon: thermodynamics and kinetics of filamentous protein aggregation. Essays Biochem 56:11–39

11. Pellarin R, Caflisch A (2006) Interpreting the aggregation kinetics of amyloid peptides. J Mol Biol 360:882–892

12. Uversky VN, Fink AL (2004) Conformational constraints for amyloid fibrillation: the importance of being unfolded. Biochim Biophys Acta 1698:131–153

13. Pimplikar SW (2009) Reassessing the amyloid cascade hypothesis of Alzheimer's disease. Int J Biochem Cell Biol 41:1261–1268

14. Liu Y, Carver JA, Calabrese AN, Pukala TL (2014) Gallic acid interacts with α-synuclein to prevent the structural collapse necessary for its aggregation. Biochim Biophys Acta Proteins Proteomics 1844:1481–1485

15. Lorenzen N et al (2014) How epigallocatechin gallate can inhibit α-synuclein oligomer toxicity in vitro. J Biol Chem 289:21299–21310

16. Ardah MT et al (2014) Structure activity relationship of phenolic acid inhibitors of α-synuclein fibril formation and toxicity. Front Aging Neurosci 6:1–17

17. Mason JM, Kokkoni N, Stott K, Doig AJ (2003) Design strategies for anti-amyloid agents. Curr Opin Struct Biol 13:526–532

18. Alvarez YD et al (2013) Influence of gold nanoparticles on the kinetics of α-synuclein aggregation. Nano Lett 13:6156–6163

19. Cabaleiro-Lago C, Szczepankiewicz O, Linse S (2012) The effect of nanoparticles on amyloid aggregation depends on the protein stability and intrinsic aggregation rate. Langmuir 28:1852–1857

20. Mirsadeghi S et al (2015) Protein corona composition of gold nanoparticles/nanorods affects amyloid beta fibrillation process. Nanoscale 7:5004–5013

21. Mahmoudi M et al (2013) The protein corona mediates the impact of nanomaterials and slows amyloid beta fibrillation. ChemBioChem 14:568–572

22. Goy-López S et al (2012) Physicochemical characteristics of protein-NP bioconjugates: the role of particle curvature and solution conditions on human serum albumin conformation and fibrillogenesis inhibition. Langmuir 28:9113–9126

23. Roberti MJ, Morgan M (2009) Quantum dots as ultrasensitive nanoactuators and sensors of amyloid aggregation in live cells. J Am Chem Soc 131:8102–8107

24. Galvagnion C et al (2015) Lipid vesicles trigger α-synuclein aggregation by stimulating primary nucleation. Nat Chem Biol 11:229–234

25. Mohammad-Beigi H et al (2015) Strong interactions with polyethylenimine-coated human serum albumin nanoparticles (PEI-HSA NPs) alter α-synuclein conformation and aggregation kinetics. Nanoscale 7:19627–19640

26. Joshi N et al (2015) Attenuation of the early events of α-synuclein aggregation: a fluorescence correlation spectroscopy and laser scanning microscopy study in the presence of surface-coated Fe_3O_4 nanoparticles. Langmuir 31:1469–1478

27. Hsieh S, Chang C, Chou H (2013) Gold nanoparticles as amyloid-like fibrillogenesis

inhibitors. Colloids Surf B Biointerfaces 112:525–529

28. Dubey K et al (2015) Tyrosine- and tryptophan-coated gold nanoparticles inhibit amyloid aggregation of insulin. Amino Acids 47:2551–2560

29. Ai Tran HN et al (2010) A novel class of potential prion drugs: preliminary in vitro and in vivo data for multilayer coated gold nanoparticles. Nanoscale 2:2724–2732

30. Linse S et al (2007) Nucleation of protein fibrillation by nanoparticles. Proc Natl Acad Sci 104:8691–8696

31. Cabaleiro-Lago C, Lynch I, Dawson KA, Linse S (2010) Inhibition of IAPP and IAPP(20-29) fibrillation by polymeric nanoparticles. Langmuir 26:3453–3461

32. Cabaleiro-Lago C, Quinlan-Pluck F, Lynch I, Dawson KA, Linse S (2010) Dual effect of amino modified polystyrene nanoparticles on amyloid β protein fibrillation. ACS Chem Nerosci 1:279–287

33. Cabaleiro-Lago C et al (2008) Inhibition of amyloid beta protein fibrillation by polymeric nanoparticles. J Am Chem Soc 130:15437–15443

34. Gao N, Sun H, Dong K, Ren J, Qu X (2015) Gold-nanoparticle-based multifunctional amyloid-β inhibitor against Alzheimer's disease. Chem A Eur J 21:829–835

35. Liao Y-H, Chang Y-J, Yoshiike Y, Chang Y-C, Chen Y-R (2012) Negatively charged gold nanoparticles inhibit Alzheimer's amyloid-β fibrillization, induce fibril dissociation, and mitigate neurotoxicity. Small 8:3631–3639

36. Pai AS, Rubinstein I, Önyüksel H (2006) PEGylated phospholipid nanomicelles interact with β-amyloid(1-42) and mitigate its β-sheet formation, aggregation and neurotoxicity in vitro. Peptides 27:2858–2866

37. Yoo S II et al (2011) Inhibition of amyloid peptide fibrillation by inorganic nanoparticles: functional similarities with proteins. Angew Chem Int Ed 50:5110–5115

38. Wu W et al (2008) TiO2 nanoparticles promote β-amyloid fibrillation in vitro. Biochem Biophys Res Commun 373:315–318

39. Xiao L, Zhao D, Chan WH, Choi MMF, Li HW (2010) Inhibition of beta 1-40 amyloid fibrillation with N-acetyl-l-cysteine capped quantum dots. Biomaterials 31:91–98

40. Ikeda K, Okada T, Sawada SI, Akiyoshi K, Matsuzaki K (2006) Inhibition of the formation of amyloid β-protein fibrils using biocompatible nanogels as artificial chaperones. FEBS Lett 580:6587–6595

41. Mirsadeghi S, Shanehsazzadeh S, Atyabi F, Dinarvand R (2016) Effect of PEGylated superparamagnetic iron oxide nanoparticles (SPIONs) under magnetic field on amyloid beta fibrillation process. Mater Sci Eng C 59:390–397

42. Coronado EA, Encina ER, Stefani FD (2011) Optical properties of metallic nanoparticles: manipulating light, heat and forces at the nanoscale. Nanoscale 3:4042–4059

43. Chithrani BD, Ghazani AA, Chan WCW (2006) Determining the size and shape dependence of gold nanoparticle uptake into mammalian cells. Nano Lett 6:662–668

44. Shukla R et al (2005) Biocompatibility of gold nanoparticles and their endocytotic fate inside the cellular compartment: a microscopic overview. Langmuir 21:10644–10654

45. Sperling RA, Rivera Gil P, Zhang F, Zanella M, Parak WJ (2008) Biological applications of gold nanoparticles. Chem Soc Rev 37:1896–1908

46. Murphy CJ et al (2008) Gold nanoparticles in biology: beyond toxicity to cellular imaging. Acc Chem Res 41:1721–1730

47. Khlebtsov N, Dykman L (2011) Biodistribution and toxicity of engineered gold nanoparticles: a review of in vitro and in vivo studies. Chem Soc Rev 40:1647–1671

48. Pham T, Jackson JB, Halas NJ, Lee TR (2002) Preparation and characterization of gold nanoshells coated with self-assembled monolayers. Langmuir 18:4915–4920

49. Oldenburg S, Averitt R, Westcott S, Halas N (1998) Nanoengineering of optical resonances. Chem Phys Lett 288:243–247

50. Lal S et al (2008) Tailoring plasmonic substrates for surface enhanced spectroscopies. Chem Soc Rev 37:898

51. Myroshnychenko V et al (2008) Modeling the optical response of highly faceted metal nanoparticles with a fully 3D boundary element method. Adv Mater 20:4288–4293

52. Huang X, Jain PK, El-Sayed IH, El-Sayed MA (2007) Gold nanoparticles: interesting optical properties and recent applications in cancer diagnostics and therapy. Nanomedicine 2:681–693

53. Jain PK, Huang X, El-Sayed IH, El-Sayed MA (2007) Review of some interesting surface plasmon resonance-enhanced properties of noble metal nanoparticles and their applications to biosystems. Plasmonics 2:107–118

54. Joh DY et al (2013) Selective targeting of brain tumors with gold nanoparticle-induced radiosensitization. PLoS One 8:e62425

55. Sonavane G, Tomoda K, Makino K (2008) Biodistribution of colloidal gold nanoparticles after intravenous administration: effect of particle size. Colloids Surf B 66:274–280

56. Prades R et al (2012) Delivery of gold nanoparticles to the brain by conjugation with a peptide that recognizes the transferrin receptor. Biomaterials 33:7194–7205

57. Kreibig U, Vollmer M (1995) Optical properties of metal clusters, vol 25. Springer, Berlin

58. Kelly KL, Coronado E, Zhao LL, Schatz GC (2003) The optical properties of metal nanoparticles: the influence of size, shape, and dielectric environment. J Phys Chem B 107:668–677

59. Pellegrotti JV et al (2014) Controlled reduction of photobleaching in DNA origami-gold nanoparticle hybrids. Nano Lett 14:2831–2836

60. Gobin AM et al (2007) Near-infrared resonant nanoshells for combined optical imaging and photothermal cancer therapy. Nano Lett 7:1929–1934

61. Wu X et al (2003) Immunofluorescent labeling of cancer marker Her2 and other cellular targets with semiconductor quantum dots. Nat Biotechnol 21:41–46

62. Pissuwan D, Niidome T, Cortie MB (2011) The forthcoming applications of gold nanoparticles in drug and gene delivery systems. J Control Release 149:65–71

63. Rotello VM (2007) Drug and gene delivery using gold nanoparticles. Drug Deliv 40–45. doi: 10.1007/s

64. Paasonen L et al (2007) Gold nanoparticles enable selective light-induced contents release from liposomes. J Control Release 122:86–93

65. Patil SD, Rhodes DG, Burgess DJ (2005) DNA-based therapeutics and DNA delivery systems: a comprehensive review. AAPS J 7:E61–E77

66. Han G, Martin CT, Rotello VM (2006) Stability of gold nanoparticle-bound DNA toward biological, physical, and chemical agents. Chem Biol Drug Des 67:78–82

67. Mayilo S et al (2009) Long-range fluorescence quenching by gold nanoparticles in a sandwich immunoassay for cardiac troponin T. Nano Lett 9:4558–4563

68. Chandrasekharan N, Kelly LA (2001) A dual fluorescence temperature sensor based on perylene/exciplex interconversion. J Am Chem Soc 123:9898–9899

69. Nath N, Chilkoti A (2002) A colorimetric gold nanoparticle sensor to interrogate biomolecular interactions in real time on a surface. Anal Chem 74:504–509

70. Zijlstra P, Paulo PMR, Orrit M (2012) Optical detection of single non-absorbing molecules using the surface plasmon resonance of a gold nanorod. Nat Nanotechnol 7:379–382

71. Mayer KM, Hafner JH (2011) Localized surface plasmon resonance sensors. Chem Rev 111:3828–3857

72. Lakhani PM, Rompicharla SVK, Ghosh B, Biswas S (2015) An overview of synthetic strategies and current applications of gold nanorods in cancer treatment. Nanotechnology 26:432001

73. Atwater HA, American S (2007) The promise of plasmonics. Sci Am 296:56–63

74. Taminiau TH, Stefani FD, van Hulst NF (2011) Optical nanorod antennas modeled as cavities for dipolar emitters: evolution of sub- and super-radiant modes. Nano Lett 11:1020–1024

75. Taminiau TH, Stefani FD, van Hulst NF (2008) Enhanced directional excitation and emission of single emitters by a nano-optical Yagi-Uda antenna. Opt Express 16:10858

76. Busson MP, Rolly B, Stout B, Bonod N, Bidault S (2012) Accelerated single photon emission from dye molecule-driven nanoantennas assembled on DNA. Nat Commun 3:962

77. Acuna GP et al (2012) Fluorescence enhancement at docking sites of DNA-directed self-assembled nanoantennas. Science 338:506–510

78. Ye X et al (2012) Improved size-tunable synthesis of monodisperse gold nanorods through the use of aromatic additives. ACS Nano 6:2804–2817

79. Kimling J et al (2006) Turkevich method for gold nanoparticle synthesis revisited. J Phys Chem B 110:15700–15707

80. Martin MN, Basham JI, Chando P, Eah S-K (2010) Charged gold nanoparticles in non-polar solvents: 10-min synthesis and 2D self-assembly. Langmuir 26:7410–7417

81. Mie G (1908) Beiträge zur Optik trüber Medien, speziell kolloidaler Metallösungen. Ann Phys 25:377–445

82. Hohenester U, Trügler A (2012) MNPBEM—a Matlab toolbox for the simulation of plasmonic nanoparticles. Comput Phys Commun 183:370–381

83. Klar T et al (1998) Surface-plasmon resonances in single metallic nanoparticles. Phys Rev Lett 80:4249–4252

84. Mock JJ, Barbic M, Smith DR, Schultz DA, Schultz S (2002) Shape effects in plasmon resonance of individual colloidal silver nanoparticles. J Chem Phys 116:6755

85. Hermanson GT (2008) Bioconjugate techniques. Elsevier, Amsterdam

86. Yushchenko DA, Fauerbach JA, Thirunavukkuarasu S, Jares-erijman EA, Jovin

TM (2010) Fluorescent ratiometric MFC probe sensitive to early stages of alpha-synuclein aggregation. J Am Chem Soc 132: 7860–7861

87. Giehm L, Otzen DE (2010) Strategies to increase the reproducibility of protein fibrillization in plate reader assays. Anal Biochem 400:270–281

88. Giehm L, Lorenzen N, Otzen DE (2011) Assays for α-synuclein aggregation. Methods 53:295–305

89. LeVine HI (1999) Quantification of beta-sheet amyloid fibril structures with thioflavin T. Methods Enzymol 309:274–284

90. Hung SC, Ju J, Mathies RA, Glazer AN (1996) Energy transfer primers with 5- or 6-carboxyrhodamine-6G as acceptor chromophores. Anal Biochem 238:165–170

91. Metzker ML, Lu J, Gibbs RA (1996) Electrophoretically uniform fluorescent dyes for automated DNA sequencing. Science 271:1420–1422

92. Koike H, Yusa T, McCormick DB, Wright LD (1970) Vitamins and coenzymes. Methods enzymology, vol 18, Elsevier, Amsterdam

93. Davis WC (1995) Monoclonal antibody protocols, vol 45. Humana Press, Totowa, NJ

94. Huang B, Wang W, Bates M, Zhuang X (2008) Three-dimensional super-resolution imaging by stochastic optical reconstruction microscopy. Science 319:810–813

Index

Fidel Santamaria and Xomalin G. Peralta (eds.), *Use of Nanoparticles in Neuroscience*, Neuromethods, vol. 135,
https://doi.org/10.1007/978-1-4939-7584-6, © Springer Science+Business Media, LLC 2018

Printed in the United States
By Bookmasters